Global Changes in the Perspective of the Past

Goal of this Dahlem Workshop:

to discuss the use of paleoclimatic data for analyzing climatic processes and for modeling the future evolution of the climate

Environmental Sciences Research Report ES 12

Held and published on behalf of the
Freie Universität Berlin

Sponsored by:
Funds from the Stiftungsfond Unilever in the
Stifterverband für die Deutsche Wissenscahaft

Global Changes in the Perspective of the Past

Edited by

J.A. EDDY and H. OESCHGER

Report of the Dahlem Workshop on
Global Changes in the Perspective of the Past

Berlin 1991, December 8–13

Program Advisory Committee:
J.A. Eddy and H. Oeschger, Chairpersons
J.-C. Duplessy, B. Frenzel, J. Imbrie, and J. Jouzel

JOHN WILEY & SONS
Chichester • New York • Brisbane • Toronto • Singapore

Library of Congress Cataloging-in-Publication Data

Dahlem Workshop on Global Changes in the Perspective of the Past (1991 : Berlin, Germany)
Global changes in the perspective of the past : report of the Dahlem Workshop on Global Changes in the Perspective of the Past, Berlin, December 8–13, 1991 / edited by J.A. Eddy and H. Oeschger.
p. cm. — (Environmental sciences research report ; ES 12) (Dahlem workshop reports)
Includes bibliographical references and indexes.
ISBN 0-471-93603-0
1. Paleoclimatology—Congresses. 2. Climatic changes—Congresses. I. Eddy, John A. II. Oeschger, H., (Hans) III. Title. IV. Series. V. Series: Dahlem workshop reports.
QC884.2.C5D34 1992
551.6—dc20 93–818
CIP

British Library Cataloguing in Publication Data

A catalogue record for this book is available from the British Library

ISBN 0-471-93603-0

Dahlem Editorial Staff: J. Lupp, C. Rued-Engel
Typeset in 10/12pt Times from author's disks by Text Processing Department, John Wiley & Sons Ltd, Chichester
Printed and bound in Great Britain by Biddles Ltd, Guildford, Surrey

Contents

The Dahlem Konferenzen

The purpose of the Dahlem Konferenzen is to promote an international interdisciplinary exchange of scientific information and ideas as well as to stimulate international cooperation in research. This is achieved by arranging discussion workshops, mainly in the life sciences and environmental sciences, organized according to a model developed and tested by the Dahlem Konferenzen.

Dahlem Konferenzen was founded in 1974 by the Stifterverband für Deutsche Wissenschaft[1] in collaboration with the Deutsche Forschungsgemeinschaft[2] to promote more effective communication between scientists. It was named after a district of Berlin, Dahlem, which has long been a home for the sciences and arts. In January, 1990, it was incorporated into the Freie Universität Berlin. Financial support comes from the Senate of Berlin, the Deutsche Forschungsgemeinschaft, and private foundations.

As scientific research has become increasingly interdisciplinary, a growing need has emerged for specialists in one field to understand the problems and work with the concepts of related fields. New insights can be gained when a problem is approached from the standpoint of another discipline, and because no existing form of scientific meeting provided the forum necessary for such exchanges, Dahlem Konferenzen created a new concept, which has since been tested and refined over the years. Now internationally recognized as the Dahlem Workshop Model, it provides a framework for coherent interdisciplinary discussion of a topic in five working days, and culminates in the draft manuscript for a book.

Dahlem Workshops provide a unique opportunity for scientists to pose questions to colleagues from different disciplines and to solicite alternative opinions on contentious issues. The aim is not to solve problems or to reach a consensus, but to identify gaps in knowledge, to find new ways to approach stubborn issues, and to define priorities for research. This approach is well summed up in the instructions given to participants: please state what you do not know, rather than what is known.

---

[1] The Donors Association for the Promotion of Sciences and Humanities, a foundation created in 1921 in Berlin and supported by German trade and industry to fund basic research in the sciences

[2] German Science Foundation

Workshop topics are proposed by leading scientists and are approved by a scientific board advised by qualified referees. Approximately one year before the workshop, an advisory committee meets to determine the scientific program and select participants, who are invited according to their scientific reputations, with the exception of a number of places reserved for junior German scientists.

The discussions at each workshop are organized around four key questions, each tackled by a group of about twelve participants with a range of expertise. There are no lectures. Instead, prior to the workshop, selected participants write background papers, which review the group's discussion topic and serve as the basis for the discussion. These papers are distributed to all participants before the meeting, with selected participants acting as referees. Based on the discussion during the week, each group prepares a report reflecting the ideas, opinions, and controversies which have emerged, as well as identifies problems areas and directions for future research.

The revised background papers and group reports are published as the Dahlem Workshop Reports. Each volume is edited by the chairperson(s) of the workshop and the Dahlem staff. The Reports provide multidisciplinary surveys by an international group of distinguished scientists, based on discussions of advanced concepts, techniques, and models. The Dahlem Workshop Reports are published in two series: Life Sciences and Environmental Sciences (formerly Physical, Chemical, and Earth Sciences).

Jennifer Altman, Acting Director
Dahlem Konferenzen der Freien Universität Berlin
Rothenburgstr. 33, 12165 Berlin, F.R. Germany

List of Participants with Fields of Research

EDOUARD BARD CEA, Laboratoire de Géosciences de l'Environment, Université d'Aix-Marseille, 13397 Marseille Cedex 13, France

Absolute geochronology, ^{14}C, U–Th by mass spectrometry, climatology of the last glaciation cycle

ANDRÉ BERGER Institut d'Astronomie et de Géophysique G. Lemaître, Université Catholique de Louvain, 2 Chemin du Cyclotron, 1348 Louvain-La-Neuve, Belgium

Modeling the response of the climate system to astronomical forcing

RAY BRADLEY Department of Geology and Geography, Morill Science Center, University of Massachusetts, Amherst, MA 01003, U.S.A.

Climatic variations in the period of instrumental and historical records

CONRADIN A. BURGA Geographisches Institut, Universität Zürich-Irchel, Winterthurerstr. 190, 8057 Zürich, Switzerland

Vegetation history, paleovegetation modeling, paleoclimatology. Area: Switzerland, Northern Italy (Alps) Fields: lake sediments, pollen, glaciers

THOMAS J. CROWLEY Applied Research Corporation, 305 Arguello Drive, College Station, TX 77840, U.S.A.

Paleoclimatology, climate modeling

JEAN-CLAUDE DUPLESSY Centre Faibles Radioactivités, CNRS, Ave. de la Terrasse, 91198 Gif-sur-Yvette, France

Paleooceanography

JOHN A. EDDY Consortium for International Earth Science Information Network, 2250 Pierce Road, University Center, MI 48710, U.S.A.

Earth system science, solar history, climate history

GRAHAM FARQUHAR Research School of Biological Sciences, Institute of Advanced Studies, Australian National University, GPO Box 475, Canberra City, ACT 2601, Australia

Carbon and oxygen isotope composition of, and discrimination by, plants

HERMANN FLOHN Meteorologisches Institut, Universität Bonn, Auf dem Hügel 20, 5300 Bonn 1, F.R.Germany

Recent changes of hydrological cycle and atmospheric circulation, natural and anthropogenic causes of climate variability, historical and palaeoclimatology

BURKHARD FRENZEL Botanisches Institut, Universität Hohenheim, Postfach 70 05 62, 7000 Stuttgart 70, F.R.Germany

Palaeoecology of the Quaternery of Eurasia, palaeoclimatology of the Quaternery, stratigraphy of the Quaternery

ANNE-KATHERIN GLIEMEROTH Botanisches Institut, Universität Hohenheim, Postfach 70 05 62, 7000 Stuttgart 70, F.R.Germany

Reconstruction of vegetation, biomass, and soil in various time-slices of the Holocene over Europe

KLAUS HERTERICH Fachbereich Geowissenschaften, Universität Bremen, Postfach 330440, 2800 Bremen 33, F.R. Germany

Ice sheet modeling

JOHN IMBRIE Department of Geological Sciences, Brown University, Providence, RI 02912-1846, U.S.A.

Paleoclimate of the past 500,000 years based on marine sediment core data

SYLVIE JOUSSAUME Laboratoire de Géochemie Isotopique, DSM/DPhG/SPER - Bat 522, CEN Saclay, 91191 Gif-sur-Yvette Cedex, France

Paleoclimate modeling with general circulation models, climatic tracer modeling

JEAN JOUZEL Laboratoire de Géochemie Isotopique, DSM DPhG/SPER, CEN Saclay, 91191 Gif-sur-Yvette Cedex, France

Paleoclimatology (from ice cores) and associated climate modeling

LAURENT LABEYRIE Centre des Faibles Radioactivités, CNRS, Ave. de la Terrasse, 91198 Gif-sur-Yvette Cedex, France

Changes in ocean circulation for the last 10^6 years using ^{18}O, ^{13}C, ^{14}C, in foraminifera and temperature of surface water with transfer functions

MICHAEL LAUTENSCHLAGER Deutsches Klimarechenzentrum GmbH, Bundesstrasse 55, 2000 Hamburg 13, F.R.Germany

Modeling the equilibrium and transient climate response in the Holocene

INGEBORG LEVIN Institut für Umweltphysik, Universität Heidelberg, Im Neuenheimer Feld 366, 6900 Heidelberg, F.R.Germany

Global and regional studies of the atmospheric CO_2 and CH_4 budgets involving isotopic species

THOMAS LITT Institut für Geologische Wissenschaften und Geiseltalmuseum der Martin-Luther-Universität, Domstrasse 5, 4020 Halle, F.R.Germany

Quaternary stratigraphy and palaeoecology interactions between prehistoric humans and environment; pollen analysis and plant macro remains

ERNST MAIER-REIMER Max-Planck-Institut für Meterologie, Bundesstrasse 55, 2000 Hamburg 13, F.R.Germany

Ocean circulation models

MICHAEL MCELROY Department of Earth and Planetary Physics, Harvard University, 29 Oxford Street, Cambridge, MA 02138, U.S.A.

Atmospheric sciences

BARBARA MOLFINO Lamont-Doherty Geological Observatory, 205 Geoscience, Palisades, N.Y. 10964, U.S.A.

Paleoestimation of upper-ocean parameters using planktonic forams and coccolithophorida: their reliability and consistency; nutricline dynamics in the equatorial Atlantic during the past 300 Ka

HANS OESCHGER Physikalisches Institut, Universität Bern, Sidlerstrasse 5, 3012 Bern, Switzerland

Earth system processes and history, carbon cycle, stable and radioactive isotopes, environmental analytics

JONATHAN T. OVERPECK Paleoclimatology Program, NOAA/National Geophysical Data Center, 325 Broadway, Boulder, CO 80303, U.S.A.

Paleoclimatology, paleoecology, global change research

W. RICHARD PELTIER Department of Physics, University of Toronto, 60 St. George Street, Toronto, Ontario M5S 1A7, Canada

Geophysical fluid dynamics

COLIN PRENTICE Department of Ecological Botany, University of Uppsala, P.O. Box 559, 751 22 Uppsala, Sweden

Modeling vegetation dynamics and impacts of climate change on vegetation from local to global scales

EHRHARD RASCHKE GKSS Research Center, Institute of Physics, 2054 Geesthacht, F.R.Germany

Atmospheric physics: radiative energy transfer, cloud physics, climate problems, energy and water cycles

DOMINIQUE RAYNAUD Laboratoire de Glaciologie et Géophysique de l'Environment, CNRS, Domaine Universitaire BP 96, 38402 Saint-Martin-d'Heres Cedex, France

Ice-core record with special emphasis on the gas record in relation to climate and biogeochemical cycles

DAVID RIND NASA, Goddard Institute of Space Studies, 2880 Broadway, New York, N.Y 10025, U.S.A.

Modeling of climate changes, both past and future

WILLIAM F. RUDDIMAN Lamont-Doherty Geological Observatory, Route 9W, Palisades, N.Y 10913, U.S.A.

Paleoclimate: marine data, GCM studies; paleoceanography: marine data; Quaternary climates: marine data, GCM studies

BENJAMIN SANTER PCMDI, Lawrence Livermore National Laboratory, P.O. Box 808, Mail Stop L-264, Livermore, CA 94550, U.S.A.

Detection of CO_2-induced climate change; statistical analysis of climate change signals in equilibrium and transient experiments performed with coupled ocean–atmosphere GCMs

MICHAEL SARNTHEIN Geologisch-Palaeontologisches Institut, Universität Kiel, Olshausenstrasse 40, 2300 Kiel, F.R.Germany

Paleooceanography, paleoclimate

STEPHEN H. SCHNEIDER Department of Biological Sciences, Stanford University, Stanford, CA 94305, U.S.A. and National Center for Atmospheric Research, 1850 Table Mesa Dr., P.O. Box 3000, Boulder, CO 80307-3000, U.S.A.

Climatology: causes and implications of climatic change

FRITZ H. SCHWEINGRUBER Swiss Federal Institute for Forest, Snow and Landscape, Research (WSL), 8903 Birmensdorf, Switzerland

Dendrochronology

NICK J. SHACKLETON The Godwin Laboratory, Cambridge University, Free School Lane, Cambridge, CB2 3RS, U.K.

Oxygen and carbon isotope analysis as tools for reconstructing past ice volume, ocean temperatures, carbon cycle, atmospheric CO_2

ULI SIEGENTHALER Physikalisches Institut, Universität Bern, Sidlerstrasse 5, 3012 Bern, Switzerland

Global carbon cycle, paleoclimate studies

FRANK SIROCKO Lamont Doherty Earth Observatory, Palisades, NY 10964, U.S.A.

Reconstruction of the SW monsoon over the Arabian Sea by means of aerosol dust flux and coastal upwelling intensity

BERNHARD STAUFFER Physikalisches Institut, Universität Bern, Sidlerstrasse 5, 3012 Bern, Switzerland

Analyses of polar ice cores, paleoclimate

MINZE STUIVER Quaternary Isotope Laboratory AK-60, University of Washington, Seattle, WA 98195, U.S.A.

Isotope geochemistry, geochronology

THOMPSON WEBB Department of Geological Sciences, Brown University, Providence, RI 02912-1846, U.S.A.

Paleoclimatology

JAMES WHITE Institute of Arctic and Alpine Research, Campus Box 450, University of Colorado, Boulder, CO 80309, U.S.A.

Stable isotope climatology, paleoclimatology; global change

ASTRID ZWICK GSF - Institut für Ökologische Chemie, Ingolstädter Landstr. 1, 8042 Neuherberg, F.R.Germany

Reconstruction of the temperature effect on the productivity of lakes by biological and chemical investigation of bioindicators (ostracodes, pigments, diatoms) in lake sediments during the late glacial

1

Introduction

J.A. EDDY
Consortium for International Earth Science Information Network,
2250 Pierce Road, University Center, MI 48710, U.S.A.

Although the work of science is often seen in terms of observations and measurements made in real time—of watching and waiting—a necessary part of the overall endeavor is concerned with looking not foward, but backward in time. The clarity of view, as a general rule, is a function of time's distance: the farther one looks, the dimmer and less distinct the scene. What can be seen, moreover, is forever restricted to what happens to have been left behind and, further, by what one can find and read of what remains of that sometimes ephemeral residuum. What one can deduce, with certainty, is always limited, often subjective, and generalizable in more than local terms by only the foolish or the bold.

Nonetheless, it is on these slender threads that hangs almost all we know of the long-term changes that have characterized the Earth, derived from the traditional labors of historians and archaeologists in the shorter term and from geologists and paleontologists in the longer. From the former came evidence of a several hundred-year Little Ice Age that had passed its peak about the time modern science began; from the latter came the surprising finding, first found in Alpine rocks but a hundred years ago, that a major Ice Age had gripped the Earth through the time of Cro-Magnon man, covering now-industrialized parts of Europe and the Americas with year-round ice a mile deep, and finally ending about ten thousand years ago. Because of the subtle and surreptitious nature of climate change, neither event was particularly evident as a distinct feature in the lifetimes of those living then; nor were there, or are there, adequate instrumental records to document even the more recent of the two events. Climate change, however severe and perhaps even today, seems visible only in retrospect—ever hidden in the present behind the masks of greater day-to-day and month-to-month variations.

Present concerns with impending climate change have increased the demand for a more complete knowledge of what has happened on the natural Earth before. Clearer pictures of the global changes of the past are needed for several reasons: They provide the only real data for testing predictive models of the coupled climate system, and our only hope of finding the limits of natural changes and of the responses and resilience

Global Changes in the Perspective of the Past
Edited by J.A. Eddy and H. Oeschger

of the biosphere to internal and external forcing. Past changes, if we can document them globally and quantitatively, can be read as grand experiments of how the system works: of what is connected to what and how.

From past records must come the answers to the most pressing questions of the immediate environmental concerns of today: How will the Earth respond, in terms of how it has behaved in the past, to vastly increased levels of atmospheric greenhouse gases? Which came first, and by how many years: the sweeps of surface temperature that marked the onset and punctuated the end of the last Ice Age, or the swings in carbon dioxide and methane that seem to have accompanied them like a shadow through their every twist and turn? How has the biosphere responded to, and perhaps ameliorated, the major climate changes of the past? And what were the initial causes of these dramatic changes?

These questions are now repeatedly asked of science, and with some urgency. The answers lie not in the conventional province of either history or geology but in an in-between domain of study that has only recently found its tools. Its primary concern is the Quaternary Era, with the desire of temporal resolution of at least centuries, or better, decades, and ultimately years and seasons. Its findings are based chiefly on breakthroughs in modern chemical analysis. It reads not so much the rocks but the softer, natural diaries written in sediments laid down in deposited silt at the bottom of still lakes and the oceans; the snows of yesteryear that lay deep as compressed ice in the glaciers of the Arctic and Antarctic and in temperate climates at the top of the Himalayas and Andes and Alps; and the dated, annual growth-rings of long-lived trees like the bristle cone pine, wherever these still stand. The code, newly broken, that links the character of these natural records to environmental change is provided by chemistry: in the case of ice, the chemical makeup of the water and entrapped air bubbles, and for trees, of the carbon, hydrogen, and oxygen of cellulose. Pollen and the remains of simple organisms found in lake and ocean sediments are similarly analyzed as telltale clues, as are coral deposits.

They tell, in the past tense, of the temperature and chemical composition of the air and water, the makeup of forests and prairies, the levels of lakes and seas, the eruption of distant volcanoes, and the surface conditions on a star, the Sun, that was then, as now, 93 million miles away. The greatest promise is through the combination of data from all of these to compile, for a given area or interval of time, quantitative pictures of the environment through past periods of significant global change.

It was toward this end that a Dahlem Workshop on *Global Changes in the Perspective of the Past* was convened in Berlin in December, 1991. The workshop brought together an international group of historians, glaciologists, dendrochronologists, limnologists, modelers, paleobotanists, geologists, geochemists, oceanographers, atmospheric scientists, and solar astronomers. The overall goal of the workshop was to discuss the use of paleoclimatic data for analyzing climatic processes and for modeling the future evolution of climate. More specific aims can be found in the questions used to structure the four discussion groups which were:

1. Are paleodata good analogs for future global changes?
2. What and where are the gaps in evaluating strategies for reconstructing global changes?
3. How can we use paleodata for evaluating the internal variability and feedbacks in the climate system?
4. How can we use paleodata for evaluating the forcing of past and future climates?

This workshop was a direct outgrowth of another Dahlem Workshop held in 1988 on *The Environmental Record in Glaciers and Ice Sheets* (Oeschger and Langway 1989). Hans Oeschger and I organized this meeting with the help of a program advisory committee comprised of Jean-Claude Duplessy, Burkhardt Frenzel, John Imbrie, and Jean Jouzel.

In the spirit of Dahlem Workshops, we hope this volume will illuminate, for a broader group of scientists, some of the burning questions yet unanswered in the field of paleostudies and, through this means, accelerate the pace of progress in this important field.

We are grateful to all those who gave their knowledge and time to both the workshop and collected papers of this volume. We particularly thank Julia Lupp, editor of Dahlem Konferenzen, as well as the support staff, Jennifer Altman, acting director, and the sponsors of the workshop, Stiftungsfond Unilever, for making this Dahlem Workshop possible.

REFERENCES

Oeschger, H. and C.C. Langway, Jr. 1989. The Environmental Record in Glaciers and Ice Sheets. Dahlem Workshop Report PC 8. Chichester: John Wiley.

2

Comparison of Interglacial Climates Regarding Space and Character

B. FRENZEL
Botanisches Institut der Universität Hohenheim,
7000 Stuttgart 70, F.R. Germany

ABSTRACT

In this chapter I will stress that while paleoclimatology probably cannot help to predict the future development of climate, it can enable us to understand the regional pattern of the climatic situation to be obtained, if the amount of future temperature increase is given. Further I emphasize that all the better known interglacials of the Quaternary were characterized by much warmer climates on the continents of the Northern Hemisphere than today. This was, in general, paralleled by an overall increase in moisture available even during the thermal optima of the interglacials considered, and holds true for today's moist and dry regions.

Although the tendency of climatic change remained roughly the same during all the interglacials, it seems likely that climate was less variable in the high and low northern latitudes than in the mid latitudes. This also holds true for all continents of the Northern Hemisphere.

In opposition to a widely held view, I suggest that the climatic history of Siberia, at least during the Upper Pleistocene, did not differ from that of other regions of the globe. Moreover, it appears that rapid changes in climate are not characteristics of possible anthropogenic changes of climate, but of the climate system itself.

INTRODUCTION

This chapter aims at contributing to the discussion of whether paleodata are good analogs for future global changes. However, several questions of general importance must be answered before interglacial climates can be discussed within the general framework of this workshop:

1. How sound is our knowledge about the quaternary stratigraphy of various regions, in view of the interglacials to be studied?

Global Changes in the Perspective of the Past
Edited by J.A. Eddy and H. Oeschger

2. How reliably are the driving factors of major changes of climate understood, and how accurately are their reactions in the Earth's system dated?
3. How fast did the climate change during the Quaternary?
4. Can the climate-controlling influences of past changes in the atmosphere's CO_2 and CH_4 contents be proven?
5. Which climate-relevant by-products, besides changes in the global mean temperatures, might have been caused by changes in the atmosphere's CO_2 contents?

Each of these problems merits a workshop of its own. Therefore, I will only discuss them briefly as they pertain to the broader topic of the importance of past climatic changes for forecasting the future.

MAIN BACKGROUND PROBLEMS

The Stratigraphical Problem

For about 2.46 million years, the Quaternary experienced several major and minor climatic changes (e.g., Hooghiemstra 1984; Dansgaard and Tauber 1969; Kukla 1987). Within this period, approximately 23 major cold and 23 warm climate periods may have occurred. This seems to be an easily accepted principle. Yet the amplitude, frequency, and timing of these major climatic changes are deduced, most of all, from fossil material only. Thus a high resolution in time and the *in situ* preservation of the fossil material are indispensable prerequisites for each paleoclimatological research project. These requirements may be fulfilled appreciably well on the deep seafloor, but not well on the continents. Here, in regard to some interglacials, the situation is even worsened by the fact that they were soon followed by very important interstadials, which may have resembled the preceding interglacials in their vegetation history. Thus traces of these interstadials may even merge with or even be confounded by those of the preceding interglacials (Frenzel 1989a). Good examples of this are the Eemian (last) interglacial and the following "interstadial" phases, St. Germain I and II, in the southwest of Central Europe, in the soil complexes s_1 to s_5 of the Chinese loesses (Kukla 1987), or in the fossil soil complexes PK II to PK V of the European loesses (Ložek 1971). However, to discuss interglacial climate history, one has to define first those times which are of interest.

Although, at present, several long continental pollen sequences are available in various regions of the world, the best coverage of these sequences, in terms of space and time, is found in Eurasia alone. Several of these interglacial sequences are botanically well defined. Yet this only holds true for Europe north of the Pyrenees. Further to the south, the warm climate organisms always lived close to their glacial refuge areas (Follieri 1962; Frenzel 1968, 1990). This caused no clear-cut sequences in their interglacial migration history. Approximately the same is true for Siberia

(Frenzel 1968). Here, the Quaternary biota were always extremely resistant to cold and drought, and several minor glacial refuge areas occurred. Due to these facts, the various interglacials of the regions mentioned are very difficult to discern on the basis of their vegetation history alone. In view of these obstacles, it seems best to try to correlate deep-sea sequences with those of the continents as precisely and as often as possible, provided there exist unequivocal marker horizons in deep-sea sediments and on the adjacent continent (e.g., Turon 1984; Heusser and Shackleton 1979). Presumably this correlation can, in general, be only done by paleomagnetics and by certain tephra layers. Much work is needed.

The Firing Problem of Incoming Solar Radiation Versus Sensible Heat

A wealth of papers have dealt with the problem of changing orbital parameters as driving factors in climate history. It is a well-known fact that this influence cannot be detected equally well in all parts of the Quaternary (Imbrie and Imbrie 1980). With respect to the interglacials, the radiation maximum seems always to have happened on the Northern Hemisphere at the very beginning of these phases, with a remarkable decrease in the incoming radiation during most of the ensuing phases of interglacial climate. This was no doubt counteracted by the final melting-off of the glacial inland ice masses, for which a huge amount of energy was needed. Thus, phases of interglacial climates in the middle and higher latitudes of the Northern Hemisphere tend to coincide with phases of an already reduced input of solar radiation. These counteracting processes can only be dated, more or less accurately, within the Holocene (^{14}C dating, dendrochronology, annualy layered lake sediments). Regarding the more remote interglacials, these two types of processes may be studied by an analysis of stable isotopes in the inland ice or in the benthic foraminifera, but with less accuracy in dating. Therefore, this important problem can only be investigated reliably for the Holocene, although the problem itself is of principal importance.

The Problem of Rapid Climatic Change and Changes in the Amount of Climate-controlling Trace Gases

It is generally believed that anthropogenic changes in climate act much faster than major natural ones ever have. Yet, due to deep-sea sediments, dendroclimatology, or in annually layered lake sediments, it can be shown that the transitions from glacials to interglacials happened only within some centuries (Sarnthein and Tiedemann 1990; Duplessy and Shackleton 1985; Becker et al. 1991; Labeyrie et al. 1987; Zolitschka 1990). On the other hand, the interglacial to glacial transition may have lasted for a much longer time, i.e., about 3 to 4 millennia (Frenzel and Bludau 1987).

During the Holocene, very rapid changes of climate occurred. According to dendroclimatology, they often lasted about 20 to 30 years, or even as brief as 2 to 3 years (Frenzel et al. 1989). Thus, even past climate changes may have happened very fast. The most extreme examples can easily be correlated with strong changes in the atmosphere's CO_2 and CH_4 contents. Yet this does not hold true for minor changes of climate. The suggested strong changes of climate that took place during the Eemian and the Holsteinian interglacials (deep-sea stages 5e and 11) deserve mention here. According to pollen evidence (Müller 1979), strong cold spells influenced the European climate, causing changes from mild interglacial to subarctic or high boreal climates and back again during some decennia. I believe that these changes were artifacts caused by a slumping in lake sediments. On the other hand, if they indeed existed, how did the atmosphere's CO_2 and CH_4 contents react? Up until now, this question could not be answered by information from Antarctic ice cores. During the Preboreal there existed a strong oscillation of the atmosphere's CO_2 contents (by about 30‰; Stauffer, pers. comm.), which does not clearly correspond to the global climate history known today.

It is therefore of utmost importance to acquire well-dated information about climatic changes and about changes in the contents of climate-controlling trace gases at an extremely high resolution in time.

How Strong Are the Ties between Changing Trace-gas Concentrations and Changes of Climate?

Although a strong parallelism exists between past major changes of climate and the contents of climate-relevant trace gases, we do not exactly understand the connections between these processes. For example, the Vostok ice cores at the very end of deep-sea stage 5e exhibit an increase in the CO_2 contents that is neither paralleled by an increase in CH_4 contents nor by an increase in global air temperatures. This CO_2 oscillation, therefore, does not seem to have influenced the global climate remarkably. Perhaps it was caused by an active humus turnover during those strong solifluction processes, which began at that time in vast areas of the Northern Hemisphere (Frenzel 1968).

Another example is the Holocene Preboreal, which occurred at about 9,000 ^{14}C B.P. At that time, the Central European and north Siberian climates experienced a pronounced cold spell (see Frenzel 1990 for a summary of the literature). Contemporaneous changes of climate have not yet been demonstrated anywhere else in the Northern Hemisphere. Thus, it seems that this climatic change was only of regional importance. The problem of whether the synchroneously occurring oscillation in the atmosphere's CO_2 contents was the cause or the consequence of this climatic change, however, remains unsolved.

Conclusions

Although a strong parallelism between changes in global climate and those of climate-relevant trace gases (e.g., CO_2 and CH_4) is evident, a linear dependence between these parameters does not exist according to our present-day knowledge. Nevertheless, the problem of strong- and rapid-acting changes of trace gases in the atmosphere and climate must be investigated thoroughly, above all in view of a possible use of paleoclimatology in forecasting climatic development. In particular, three problems should be intensively investigated:

1. The increasing contents of climate-relevant trace gases might eventually counteract the decrease in solar radiation incidence to the globe. With respect to this, it should be noted that since about 4,000–5,000 B.P. (^{14}C dating), the polar tree limit has slowly retreated to the south while the frequency of glacier advances or retreats and of phases of enhanced solifluction processes (Mathews et al. 1993) has increased. This points to a higher climatic variability during the last 5,000 years.
2. It might be argued that both the final melting-off of the late Pleistocene inland ice masses and the increasing contents of climate-relevant trace gases may lead to an increase in warmth available on the Earth's surface. Yet the combined CO_2 and CH_4 increase is a global phenomenon; the melting-off of the last inland ice masses was only a regional phenomenon, with presumably other consequences on the global atmospheric circulation pattern. Nevertheless, both processes should be investigated for their mutual dependence and interactions.
3. An increase in the CO_2 contents, even within the framework of the expected results, means a stimulation of photosynthesis and bioproduction (Allen 1990; Bazzaz 1990) and a gentle reduction in the degree of the stomate opening. Both effects might, directly or indirectly, improve the continental water budget, most of all in the southern temperate or even within the seasonally dry regions of the globe. The interglacial Mediterranean moist forest of Colchidean type (Follieri 1979) are good examples of this. For the boreal coniferous forest, an increase in bioproduction may mean an increase in the thickness of the humus layer. This in turn would influence the heat and moisture fluxes. Depending on the local situation, this could cause an increase of paludification or of seasonally or even perennially frozen ground. These possible changes in the ecological situation, due to an increase in CO_2, should be investigated intensively.

Thus, one has the impression that important consequences of an anthropogenic increase in the concentration of climate-relevant trace gases are not well understood, although they might form a new element in climate.

Based on what has been discussed up until now, it follows that paleoclimatology cannot presumably be used directly as a firm basis for forecasting the future climate. On the other hand, paleoclimatology can help us understand general trends of the global climate system as well as characteristics and thresholds of changes in the geosphere and biosphere. Moreover, a sound reconstruction of past climates favors the improvement of climate models and should enable a more reliable differentiation between spontaneously occurring and human-made changes of climate to be done, focusing on the last 7,000 years, or so. Much more research is needed, most of all concerning this last problem.

COMPARISON OF INTERGLACIAL CLIMATES

In view of the difficulties mentioned, a study regarding interglacial climates should focus on the following problems:

— the definition of common trends in interglacial climates on the Northern Hemisphere,
— the study of general or regional differences in climate history of various interglacials,
— the search for general principles in differences between the interglacial climates of various regions on the Northern Hemisphere.

Common Trends in the Development of Interglacial Climates in the Northern Hemisphere

It is evident that at the very beginning of the interglacial climates, two types of climatic changes occurred: (a) a continuous increase in temperature and moisture, e.g., the beginning of the Holsteinian interglacial (Frenzel 1967; Sarnthein and Tiedemann 1990), and (b) a discontinuous oscillating increase in temperature and moisture, as was evidenced during the younger Quaternary interglacials (Sarnthein and Thiedemann 1990; Frenzel 1967, 1973, 1982). At any rate, it seems that the rise of the sea level occurred very fast during the onsets of the Eemian interglacial and the Holocene (Streif 1991 and references therein).

During the interglacials, the higher latitudes experienced a much stronger warming in comparison with the present-day situation (by about 3° to 6°C) than did the middle and lower latitudes (by about 2° to 3°C). Soviet scientists even stress that during interglacial times, the mid latitudes may have experienced a somewhat cooler climate than at the present (Frenzel et al. 1991). However, to me it seems that this deviation in reconstructed temperatures was caused by the divergent character of vegetation, most of all in the Mediterranean area, not by climate itself.

During all interglacials, the water supply in the Northern Hemisphere was much better than it is today. The only exception seems to have been central North America, where the moisture supply seems to have been less during the Holocene and the Sangamonian than today (Frenzel et al. 1992). Evidently the better water supply within the interglacial mid latitudes caused a remarkable transgression of the Caspian Sea, at least during the last interglacial and during some phases of the Holocene (e.g., Fedorov 1978; Varushchenko et al. 1980).

Differences between Climates of Various Interglacials

The differences in the development of the early Holsteinian interglacial climate in comparison with those of the interglacials of the younger Pleistocene have already been stressed. Moreover, in Eurasia the entire Holsteinian interglacial was characterized by extremely moist climates. This even holds true for Central Asia and the loess regions of northern China (Lazarenko 1980; Lazarenko et al. 1980; Dodonov 1988; Kukla 1987; Liu Tungsheng and Yuan Baoyin 1987).

Moreover, during the Holsteinian the relative temperature increase, in comparison to today's conditions, was much stronger all over Europe and western Siberia than during any other interglacial period (Frenzel 1967).

It seems that the general tendency of the water supply in the Great Plains and Central Asia differed from each other during the last interglacial and the Holocene climatic optimum, respectively: the climate in the Great Plains was drier than it is today, whereas it was, at the same time, more moist in Central Asia. In contrast, during the mid-Holocene, the climate was drier in Central Asia and in the Great Plains than it is today. However, Soviet reconstructions of mid-Holocene climates of Central Asia indicate moister conditions than today. What has caused these discrepancies in the reconstructed water budgets is unknown (Frenzel et al. 1991).

Another point of interest is the medieval European climatic optimum (A.D. 1080 to approx. 1360). At that time, in contrast to former warm climate periods, temperatures seem to have been generally higher by about 0.5°C to 1.5°C all over Europe, including the area northwest of Siberia, without any regional differences. This coincided with an amelioration of the water supply nearly everywhere, with the exception of the northern part of the Lake Caspian basin (Varushchenko et al. 1980).

General Tendencies in the Differences between Various Interglacial Climates

Irrespective of the approaches used to reconstruct former interglacial climates, it is quite obvious that the mid latitudes of the Northern Hemisphere were those regions in which more remarkable differences in climate existed from interglacial to interglacial than further to the north or to the south. On the other hand, interglacial climates of the higher and lower latitudes strongly resembled one another, although quantitative differences existed among them. We should question whether or not these differences were caused by tectonic movements. It seems more likely to me that the mid latitudes simply experienced a much stronger interglacial climatic variability than those further to the north or south. If the end of the Pliocene is included (Frenzel 1967, 1968), this tendency is even more pronounced.

A Special Siberian Young Pleistocene Interglacial?

As has already been stressed, two strong interglacials (St. Germain I and II) occurred during the first part of the last European glaciation. At that time, the vegetation of central and southeastern France strongly resembled that of today (de Beaulieu and Reille 1984; Woillard 1978). These phases are hypothetically included in the deep-sea stages 5a and 5c. The general view (Behre 1989, 1992) is that these warm climate phases were the equivalent of the Brørup and the Odderade interstadials, respectively, which were characterized by typical boreal forests in the northern part of central Europe. I do not share this view since this would mean an extremely steepclimatological gradient from eastern France to southern and eastern Germany or to northern Switzerland. Thus I would not be surprised if it should definitely turn out that the St. Germain I and II interstadials were not the homologue of the Brørup and Odderade interstadials (Frenzel 1988).

Soviet scientists, on the other hand, have stressed that in Siberia, the first part of the last glaciation was interrupted by an extremely warm and moist interglacial, during which the polar timberline lay beyond its natural position of today (Kind 1974; Kind and Leonov 1982; Alekseev 1978). This so-called Karga interglacial is believed to have lasted from approximately 60,000 or 55,000 B.P. to approximately 30,000 or 25,000 B.P. In other regions of the world, no indications of a comparable interglacial are known. If the Soviet view is correct, this would mean a remarkable independence of the Siberian climate system, at least during the younger Pleistocene. A wealth of pollen analytical and mollusc data from the Arctic Ocean seems to favor the Soviet view. Consequently, it is believed that the Arctic Ocean of that time invaded the lowlands of north Siberia.

I doubt the correctness of this interpretation (see Frenzel 1991 and references therein). Most of the ^{14}C data used were either indefinite or they became older and older, if repeated. Also, the marine mollusc fauna in the Arctic Ocean during the Karga interglacial strongly resembled that of the classical Kazantsevo interglacial, which is the equivalent of the Eemian interglacial. Therefore I believe that most sites of the Karga interglacial date from Kazantsevo times, and it is possible that traces of the equivalents of St. Germain I and II or the St. Pierre interstadials were included into the so-called Karga interglacial. If this assumption is correct, it would mean that the St. Germain I and II interstadials are of much more general importance than originally suggested and would necessitate a reinterpretation of some old paleobotanical and paleoecological findings (Frenzel 1988).

CONCLUDING REMARKS

In most regions of the Northern Hemisphere, all the interglacials mentioned experienced a strong expansion of the forest belt onto the present-day tundra zone as well as on the recent steppe regions, although during their optima, climate was much warmer than it is today. Evidently it was also much moister. From interglacial to interglacial, the structure of the forest types changed. The general distribution patterns of the plant taxa involved, however, remained broadly the same in the middle and higher latitudes of the Northern Hemisphere. Thus, warm interglacial climates of these types, with a velocity of climatic change resembling that discussed here, are of no harm to the biosphere. Most of the currently extinct taxa died out during cold climate phases or at the transition from the last glacial to the Holocene, but not during the interglacials (see Frenzel 1989b and references therein).

However, we still do not know how much additional stress on the geosphere and biosphere has been caused by the onset of agriculture throughout the world. Much more research is needed to determine this.

LIST OF OPEN QUESTIONS OR UNSOLVED PROBLEMS

1. Up to now, most of the deep-sea stages ($\delta^{18}O$ stages) have not been definitely correlated with events on the continents, although this exact correlation is of crucial importance for further palaeoclimatological research. This correlation could be best done by tephra layers, but where are the best regions to look for stratigraphically well-situated tephra layers?
2. An exact measuring of time needs to be done to facilitate the study of past climatic changes. The further back into the past, the worse the general situation of radiometric or other physical time-measuring methods becomes. Therefore, dendrochronology and the study of annually layered lake sediments seems to be the only solution. Where on the Earth are fossil tree trunks available that cover a longer span of time than the Holocene? Only in Siberia? Could, e.g., layers of corals also be used for this type of exact chronology? Is the reliability of this method using cross correlations between corals of different ages already proven? Are annually layered deep-sea sediments available and can they be used for this purpose?
3. Major changes of climate and of changes in the atmosphere's contents of CO_2 and CH_4 seem to parallel each other broadly. Which efforts should be done to solve the important problem of interdependence or of lead and lag, respectively, between climate and the changing contents of climatically effective trace gases? Where on the Earth, apart from the huge ice masses of Greenland and of Antarctica, are cold glaciers in mid latitudes available to study this problem? Only in the Dunde ice cap?
4. It can be shown that the character of interglacial climates on the Northern Hemisphere resembled each other in the high as well as in the low latitudes. Yet within the mid latitudes, the interglacial climates are characterized by a remarkable variability from interglacial to interglacial. How may this phenomenon be explained? Is it possible that, in general, climate in the mid latitudes of the Northern Hemisphere is much more unstable than that of the other latitudes mentioned or is it much more sensitive to changing boundary conditions?
5. How might the differences in climatic change at the very beginning of the Holsteinian interglacial (deep-sea stage 11) be explained in comparison with other interglacials?
6. How can the strange climatic situation of the Holsteinian interglacial be explained? At first it was looked upon as the "big interglacial" although it was only a short one (approx. 15,000 years; Penck and Brücker 1909). Later it was held to have been a cold interglacial (Moskvitin 1960a,b) although it was an extremely moist and warm one. Is it necessary to think about epirogenetic movements?
7. How strong was the interglacial deglaciation in Greenland and in Antarctica?

8. The last interglacial inland ice masses seem to have already accumulated, when on the European continent optimal climatic conditions still existed. Is this a general tendency of the late Pleistocene interglacial climates and, if so, which type of general atmospheric circulation was responsible for this?

REFERENCES

Alekseev, M.N. 1978. Anthropogene of East Asia. Stratigraphy and correlation (in Russian with an English summary). Moscow: Nauka.

Allen, L.H. 1990. Plant responses to rising carbon dioxide and potential interactions with air pollutants. *J. Environ. Qual.* **19:**15–34.

Bazzaz, F.A. 1990. The response of natural ecosystems to the rising global CO_2 levels. *Ann. Rev. Ecol. Syst.* **21:**167–196.

Beaulieu, J.-L. de, and M. Reille, 1984. A long Upper Pleistocene pollen record from Les Echets, near Lyon, France. *Boreas* **13:**111–132.

Becker, B., B. Kromer, and P. Trimborn. 1991. A stable-isotope tree ring timescale of the Late Glacial/Holocene boundary. *Nature* **353**:647–649.

Behre, K.-E. 1989. Biostratigraphy of the last glacial period in Europe. *Quat. Sci. Rev.* **8:**25–44.

Behre, K.-E., and J. van der Pflicht. 1992. Towards an absolute chronology for the last glacial period in Eruope: Radiocarbon dates from Oerel, northern Germany. *Veg. Hist. Archaeobot.* **1:**111–117.

Dansgaard, W., and H. Tauber. 1969. Glacier oxygen-18 content and pleistocene ocean temperatures. *Science* **166:**499–502.

Dodonov, A.E. 1988. Voprosy regional 'noj i mezhregional 'noj korreljacii verchnepliocen-chetvertichnykh otlozhenij juzhnogo Tadzhikistana. *Bjull. Kom. po izuch. chetvert. perioda* **57**:18–27.

Duplessy, J.-C. and N.I. Shackleton. 1985. Response of global deep-water circulation to Earth's climatic change 135,000–107,000 years ago. *Nature* **316**:500–507.

Fedorov, P.V. 1978. Pleistocen Ponto-Kaspiya. *Trudy Geol. In-ta Akad. Nauk SSSR* **310**:116.

Follieri, M. 1962. La foresta colchica fossile di Riano Romano. Il Analisi pollinichi. *Ann. di Bot.* **27**:245–280.

Follieri, M. 1979. Late pleistocene floristic evolution near Rome. *Pollen et Spores* **21:**135–148.

Frenzel, B. 1967. Die Klimaschwankungen des Eiszeitalters. Braunschweig: Vieweg, 296 pp.

Frenzel, B. 1968. Grundzüge der pleistozänen Vegetationsgeschichte Nord-Eurasiens. Erdwissenschaftliche Forschung, vol. 1. Wiesbaden: Steiner, 326 pp.

Frenzel, B. 1973. Climatic fluctuations of the Ice Age. Cleveland: Case Western Reserve Univ. Press.

Frenzel, B. 1982. The history of flora and vegetation during the Quaternary. *Prog. Bot.* **44**:406–417.

Frenzel, B. 1988. Forschungen zur Geographie und Geschichte des Eiszeitalters (Pleistozän) und der Nacheiszeit (Holozän). In: Jahrbuch der Akademie der Wissenschaften und der Literatur zu Mainz, pp. 171–177. Stuttgart: Franz Steiner.

Frenzel, B. 1989a. The history of flora and vegetation during the Quaternary. *Prog. Bot.* **50**:327–340.

Frenzel, B. 1989b. Theoretische Grundprobleme der botanischen Biostratigraphie des Eiszeitalters. In: Quaternary Type Sections: Imaginations or Reality, ed. R. Schlüchter et al., pp. 33–39. Rotterdam: Balkema.

Frenzel, B. 1990. Die Vegetationsentwicklung im Eiszeitalter. In: Eiszeitforschung, ed. H. Liedtke, pp. 69–90. Darmstadt: Wiss. Buchges.

Frenzel, B. 1991. The history of flora and vegetation during the Quaternary. *Prog. Botany* **52**:359–380.

Frenzel, B. and W. Bludau. 1987. On the duration of the interglacial to glacial transition at the end of the Eemian interglacial (deep-sea stage 5e): Botanical and sedimentological evidence. In: Abrupt Climatic Change. Evidence and Implications, ed. W.H. Berger and L.D. Labeyrie, pp. 151–162. Dordrecht: Reidel.

Frenzel, B., G. Furrer, and W. Lauer. 1989. Zur Paläoklimatologie der letzten 2,500 Jahre in Mitteleuropa. In: Akademie der Wissenschaften und der Literatur, 1949–1989, pp. 143–182. Stuttgart: Franz Steiner.

Frenzel, B., M. Pécsi, and A.A. Velichko. 1991. Atlas of paleoclimates and paleoenvironments of the Northern Hemisphere. Late Pleistocene-Holocene. Stuttgart: Gustav Fischer.

Heusser, L.E., and N.J. Shackleton. 1979. Direct marine-continental correlation: 150,000-year oxygen and isotope-pollen record from the North Pacific. *Science* **204:**837–839.

Hooghiemstra, H. 1984. Vegetational and climatic history of the high plain of Bogota, Colombia: A continuous record of the last 3.5 million years. *Diss. Botanicae*, vol. 79. Vaduz: Cramer, 368pp.

Imbrie, J., and I.Z. Imbrie. 1980. Modeling the Climatic Response to Orbital Variations. *Science* **207**:943–953.

Kind, N.V. 1974. Geokhronologiya pozdnego antropogena po izotopnym dannym. *Trudy Geol. In-ta Akad. Nauk SSSR* **257**:255–257.

Kind, N.V. and B.N. Leonov. 1982. Antropogen Tajmyra. Moskva: Naukal, 184 pp.

Kukla, G. 1987. Loess stratigraphy in Central China. *Quat. Sci. Rev.* **6**:191–219.

Labeyrie, L.D., J.-C. Duplessy, and P.L. Blanc. 1987. Variations in mode of formation and temperature of oceanic deep waters over the past 125,000 years. *Nature* **327**:477–482.

Lazarenko, A.A. 1980. Pogrebjonnye pochvy ljossovoj formacii Srednej Azii i ikh paleogeograficheskoe znachenie. *Doklady Akad. Nauk SSSR* **252**:181–185.

Lazarenko, A.A., N.S. Bolikhovskaja, and V.V. Semenov. 1980. Opyt drobnogo stratigraficheskogo raschlenenija ljossovoj formacii Pritashkentskogo rajona. *Izv. Akad. Nauk SSSR, ser. geol.* **5**:52–66.

Liu Tungsheng and Yuan Baoyin. 1987. Paleoclimatic cycles in northern China: Luochuan loess section and its environmental implications. In: Aspects of Loess Research, ed. Liu Tungsheng, pp. 3–26. Beijing: China Ocean Press.

Ložek, V. 1971. Problem der Zahl der quartären Klimaschwankungen. *Quartär* **22**:11–16.

Mathews, J., B. Frenzel, and B. Gläser, eds. 1993. Solifluction and climatic variation in the Holocene. *Paläoklimafor.* **11**, in press.

Moskvitin, A.J. 1960a. Paleogeografija jugo-vostoka Evropy v plejstocene (with English summary). *Dokl. sov. geol. Meždunarodn. Geol. Kongr.* **21. sessija:**41–47.

Moskvitin, A.J. 1960b. Über wärmere und Kühlere Interglaziale in der UdSSR. *Ber. der geolog. Ges. in der DDR. Gesamtgeb. Geolwiss.* **5:**5–20.

Müller, H. 1979. Climatic changes during the last three interglacials. In: Man's Impact on Climate Developments in Atmospheric Scinece. Proc. Int. Conf. in Berlin, ed. Bach, Pankrath, and Kellog, vol. 10. New York.

Penck, A. and E. Brucker. 1909. Die Alpen im Eiszeitalter. 1. Band: Die Eiszeit in den nördl. Ostalpen. Band 2: Die Eiszeit in den nördl. Westalpen. Leipzig: Tauchnitz.

Sarnthein, M. and R. Tiedemann. 1990. Younger Dryas-style cooling events at glacial terminations I–VI at ODP site 658: Associated benthic $\delta^{13}C$ anomalies constrain meltwater hypothesis. *Palaeooceanog.* **56**:1041–1055.

Streif, H.-J. 1991. Zum Ausmaß und Ablauf eustatischer Meeresspiegelschwankungen im südlichen Nordseegebiet seit Beginn des letzten Interglazials. In: Klimageschichtliche Probleme der letzten 130,000 Jahre, ed. B. Frenzel, Paläoklimaforschung, vol. 1, pp. 231–249. Stuttgart: Gustav Fischer.

Turon, J.-L. 1984. Direct land/sea correlations in the last interglacial complex. *Nature* **309:**673–676.

Varushchenko, A.N., S.I. Varushchenko, and R.K. Klige. 1980. Izmenenie urovnja Kaspijskogo morja v pozdnem plejstocene-golocene. In: Kolebanija uvlazhnennosti Aralo-Kaspijskogo regiona v golocene (Oscillations of moisture in the Aral-Kaspi regions during the Holocene), ed. B.V. Andrianov, L.V. Zorin, and R.V. Nikolaeva, pp. 79–90. Moskva: Nauka.

Woillard, G.M. 1978. Grande Pile peat bog: A continuous pollen record for the last 140,000 years. *Quat. Res.* **9:**1–21.

Zolitschka, B. 1990. Spätquartäre jahreszeitlich geschichtete Seesedimente ausgewählter Eifelmaare. Paläolimnologische Untersuchungen als Beitrag zur spät- und postglazialen Klima- und Besiedelungsgeschichte. *Documenta naturae* **60**:1–226.

3

Use and Misuse of the Geologic "Analogs" Concept

T.J. CROWLEY
Applied Research Corporation, 305 Arguello Drive,
College Station, TX 77840, U.S.A.

ABSTRACT

The concept of a geologic analog for a future greenhouse warming needs clarification. On one hand, it is unlikely that any past time period would be a satisfactory analog: the rate of change in the future will be unprecedentedly rapid, past tectonic changes would have resulted in regional climates different from the present, and the nature of past forcing was not always the same. Nevertheless, the geologic record can provide valuable information about processes operating within the climate system. These include the general correspondence of warm periods with times of high atmospheric CO_2 levels, sudden variations of North Atlantic Deep Water production, sea-level changes potentially associated with warmer climates, and the possibility of a climate-induced extinction event. If geologic studies are to enhance our understanding of future climate change, we should abandon the "time-slice analog" concept and focus on the "process analog" approach.

INTRODUCTION

Future variations in temperature due to greenhouse-gas increases will probably result in a climate change large even on geologic scales and rapid beyond precedent. A schematic comparison (Figure 3.1) of the magnitude and rate of future global temperature increase suggests that, within the next two centuries, temperatures may reach their highest level since the thermal maximum of the mid-Cretaceous, 100 million years ago (100 Ma).

Some time ago, in the face of such potential changes, a search began for possible geologic analogs for a future warming, i.e., past warm periods that can show us what the future will be like. Although the geologic record can provide a valuable perspective for future climate change, I believe that the reasoning underlying this search is flawed (Crowley 1990). This chapter will summarize what I consider to be both invalid and valid approaches to the use of the past as a key to the future.

Global Changes in the Perspective of the Past
Edited by J.A. Eddy and H. Oeschger

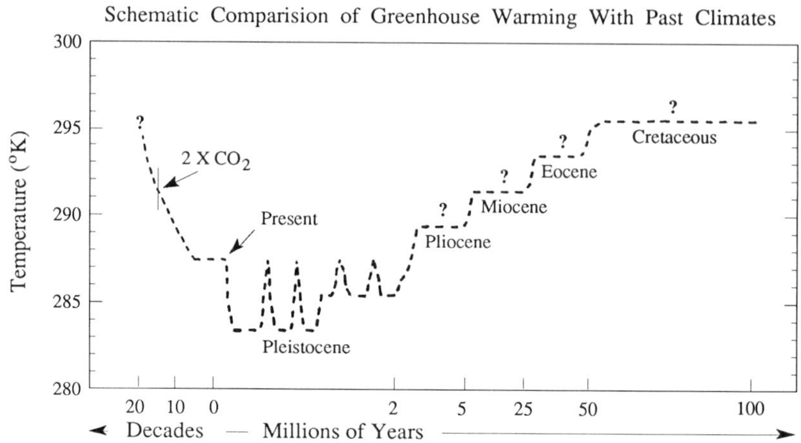

Figure 3.1 Schematic comparison of future greenhouse warming with past changes in temperature. Future changes are based on results reported in Schlesinger (1989). Note that pre-Pleistocene changes are not well fixed in absolute magnitude, but the relative warmth of the intervals is approximately correct. After Crowley (1990).

MISUSE OF THE ANALOG CONCEPT: "TIME-SLICE ANALOGS"

A number of authors have cited various past warm periods as possible analogs for the future (Kellogg 1977; Owen and Rea 1985; Budyko and Izrael 1987; Zubakov and Borzenkova 1988). The most commonly cited intervals are the early Holocene (6000–9000 B.P.) and the last interglacial (120,000 B.P.), although some claims have also been made for the Pliocene (3–5 Ma), the early Eocene (55 Ma), and the mid-Cretaceous (100 Ma). There are serious problems in utilizing any of these time periods as analogs (Crowley 1990; cf. Webb and Wigley 1985). The Quaternary warm intervals probably represented only seasonal changes or mean annual changes that were not globally synchronous. Seasonal changes on land probably reflected orbital insolation differences during these time periods (e.g., Prell and Kutzbach 1987; COHMAP 1988). Mean annual changes occurred in some areas; however, in some cases these were not synchronous globally (Figure 3.2). For example, Figure 3.3 illustrates sea-surface temperatures (SSTs) during the last interglacial. For the sites illustrated, the average difference in SSTs (present minus 120,000 B.P.) is -0.05 ± 2.06°C ($n = 52$).

Pre-Pleistocene warm intervals probably had mean annual changes in global temperatures; however, for two reasons, the use of these time periods as analogs is unjustifiable: First, the solid boundary conditions of the Earth (mountains, ocean gateways, continental positions) were sometimes very different from those at the present (Figure 3.4). These differences might have more effect on regional and

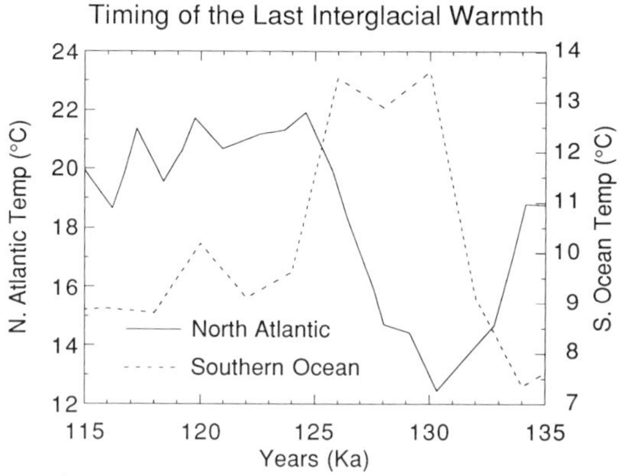

Figure 3.2 Comparison of the timing of peak warmth during the last interglacial in the North Atlantic and Southern Hemisphere. North Atlantic record from Ruddiman and McIntyre (1981); Southern Hemisphere record from Martinson et al. (1987). Peak warmth does not occur synchronously in the two cores, a response that may reflect redistributions of heat by the ocean–atmosphere system. Note that this separation in timing may not be as dramatic in all records. From Crowley (1990), reprinted from *J. Climate* with permission of the American Meteorological Society.

seasonal variations than on the global mean; however, they are still important. Second, the future rate of CO_2 (and temperature?) change will be very rapid, whereas past warm periods presumably represented an equilibrium response to changes in boundary conditions at a slower rate than will occur in the future.

The argument in favor of a pre-Pleistocene analog would be strengthened if future changes resulting from a transient CO_2 increase resemble those of an equilibrium response. Some similarities between the transient and equilibrium states of the ocean have been observed in one general circulation model (GCM) run (Mikolajewicz et al. 1990; Cubasch et al. 1991). However, it is too early to assume that all aspects of transient and equilibrium runs would be the same. Furthermore, since potential pre-Pleistocene cases have different geographic configurations, the breakdown of the analog concept still holds.

USE OF THE ANALOG CONCEPT: "PROCESS ANALOGS"

If we shift our perspective from a narrow "time-slice analog" to that of a "process analog," then the geologic record can provide valuable information about the future. By process analogs I mean either the components or processes operating in a climate system. The study of a particular process may enable us to assess the believability of the same process as it is forecast to change in the future. As an example, the section below lists four areas where the study of process analogs is enlightening.

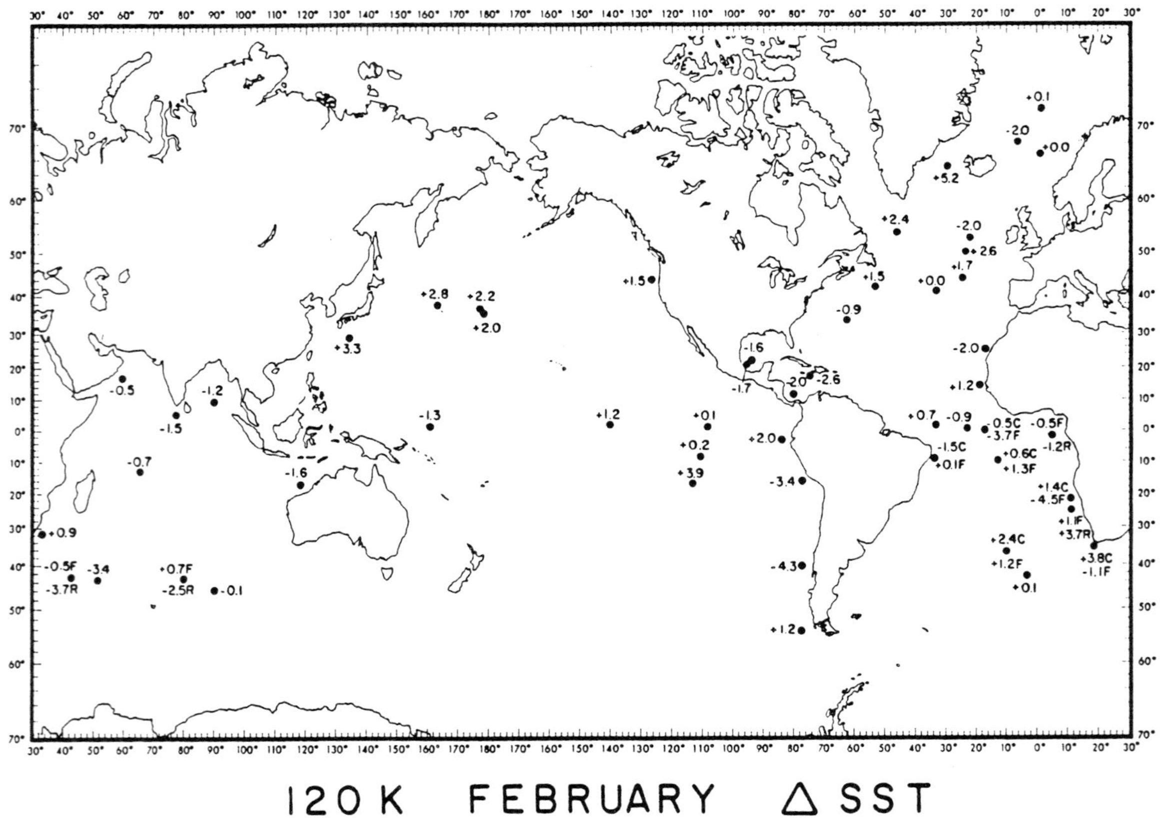

Figure 3.3 Difference between modern sea-surface temperature (SST) and estimated February SST (in °C) at the last interglaciation 120,000 years ago. Negative values mean that the last interglacial ocean was colder than today. Most SST values are similar to present; any spatially coherent increases in one region are more or less offset by equivalent decreases in other areas. From CLIMAP (1984), reprinted with permission from *Quaternary Research*.

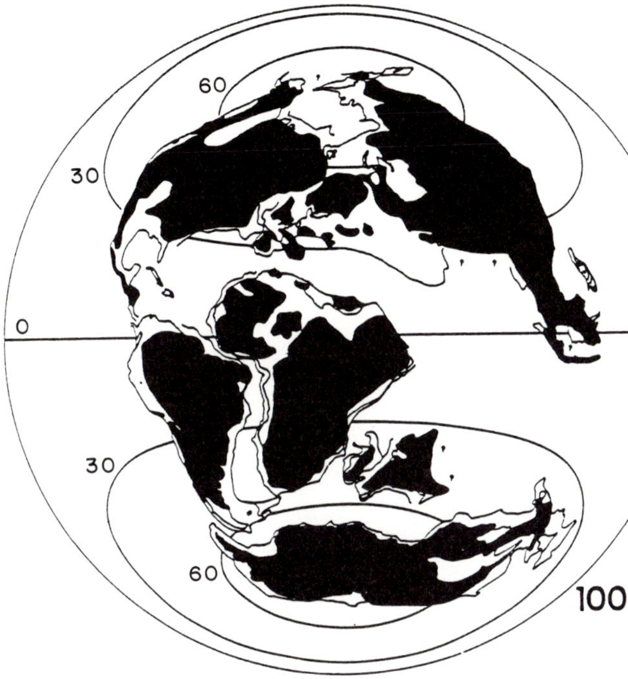

Figure 3.4 Paleogeographic reconstruction at 100 Ma. Light areas on continents indicate regions flooded by shallow seas (maximum depth 100–200 m). After Barron et al. (1980).

Instabilities in the Climate System

Geologic studies suggest that there may have been relatively abrupt changes in North Atlantic Deep Water (NADW) formation during the Pleistocene (Broecker et al. 1985). These changes would presumably reflect variations in the salt content of the surface waters of the ocean. An ocean modeling study for the Younger Dryas (10,500 B.P.) supports Broecker's proposal that variations in freshwater flux to the subpolar North Atlantic could trigger such an abrupt NADW reduction (Figure 3.5). Since almost all CO_2 doubling experiments yield higher precipitation in mid-latitudes (due to higher moisture content of the warm air), one might conjecture that a freshening in the subpolar North Atlantic might also occur in the future. Ocean GCMs support such a response (Washington and Meehl 1989; Mikolajewicz et al. 1990; Cubasch et al. 1991). In this case, observations of past climate variations provided the first indications that such changes could occur. This is a good example of the process analog approach at work.

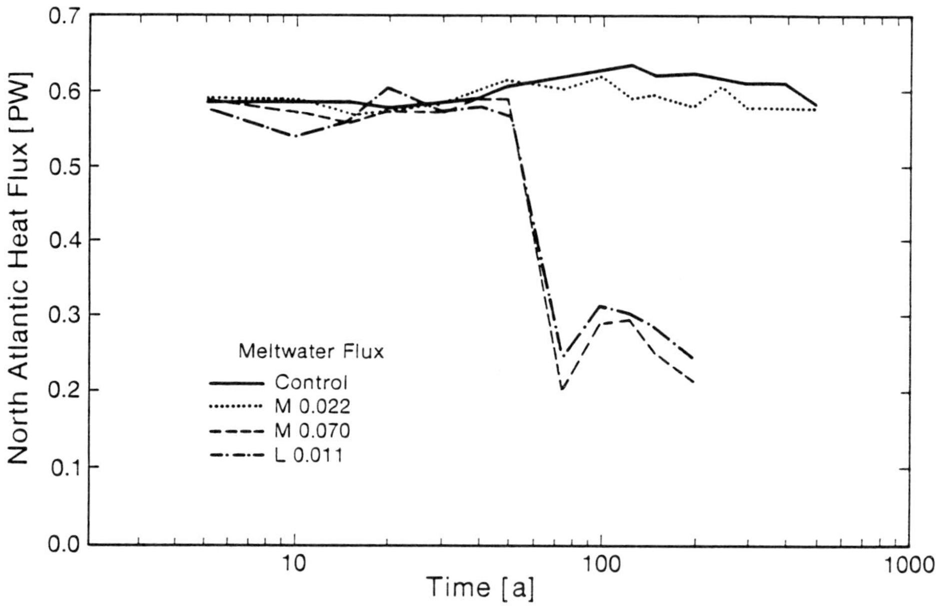

Figure 3.5 Effect of low salinity meltwater plumes on North Atlantic heat transport. Heat transport varies as a function of North Atlantic Deep Water (NADW) production: as NADW production decreases, so too does heat transport. Figure illustrates results from several sensitivity experiments with an ocean general circulation model (GCM), with time indicating model years from beginning of the run. "L" refers to outflow from the St. Lawrence Rivers (in Sverdrups; 1 Sv = 1×10^6 m^3s^{-1}), "M" refers to outflow from the Mississippi. Note that low values of outflow are insufficient to reduce NADW production and that larger values of Mississippi outflow are required to achieve the same level of reduction as occurs with the St. Lawrence outflow. After Maier-Reimer and Mikolajewicz (1989).

Sea-level Variations

The Quaternary record also provides some constraints on models for future sea-level variations. Estimated future increases are on the order of 0.6 mm yr^{-1} (Meier 1990). These numbers are a factor of three smaller than some earlier estimates and reflect, in part, the possibility of increased accumulation rates on the Antarctic and perhaps Greenland ice caps as a result of greater moisture content of the air (Oerlemans 1982). Satellite observations provide some support for this trend for part of Greenland (Zwally et al. 1989), and geologic observations suggest that it may also hold true for Antarctica. For example, estimated accumulation rates for the Antarctic ice sheet were only about one-half that of the present during the last glacial maximum (Lorius et al. 1985). Thus, the conjectured relationships between increased temperature and greater accumulation is supported by the geologic record.

Geologic studies have also modified the picture of a future sea-level "catastrophe" as a result of sudden melting of the West Antarctic ice sheet (e.g., Stuiver et al. 1981). Such a scenario was motivated in part by observations that sea level during the last interglacial was about 6 m higher than present, and that such changes could potentially be achieved by small increases in temperature. Although such a change is still possible, the sea-level history for the last interglacial has been complicated by alternate hypotheses as to the melting of ice on Greenland (Koerner 1989) and the actual level of the East Antarctic ice sheet (Robin 1985). Both of these ice sheets would be less prone to instabilities than the West Antarctic ice sheet.

General Relationship between CO_2 and Warm Climates

Even though pre-Pleistocene warm periods cannot be used as time-slice analogs, they still can provide useful insights into how the system operates. For example, there is a generally good relationship for both the late Pleistocene (Barnola et al. 1987) and the last 600 million years between times of past increased CO_2 levels and warmer climates (Figure 3.6). One exception involves the late Ordovician glaciation (440 Ma), which may have occurred at a time of high CO_2. This glaciation might be explained by decreased solar output and the unique geographic configuration of the continents at that time (Crowley and Baum 1991). Thus, the geologic record in general supports a positive relation between CO_2 and climate. Budyko et al. (1985; see also MacCracken et al. 1990) estimate that past changes are consistent with ~3°C temperature increase for a doubling of CO_2. However, given the uncertainties in estimates of past CO_2 and global average temperatures, this number is probably uncertain by at least 50%. The net result is that the paleo-estimate of climate sensitivity for a CO_2 increase is within the same range of uncertainty as that from the present generation of climate models (Houghton et al. 1990).

Climate Change and Extinctions

Future changes in climate may also affect extinction rates. The geologic record provides the only perspective on how biota respond to climate change. Although there has been much anecdotal support for a significant climate-extinction link (see review in Crowley and North 1988), there is surprisingly little in the line of rigorous statistical analysis of this alleged connection. If it exists at all, some evidence suggests that the biosphere may be most sensitive to climate change during the initial stage of a perturbation (Crowley and North 1988). This subject warrants more study.

POTENTIAL CONTRIBUTIONS FROM THE GEOLOGIC RECORD

Although the above cases provide examples of how the geologic record can aid our understanding of future climate change, there are several significant aspects of

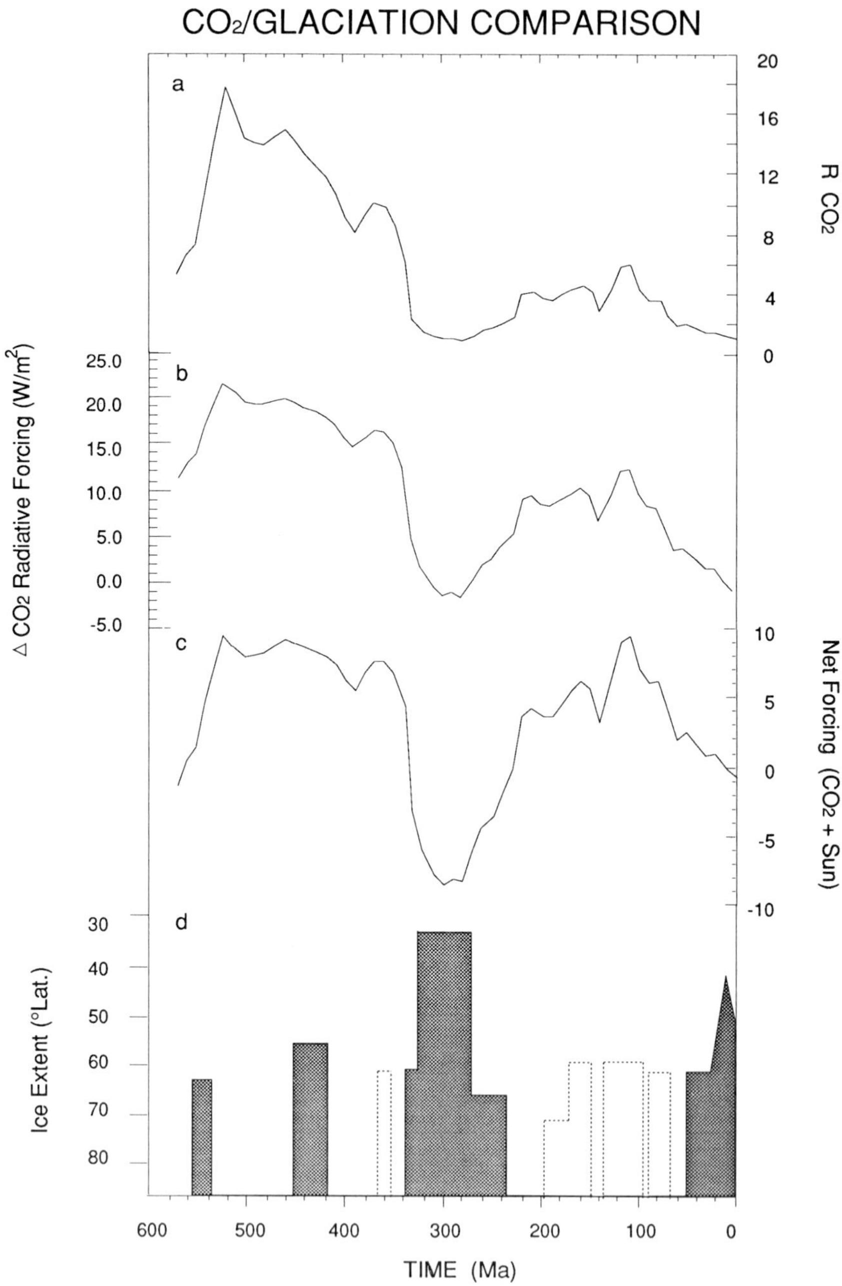
CO2/GLACIATION COMPARISON
a
b
c
d
R CO2
20
16
12
8
4
0
Δ CO2 Radiative Forcing (W/m2)
25.0
20.0
15.0
10.0
5.0
0.0
-5.0
Net Forcing (CO2 + Sun)
10
5
0
-5
-10
Ice Extent (°Lat.)
30
40
50
60
70
80
600
500
400
300
200
100
0
TIME (Ma)

greenhouse predictions that either do not receive support from the geologic record or which have not yet been adequately tested. For example, almost all models predict higher SSTs in the Tropics with increased CO_2 levels (e.g., Manabe and Bryan 1985; Schlesinger 1989). Such temperatures are not found during past times of higher CO_2 levels (Crowley 1991). Models also make predictions (Manabe and Bryan 1985; Manabe and Wetherald 1986) about zonal or regional changes in precipitation and soil moisture (in North America and Eurasia), which so far have not been adequately tested against the geologic record. One recent simulation with a coupled ocean–atmosphere GCM (Cubasch et al. 1991) suggests that land temperatures should increase more rapidly than ocean temperatures. Modeling and geologic studies suggest that when such changes occurred during the early Holocene, they were associated with increased intensity of the African–Asian monsoon system (COHMAP 1988).

CONCLUSIONS

I have argued that use of the "time-slice analog" approach to future climate change is misleading. However, the study of certain processes operating in the climate system can provide valuable insight into future climate change. A topic not addressed herein—validation of climate models with the geologic record (Crowley and North 1991)—also represents a contribution of geologic studies towards understanding the future. However, there are a number of important climate predictions about the future (tropical SSTs and soil moisture variations) that presently receive either little support or have not been adequately tested. Hopefully, future studies will shed more light on these areas of uncertainty.

ACKNOWLEDGEMENTS

Supported by NSF grant ATM90–02808 to T.J.C. I thank A. Berger, J.-C. Duplessy, and T. Webb for comments.

Figure 3.6 Comparison of estimated CO_2 variations with geologic evidence for glaciation. (a) Phanerozoic CO_2 variations as calculated by a geochemical model (from Berner 1990); (b) radiative departures from the present of Figure 3.6a; (c) combined effect of radiative forcing (Figure 3.6b) and solar luminosity variations vs. time; (d) area of glaciation vs. time as estimated by Frakes and Francis (1988). Dashed areas represent evidence for ice but not necessarily continental glaciation; hachured areas represent times of continental glaciation. Note the coincidence of low "net CO_2" and glaciation for the Carboniferous and late Cenozoic, but high CO_2 levels for the Ordovician glaciation. After Crowley and Baum (1991).

REFERENCES

Barnola, J., D. Raynaud, Y. Korotkevich, and C. Lorius. 1987. Vostok ice core provides 160,000-year record of atmospheric CO_2. *Nature* **329**:408–414.

Barron, E., J. Sloan, and C. Harrison. 1980. Potential significance of land-sea distribution and surface albedo variations as a climatic forcing factor: 180 M.y. to the present. *Palaeogeog. Palaeoclim. Palaeoecol.* **30**:17–40.

Berner, R. 1990. Atmospheric carbon dioxide over Phanerozoic time. *Science* **249**:1382–1386.

Broecker, W., D. Peteet, and D. Rind. 1985. Does the ocean–atmosphere system have more than one stable mode of operation? *Nature* **315**:21–26.

Budyko, M., and Y. Izrael. 1987. Anthropogenic Climate Changes. Leningrad: Gidrometeoizdat. (in Russian) (English translation: Arizona Univ. Press, 1990).

Budyko, M., A. Ronov, and A. Yanshin. 1985. The History of the Earth's Atmosphere. Leningrad: Gidrometeoizdat. (in Russian) (English translation: Springer-Verlag, Berlin, 1987.)

CLIMAP. 1984. The last interglacial ocean. *Quat. Res.* **21**:123–224.

COHMAP. 1988. Climatic changes of the last 18,000 years: Observations and model simulations. *Science* **241**:1043–1052.

Crowley, T. 1990. Are there any satisfactory geologic analogs for a future greenhouse warming? *J. Climate* **3**:1282–1292.

Crowley, T. 1991. Past CO_2 changes and tropical sea surface temperatures. *Paleoceanog.* **6**:387–394.

Crowley, T., and S. Baum. 1991. Towards reconciliation of late Ordovician (~ 440 Ma) glaciation with very high CO_2 levels. *J. Geophys. Res.* **96**:22597–22610.

Crowley, T., and G. North. 1988. Abrupt climate change and extinction events in earth history. *Science* **240**:996–1002.

Crowley, T., and G. North. 1991. Paleoclimatology. New York: Oxford Univ. Press.

Cubasch, U., K. Hasselmann, H. Höck, E. Maier-Reimer, U. Mikolajewicz, B. Santer, and R. Sausen. 1991. Time-dependent greenhouse warming computations with a coupled ocean–atmosphere model. Max-Planck-Institut für Meteorologie Rep. No. 67, Hamburg.

Frakes, L., and J. Francis. 1988. A guide to Phanerozoic cold polar climates from high-latitude ice-rafting in the Cretaceous. *Nature* **333**:547–549.

Houghton, J., G. Jenkins, and J. Ephraums. 1990. Climate Change: The IPCC Scientific Assessment. Cambridge: Cambridge Univ. Press.

Kellogg, W. 1977. Effects of human activities on global climate. WMO Report No. 486. Geneva, Switzerland: World Meteorological Organization.

Koerner, R. 1989. Ice core evidence for extensive melting of the Greenland ice sheet in the last interglacial. *Science* **244**:964–968.

Lorius, C., J. Jouzel, C. Ritz, L. Merlivat, N. Barkov, Y. Korotkevich, and V. Kotlyakov. 1985. A 150,000-year climatic record from Antarctic ice. *Nature* **31**:591–596.

MacCracken, M., M. Budyko, A. Hecht, and Y. Izrael. 1990. Prospects for Future Climate: A Special US/USSR Report on Climate and Climate Change. Chelsea, MI: Lewis.

Maier-Reimer, E., and U. Mikolajewicz. 1989. Experiments with an OGCM on the cause of the Younger Dryas. In: Oceanography 1988, ed. A. Ayala-Castanares, W. Wooster, and A. Yanel-Arancibia, pp. 87–100. Mexico D.F.: UNAM.

Manabe, S., and K. Bryan. 1985. CO_2-induced change in a coupled ocean–atmosphere model and its paleoclimatic implications. *J. Geophys. Res.* **90**:11689–11708.

Manabe, S., and R. Wetherald. 1986. Reduction in summer soil wetness induced by an increase in atmospheric carbon dioxide. *Science* **232**:626–628.

Martinson, D., N. Pisias , J. Hays, J. Imbrie, T. Moore, and N. Shackleton. 1987. Age dating and the orbital theory of the ice ages: Development of a high-resolution 0–300,000-year chronostratigraphy. *Quat. Res.* **27**:1–29.

Meier, M. 1990. Reduced rise in sea level. *Nature* **343**:115–116.

Mikolajewicz, U., B. Santer, and E. Maier-Reimer. 1990. Ocean response to greenhouse warming. *Nature* **345**:589–593.

Oerlemans, J. 1982. Response of the Antarctic ice sheet to a climatic warming: A model study. *J. Climatol.* **2**:1–11.

Owen, R., and D. Rea. 1985. Seafloor hydrothermal activity links climate to tectonics: The Eocene carbon dioxide greenhouse. *Science* **277**:166–169.

Prell, W., and J. Kutzbach. 1987. Monsoon variability over the past 150,000 years. *J. Geophys. Res.* **92**:8411–8425.

Robin, G. 1985. Contrasts in Vostok ice core: Changes in climate or ice volume? *Nature* **316**:578–579.

Ruddiman, W., and A. McIntyre. 1981. Oceanic mechanisms for amplification of the 23,000-year ice-volume cycle. *Science* **212**:617–627.

Schlesinger, M. 1989. Model projections of the climatic changes induced by increased atmospheric CO_2. In: Climate and Geo-Sciences, ed. A. Berger, S. Schneider, and J. Duplessy, pp. 375–415. Dordrecht: Kluwer.

Stuiver, M., G. Denton, T. Hughes, and J. Fastook. 1981. History of marine ice sheet in West Antarctica during the last glaciation: A working hypothesis. In: The Last Great Ice Sheets, ed. G. Denton and T. Hughes, pp. 319–436. New York: Wiley.

Washington, W., and G. Meehl. 1989. Climate sensitivity due to increased CO_2: Experiments with a coupled atmosphere and ocean general circulation model. *Clim. Dynam.* **4**:1–38.

Webb, T., and T. Wigley. 1985. What past climates can indicate about a warmer world. In: The Potential Climatic Effects of Increasing Carbon Dioxide, ed. M. MacCracken and F. Luther, pp. 239–257. Report DOE/ER–0237. Washington, D.C.: U.S. Dept. of Energy.

Zubakov, V., and I. Borzenkova. 1988. Pliocene palaeoclimates: Past climates as possible analogues of mid-twenty-first century climate. *Palaeogeog. Palaeoclim. Palaeoecol.* **65**:35–49.

Zwally, H., A. Brenner, J. Major, R. Bindschadler, and J. Marsh. 1989. Growth of Greenland ice sheet: Measurement. *Science* **246**:1587–1591.

4

Self-regulatory Processes in the Biosphere in the Face of Climate Change

I.C. PRENTICE[1] and M. SARNTHEIN[2]
[1]Department of Plant Ecology, Lund University, Östra Vallgatan 14, 223 61 Lund, Sweden
[2]Geologisch–Palaeontologisches Institut der Universität Kiel, Olshausenstrasse 40, 2300 Kiel, F.R. Germany

ABSTRACT

This chapter aims to provoke discussion about important but incompletely understood self-regulatory processes in the biosphere–ocean–atmosphere system. Atmospheric dynamics and climatic seasonality are seen as key intermediaries in the control of the hydrological and carbon cycles, producing fast responses in terrestrial and oceanic biota. Changes in the hydrological cycle are driven primarily by external physical forcing through changes in atmospheric circulation, although hydrological feedbacks may modify vegetation distribution and climate at a subcontinental scale. Variations in nutrient supply to the low- and mid-latitude surface water, producing changes in plankton productivity in the crucial upwelling zones, are an important forcing factor for glacial–interglacial changes in oceanic carbon storage and atmospheric CO_2. Effects of climatic change on biome distributions since the last glacial maximum are more than sufficient to explain the increase in terrestrial carbon storage inferred from global changes in benthic $\delta^{13}C$.

INTRODUCTION AND DEFINITIONS

In this chapter, we discuss some of the self-regulatory processes in the biosphere–ocean–atmosphere system that are suggested by analysis of the palaeoenvironmental record and which are relevant to future global changes. Time scales of these processes span years to centuries (Table 4.1). These time scales are relevant to understanding possible changes in the climate system during the next 200 years.

Global Changes in the Perspective of the Past
Edited by J.A. Eddy and H. Oeschger

Table 4.1 Time constants of some interactive processes in the biosphere–ocean–atmosphere system.

Upwelling/plankton production/atmospheric CO_2	1 yr
Dust transport/plankton production/CO_2	1 yr
Sea-ice distribution/atmospheric circulation	1–10 yr
Vegetation structure/continental hydrology	1–100 yr
Ocean circulation/atmospheric CO_2	100–1000 yr
Biomass/atmospheric CO_2	100–1000 yr
Soil carbon, peat/atmospheric CO_2, CH_4	>1000 yr

Information on past activities of these processes can be obtained from records preserved in terrestrial sediments, with time resolution down to one year in varved sediments. In ocean sediments, the resolution is usually worse (no better than 50–200 years, except in a few places such as the Gulf of California, where fast-accumulating varved sediments have been found); however, records of longer time spans are readily obtainable and can be directly linked to global climate changes. We focus on records from the last three major climatic cycles (the past 350,000 years), while recognizing that relevant information also resides in comparable records covering longer time spans, e.g., from the Pliocene onwards, or during the Cretaceous.

Major self-regulatory processes in the biosphere–ocean–atmosphere system include changes in (a) the species composition of terrestrial and marine biota, (b) ecosystem structure, (c) water and CO_2 exchanges among the atmosphere, land, and oceans, which produce changes in precipitation and the greenhouse effect, and (d) nutrient generation in the surface ocean. Others include the global cycles of CH_4 and dimethyl sulfide and their implied effects on sea-surface temperature as well as speciation and extinction, which are not considered here.

OLD AND NEW STATEMENTS

Our understanding of the nature of the interactions between climate and biosphere is clouded by the persistence of a number of overgeneralizations and misconceptions. In some cases, the traditional view involves some kind of biosphere feedback mechanism as an overriding control, whereas in other cases it discounts biosphere feedback. In this section, we mention some persistent oversimplifications of both types and suggest alternative models (which may or may not be correct). In the next section, we summarize the evidence that supports our suggestions, based on paleoenvironmental records and model studies of various kinds.

- A misconception that is still widespread (especially among ecologists) concerns the magnitude of the possible long-term influence of deforestation on precipitation. Classic examples include the blame attached to Roman and

Greek shipbuilders in causing widespread increases in aridity 2000–3000 years ago in the Mediterranean region and to Neolithic grazing for the desertification of the Sahara. These ideas are inconsistent with the paleoclimatological evidence that such large-scale changes in vegetation and the hydrological cycle are externally forced, particularly by orbital changes which, in turn, produce changes in atmospheric circulation. Feedbacks from vegetation to climate exist, but they are not the main cause of such changes.

- Changes in global biogeochemical cycles are often perceived (especially by physical scientists) as being determined exclusively by long-term changes in physicochemical processes and as being capable of reconstruction from changes in the annual means of relevant environmental variables. Paleobiological evidence suggests, on the contrary, that the abundance and distributions of both terrestrial and marine organisms are (a) important in determining global biogeochemical processes and (b) controlled by the seasonality and spatial variability of environmental variables. The relevant scales of spatial variability are, in some cases, beyond the limit of resolution of current atmospheric and ocean circulation models.
- In an excellent illustration of the above, variations in deep-water formation in high latitudes have often been considered as the single key agent in determining deep-water ventilation and ocean carbon storage. Paleoproductivity data, however, suggest that changes in the vertical transfer of organic carbon in the ocean, induced by plankton production in the spatially very restricted upwelling zones of middle and low latitudes, are a major agent. Such changes are considered to be mediated by atmospheric dynamics, which control the seasonally changing availability of nutrients in the surface ocean layers.
- Terrestrial vegetation and soils are often assumed (for example, in studies of recent changes in the carbon cycle) to provide a negative feedback opposing CO_2 increases through physiological mechanisms. However, paleovegetation data suggest that changes in biome distributions since the last glacial maximum are at least large enough to explain the increase in terrestrial carbon storage implied by $\delta^{13}C$ changes in benthic foraminifera, and that these changes can largely be explained as responses of the biota to changing regional climates.

EXTERNAL FORCINGS AND FEEDBACKS IN THE TERRESTRIAL AND OCEANIC BIOSPHERES

The Hydrological Cycle on Land

Based on lake-level variations, pollen evidence, and changes in the distribution of sand dunes, it has been established that the subtropical arid belt of Africa and Eurasia expanded towards a maximum at the last glacial maximum (Sarnthein 1978; Street-Perrott and Harrison 1985) and contracted almost to zero during the Holocene around

6000 years ago (Petit-Maire et al. 1990). The late Holocene readvance of the deserts to approach their glacial extents and the associated decrease in biomass gave rise to the common idea that human activities were somehow responsible for the change in climate. The paleoclimatic records, however, show that (a) these changes took place in parallel over a very large area and (b) they were not entirely unidirectional but instead showed (regionally synchronous) reversals at various times.

A natural mechanism for these changes explains the late Holocene decreasing trend in the moisture budget of the northern subtropical continents in terms of variations in summer insolation north of the equator (Kutzbach and Street-Perrott 1985; COHMAP 1988). Northern Hemisphere summer insolation was at a maximum and winter insolation at a minimum during the early Holocene. The result was a stronger-than-present land–sea pressure contrast and enhanced monsoons. The insolation anomaly has decayed over the past 6000 years, producing a return to aridity. Additional contributory causes of decreasing moisture include the observed decrease in equatorial sea-surface temperatures and an increase in coastal upwelling off the western margins of the continents (Tetzlaff et al. 1989; Henning and Flohn 1981).

These observations lead us to reject the idea of a dominant human role in the progressive drying of Africa and southern Europe since the mid-Holocene. However, model experiments have shown that the presence of forest vegetation in runoff areas can increase the penetration depth of oceanic moisture transport into the continents (e.g., Tetzlaff et al. 1988; Xue et al. 1990). This is due, in part, to increased evapotranspiration from forests, which have a higher leaf-area index and greater rooting depth than nonforest vegetation (part of the transpired water is transferred further inland and precipitated again), and in part to more complex influences of changes in surface roughness and albedo on large-scale moisture convergence (Charney et al. 1977; Xue et al. 1991; Nobre et al. 1991). Thus, there are positive feedbacks at work that may amplify the natural forcing. Paleoenvironmental data, however, imply that these are neither the primary nor the main causes of the observed changes during the Holocene.

Mechanisms that can explain the retreat of Mediterranean rainfall and of north Saharan vegetation during the past 6000 years have yet to be discovered. Nevertheless, these changes occurred in phase with the changes further south, which implies that they, too, were externally forced.

Seasonality and the Biosphere

Climatic seasonality is primarily controlled by insolation and atmospheric dynamics. This is true both in the oceans and on land. For example, the seasonal shift of the trade winds controls seasonal variations in sea-surface temperatures in oceanic upwelling zones, and these variations in turn affect the seasonal moisture supply to the continents. In high latitudes on the continents, temperature seasonality is partly controlled by the strength of onshore winds and local orography. The formation of sea ice in the high-latitude ocean represents a particularly important threshold mechanism that can enhance seasonality over large parts of the hemisphere within decades. In the Southern Ocean, seasonal storm tracks may be responsible for low ocean productivity despite high nutrient availability, because no full bloom can develop under highly turbulent conditions (Wefer 1989).

Both ocean and land biota respond to seasonality. On land, many types of plants are restricted in their distribution by low winter temperatures, and this has a large effect on the structure of vegetation and in the seasonal pattern of leaf display. For example, the boundary between broad-leaved evergreen and summer-green deciduous forests corresponds to the –15°C isotherm of absolute minimum temperature (Woodward 1987). All trees have a certain requirement for growing-season warmth, which can be quantified as the annual average time integral of temperatures above 5°C or "growing degree days." Boreal conifers require about 350 growing degree days (Prentice et al. 1992). Below this limit, boreal forest is replaced by tundra, with considerably lower biomass. A striking example of the effect of seasonality is Mediterranean sclerophyll vegetation, which is found only in regions with winter rain and summer drought, and where winter temperatures are adequate for broad-leaved evergreen trees or shrubs (Prentice et al. 1992). Mediterranean vegetation has less than half the biomass and carbon storage of the temperate broad-leaved forests that are found in climates with similar temperatures and total annual rainfall, but with a more even seasonal distribution of rainfall (Olson et al. 1983).

In the oceans, a seasonal concentration of upwelling, nutrient supply, and plankton blooms results in a much higher export production than the same annual nutrient supply spread over the whole year (Eppley and Peterson 1979; Berger et al. 1989). The main reason for this is that organic matter in high-productivity areas becomes flocculated into heavy particles that sink rapidly. This process can be enhanced by the seasonal addition of dust-derived mineral grains, which act as carriers and add weight to the organic particles. In general, export production increases with the square of primary production, so that a modest increase in primary production causes a significant increase in the rate of carbon transfer from the sea surface to the deep ocean.

Linked to this increase is a shift in dominance from secondary producers to primary producers, to noncarbonate-producing organisms (such as diatoms and dinoflagellates) from organisms with carbonate tests (such as foraminifera and coccolithophorids). These shifts produce an increase in the ratio of organic carbon to carbonate carbon in the particles that settle to the ocean floor (the "rain ratio") and thereby lead to an increased dissolution of calcium carbonate by the remineralized organic carbon (Berger and Keir 1984). This, in turn, results in higher alkalinity, which—if it occurs over large areas of the ocean—implies a greater potential to sequester dissolved CO_2 at depth (Boyle 1989). Periodic phases of intermittent exhaustion of the nitrogen reservoir in the surface ocean result in a further species shift towards dinoflagellate blooms (red tides), such as occur during El Niño regimes (Codispoti et al. 1981).

The Carbon Cycle in the Ocean

There is increasing evidence that narrow zones of high productivity along the continental margins play a major role in the vertical transfer of carbon from the atmosphere and surface ocean to the deep ocean. Together with the equatorial upwelling zones, these coastal upwelling zones appear to account for more organic carbon flux than the rest of the ocean put together (Koblentz-Mishke et al. 1970; Berger et al. 1989; Sarnthein et al. 1988; Ross et al. 1991). General circulation models (GCMs) of the ocean have, so far, been unable to simulate these highly localized fluxes (Heinze et al. 1991).

A major unsolved problem is whether the glacial–interglacial changes in atmospheric CO_2 were primarily caused by changes in the horizontal advection of water masses in the deep ocean, or by changes in the vertical chemical fluxes in the ocean. A first global balance-sheet of changes in export production from glacial to interglacial times (Sarnthein et al. 1988), however, indicates that the downward flux of particulate carbon decreased by 2–4 Pg C yr^{-1}. Model studies (Michel 1991; Ross et al. 1991) suggest that changes in the rain ratio below upwelling zones may be responsible for a significant fraction of the total change in atmospheric CO_2, in part via changes in deep-water alkalinity. Moreover, detailed time-series studies of ocean sediments have shown a consistent phase relationship with the decrease in production which leads the increase in deep-water ventilation by up to several thousand years (Sarnthein et al. 1988).

From $\delta^{13}C$ values in near-surface dwelling planktonic foraminifera, we also know that the nutrient reservoirs in the low- and mid-latitude high-productivity zones were greater during glacial times. The additional nutrients may have been derived in part from an increase in the nutrient content of the upwelling source waters (for example, through enhanced advection of nutrient-enriched South Atlantic Central Water to the equatorial Atlantic upwelling zone and northwestern Africa). In parts of the Pacific, the increased nutrient supply during glacial times is due, at least in part, to increased dust discharge (in particular iron) from the continents (Lyle et al. 1992). These mechanisms make it plausible that the productivity changes observed in low- and mid-latitudes were related to increased nutrient supply and that they contributed to the glacial–interglacial changes in the total oceanic uptake of CO_2.

Thus, we propose that the intensity of atmospheric circulation is the link between high-latitude glacial boundary conditions and low- and mid-latitude oceanic nutrient distribution and CO_2 uptake by the oceanic biosphere.

Glacial–Interglacial Carbon Transfer between Ocean and Land

The ice-core record shows that atmospheric CO_2 concentration increased from 190 ppmv to 275 ppmv from the last glacial maximum to preindustrial times, an increase of nearly 200 Pg C (Barnola et al. 1987). Meanwhile, the average global ocean $\delta^{13}C$, deduced from measurements on benthic foraminifera, is estimated to have increased by 0.32‰ (Duplessy et al. 1988). A standard calculation, assuming a uniform carbon isotopic ratio for terrestrial organic carbon and neglecting atmospheric changes, attributes this isotopic shift to a further transfer of 450 Pg C from the deep ocean (via the atmosphere) to vegetation and soils.

To explain these global changes, we require an analysis of what happened on land as well as in the ocean. Estimates of the increase in terrestrial vegetation and soil carbon storage from the glacial maximum to recent times have been made, based on simulated or reconstructed changes in biome distributions and assuming that carbon storages are fixed properties of each biome. These estimates range widely, from around zero (Prentice and Fung 1990) to 1350 Pg C (Adams et al. 1990). The zero estimate is almost certainly wrong. It was obtained from an experiment with an atmospheric GCM that produced wetter-than-present glacial maximum conditions in the Tropics and Subtropics, at variance with the terrestrial evidence and with other atmospheric GCM experiments (e.g., COHMAP 1988). The 1350 Pg estimate, on the other hand, represents an upper bound. Its authors assumed the maximum plausible reductions in forest area in key regions (such as the Sunda shelf and the Amazon lowlands), where paleovegetation data from the glacial maximum are sparse or absent.

Disregarding both of these estimates, a rough calculation suggests that terrestrial carbon storage did increase by at least 450 Pg C between the glacial maximum and preindustrial times. Tundra and forest soils in the formerly glaciated regions alone should account for a 400 Pg C increase (Schlesinger 1990). Peatlands of Holocene age account for 280–450 Pg C (Adams et al. 1990; Gorham 1991), including substantial volumes in the unglaciated regions. Widespread evidence for glacial-age low lake levels and extensive dunes in the Tropics suggest further increases (hundreds of Pg C?) in tropical biomass.

The benthic $\delta^{13}C$ calculation is not unambiguous, and Crowley (1991) suggested three ways in which it might underestimate the increase in terrestrial carbon storage. The first explanation effectively requires tropical biomass to have been significantly greater at the glacial maximum than it is at present; however, the terrestrial evidence does not support this view. The second involves a major sequestration of organic carbon from the land into continental slope sediments. Organic carbon accumulation rates in these sediments were indeed higher than they are today; however, this storage can only amount to about 1–5% of the local primary production and therefore could only account for a part of the "missing" carbon. The third invokes a greater abundance of C_4 plants at the glacial maximum. C_4 plants have a lower CO_2 compensation point than C_3 plants, which presumably would have been an advantage when atmospheric CO_2 was low. Carbon assimilated by the C_4 pathway has a less negative $\delta^{13}C$, so the ocean $\delta^{13}C$ would have been more negative; however, this means that the standard calculation would overestimate the change in carbon storage. So, it is not easy to explain an underestimate by any of these effects. Yet we note that the calculation would be sensitive to any change in atmospheric and biospheric $\delta^{13}C$.

We conclude that there is as yet no firmly based quantitative estimate of the change in terrestrial carbon storage since the last glacial maximum, but that changes in biome distributions were large enough to explain the observed change in benthic $\delta^{13}C$. Our calculation neglects possible further effects of CO_2 on carbon storage within biomes, which would imply a larger change that would be harder to reconcile with the benthic $\delta^{13}C$ record. Low CO_2 at the glacial maximum might have increased the area of arid biomes by decreasing plant water-use efficiency; however, this would not explain the widespread abiotic evidence of aridity. Thus it appears that the major changes in terrestrial carbon storage were driven primarily by climate changes (effects of ice-sheet configurations and atmospheric CO_2 concentrations on atmospheric circulation; COHMAP 1988), mediated by the responses of the terrestrial biota to changing regional climates.

IMPLICATIONS FOR FUTURE CHANGE

The findings summarized in this chapter illustrate how paleorecords may contribute to an understanding of how and to what extent self-regulating processes in the biosphere interact with climatic change. In each of the processes we have considered, atmospheric dynamics has emerged as the primary link between external forcing and biospheric response. Variations in sea-ice cover in the high-latitude oceans may act as a major trigger for rapid changes in atmospheric dynamics. Changes in atmospheric circulation and sea-ice cover can happen in a few years and hence we may expect fast changes in the future. This conclusion contrasts with prognoses based on the much slower turnover rates of the deep ocean.

Self-regulatory processes alone cannot explain observed past changes in the continental hydrological cycle because the major control of regional aridity is external to the biosphere. One implication of this finding is that although deforestation on humid areas may secondarily reduce the penetration of precipitation into the continents, reforestation of semi-arid areas cannot compensate for a lack of moisture advection determined by the atmospheric circulation.

Continental desertification may, however, have the unexpected secondary effect of increasing the flux of iron-rich aeolian dust into the ocean. The paleorecord suggests that this flux of iron forms a link between the hydrological and carbon cycles: increased dust flux produces increased plankton productivity and increased CO_2 extraction from the atmosphere.

Carbon sequestration by the terrestrial biosphere is a potential self-regulating mechanism in CO_2-induced global warming. On a time scale of years, the terrestrial biosphere may be a carbon sink because increasing CO_2 implies increasing net primary production, especially in water-limited biomes where water-use efficiency is critical, and this may translate into increasing biomass and soil organic matter storage. In the long run, however, increased production in any one biome may be counteracted to an unknown extent by increased disturbance, decomposition, and erosion, while climatic changes will force more complex regional responses dependent on the temperature and moisture requirements of different types of plant (Prentice 1992).

ACKNOWLEDGEMENTS

We thank Tom Crowley, Jonathan Overpeck, and Uli Siegenthaler for their helpful comments.

REFERENCES

Adams, J.M., H. Faure, L. Faure-Denard, J.M. McGlade, and F.I. Woodward. 1990. Increases in terrestrial carbon storage from the Last Glacial Maximum to the present. *Nature* **348**:711–714.

Barnola, J.M., D. Raynaud, Y.S. Korotkevitch, and C. Lorius. 1987. Vostok ice core provides 160,000-year record of atmospheric CO_2. *Nature* **329**:408–414.

Berger, W.H., and R.S. Keir. 1984. Glacial–Holocene changes in atmospheric CO_2 and the deep-sea record. In: Climate Processes and Climate Sensitivity, ed. J.E. Hansen and T. Takahashi. *Geophys. Monog.* **29**:337–351. Washington, D.C.: American Geophysical Union.

Berger, W.H., V.S. Smetacek, and G. Wefer. 1989. Ocean productivity and paleoproductivity: An overview. In: Productivity of the Ocean: Present and Past, ed. W.H. Berger, V.S. Smetacek, and G. Wefer, pp. 1–34. Dahlem Workshop Report PC 8. Chichester: Wiley.

Boyle, E.A. 1989. The role of vertical chemical fractionation in controlling Late Quaternary atmospheric carbon dioxide. *J. Geophys. Res.* **93**:15701–15714.

Charney, J.G., W.J. Quirk, S.H. Chow, and T. Kornfield. 1977. A comparative study of the effects of albedo change on drought in semi-arid regions. *J. Atmos. Sci.* **34**:1366–1388.

Codispoti, L.A. 1981. Temporal nutrient variability in three different upwelling regions. In: Coastal Upwelling. *Coast. Estuar. Sci.* **1**:209–220.

COHMAP. 1988. Climatic changes of the last 18,000 years: Observations and model simulations. *Science* **241**:1043–1052.

Crowley, T.J. 1991. Ice age carbon. *Nature* **352**:575–576.

Duplessy, J.C., N.J. Shackleton, R.G. Fairbanks, L. Labeyrie, D. Oppo, and N. Kallel. 1988. Deepwater source variations during the last climatic cycle and their impact on the global deepwater circulation. *Paleoceanog.* **3**:343–360.

Eppley, R., and B.J. Peterson. 1979. Particulate organic matter flux and plankton new production in the deep ocean. *Nature* **282**:677–680.

Gorham, E. 1991. Northern peatlands: Role in the carbon cycle and probable responses to climatic warming. *Ecol. Applic.* **1**:182–195.

Heinze, C., E. Maier-Reimer, and K. Winn. 1991. Glacial pCO_2 reduction by the world ocean: Experiments with the Hamburg carbon cycle model. *Paleoceanog.* **6**:395–430.

Henning, D., and H. Flohn. 1981. Some aspects of evaporation and sensible heat flux of the tropical Atlantic. *Contr. Atmos. Phys.* **53**:430–441.

Koblentz-Mishke, O.J., V.V. Volkowinski, and J.G. Kabanova. 1970. Plankton primary productivity of the world ocean. In: Scientific Exploration of the South Pacific, ed. W.S. Wooster, pp. 183–193. Washington, D.C.: National Academy of Science.

Kutzbach, J.E., and F.A. Street-Perrott. 1985. Milankovitch forcing of fluctuations in the level of tropical lakes from 18 to 0 kyr B.P. *Nature* **317**:130–134.

Lyle, M.W., F.G. Prahl, and M.A. Sparrow. 1992. Ice age paleotemperatures, upwelling and productivity from a temperature record in the equatorial Pacific Ocean. *Nature* **355:**812–815.

Michel, E. 1991. L'océan au dernier maximum glaciaire: Le cycle du carbone et la circulation. Constraintes isotopiques et modélisation. Thése, Université de Paris-Sud, Centre d'Orsay.

Nobre, C.A., P.J. Sellers, and J. Shukla. 1991. Amazonian deforestation and regional climate change. *J. Climate* **4**:957–988.

Olson, J.S., J.A. Watts, and L.J. Allison. 1983. Carbon in live vegetation of major world ecosystems. Rep. ORNL–5862. Oak Ridge, TN: Oak Ridge National. Laboratory.

Petit-Maire, N., D. Commelin, J. Fabre, and M. Fontugne. 1990. First evidence for Holocene rainfall in the Tanezrouft hyperdesert and its margins. *Palaeogeog. Palaeoclim. Palaeoecol.* **79**:333–338.

Prentice, I.C. 1992. Biome modeling and the carbon cycle. In: The Global Carbon Cycle, ed. M. Heimann. NATO ASI Series, in press.

Prentice, I.C., W. Cramer, S.P. Harrison, R. Leemans, R.A. Monserud, and A.M. Solomon. 1992. A global biome model based on plant physiology and dominance, soil properties and climate. *J. Biogeog.* **19:**117–134.

Prentice, K.C., and I.Y. Fung. 1990. The sensitivity of terrestrial carbon storage to climate change. *Nature* **346**:48–51.

Ross, R.M., M. Sarnthein, and K. Winn. 1991. Organic carbon burial in regions of coastal upwelling: Application of a model to the northwest coast of Africa. *Geol. Soc. Am. Ann. Meet. Abst.* **23/5**:A339.

Sarnthein, M. 1978. Sand deserts during glacial maximum and climatic optimum. *Nature* **271**:43–46.

Sarnthein, M., K. Winn, J.C. Duplessy, and M.R. Fontugne. 1988. Global variations of surface ocean productivity in low and mid-latitudes: Influences on CO_2 reservoirs of the deep ocean and atmosphere during the last 21,000 years. *Paleoceanog.* **3**:361–389.

Schlesinger, W.H. 1990. Evidence from chronosequence studies for a low carbon-storage potential of soils. *Nature* **348**:232–234.

Street-Perrott, F.A., and S.P. Harrison. 1985. Lake levels and climate reconstruction. In: Paleoclimate Analysis and Modeling, ed. A.D. Hecht, pp. 291–340. Chichester: Wiley.

Tetzlaff, G., M. Peters, L.J. Adams, and W. Janssen. 1988. Wasserhaushalt im Sahel. *Die Erde* **119**:163–170.

Tetzlaff, G., M. Peters, W. Janssen, and L.J. Adams. 1989. Aeolian dust transport in West Africa. In: Paleoclimatology and Paleometeorology: Modern and Past Patterns of Global Atmospheric Transport, ed. M. Leinen and M. Sarnthein, pp. 185–203. Dordrecht: Kluwer.

Wefer, G. 1989. Particle flux in the ocean: Effects of episodic production. In: Productivity of the Ocean: Present and Past, ed. W.H. Berger, V.S. Smetacek, and G. Wefer, pp. 139–153. Dahlem Workshop Report PC 8. Chichester: Wiley.

Woodward, F.I. 1987. Climate and Plant Distribution. Cambridge: Cambridge Univ. Press.

Xue, Y., K.-N. Liou, and A. Kasahara. 1990. Investigation of the biogeophysical feedback on the African climate using a two-dimensional model. *J. Climate* **3**:337–352.

Xue, Y., P.J. Sellers, J.L. Kinetr, and J. Shukla. 1991. A simplified biosphere model for climate studies. *J. Climate* **4**:345–364.

5

How Will Future Climate Changes Differ from Those of the Past?

D. RIND
NASA, Goddard Institute for Space Studies, 2880 Broadway,
New York, NY 10025, U.S.A.

ABSTRACT

Expected future climate changes are compared with past changes in the areas of forcing, timing of change, and background conditions. The rapid warming anticipated from anthropogenic greenhouse gases has no obvious precedents, and the anticipated warmth has probably never existed with an ocean/cryosphere/vegetative distribution similar to today's. The most relevant use of paleoclimate data is to test current models and physical processes, which may then help these models forecast future changes.

INTRODUCTION

The idea that knowledge of the past is central to predicting the future is prevalent in many of the social sciences (e.g., "those who cannot remember the past are condemned to repeat it," George Santayana in *The Life of Reason*). For some of the physical sciences, however, this concept is less relevant: modern physics has moved far beyond the formulations of even the last century, although many of the fundamental principles still apply. Exactly where in this spectrum does climate change fall? Has the anthropogenic influence so perturbed the system that, aside from the responses embodied in the conservation relationships, the past has little to offer us in the way of future perspectives? Or do "natural" laws still dominate the system, to the extent that past climate changes are suitable analogs for predicting characteristics of future climate? This issue will be explored with respect to the following categories: (a) the forcing of future climate changes, (b) the expected timing of the changes, and (c) the background in which the changes will be happening. The following conclusion will be offered: the uniqueness of the expected climate change, in combination with our lack of certainty concerning the actual events of the past, make it highly likely that future

Global Changes in the Perspective of the Past
Edited by J.A. Eddy and H. Oeschger

climate changes will be substantially different from changes in the past. Nevertheless, paleoclimate data should help us understand the sensitivity of different components of the climate system, which may then improve our forecasts of the future climate.

CLIMATE FORCING

The forcing of future climate changes, for at least the next several centuries, should be dominated by the accumulation of anthropogenically generated greenhouse gases (e.g., IPCC 1990). Additional climate forcings may arise, such as volcanic eruptions (e.g., the recent Mt. Pinatubo event; EOS 1991) or solar irradiance variations, like those associated with the recent solar cycle (Hickey et al. 1988). However, these are likely to be either transient or of small amplitude compared with the 4 Wm^{-2} net radiation imbalance at the tropopause expected from the trace-gas increases over the next fifty years.

What about the forcing of past climate changes? The most obvious forcing has been that associated with orbital mechanics, by which variations in the Earth's orbital precession, tilt and eccentricity vary the insolation received at the top of the atmosphere. Figure 5.1 shows the calculated July insolation at 65°N, from the formulae of Berger (1978), for the past eight hundred thousand years (800 Ka), and the next 25 Ka

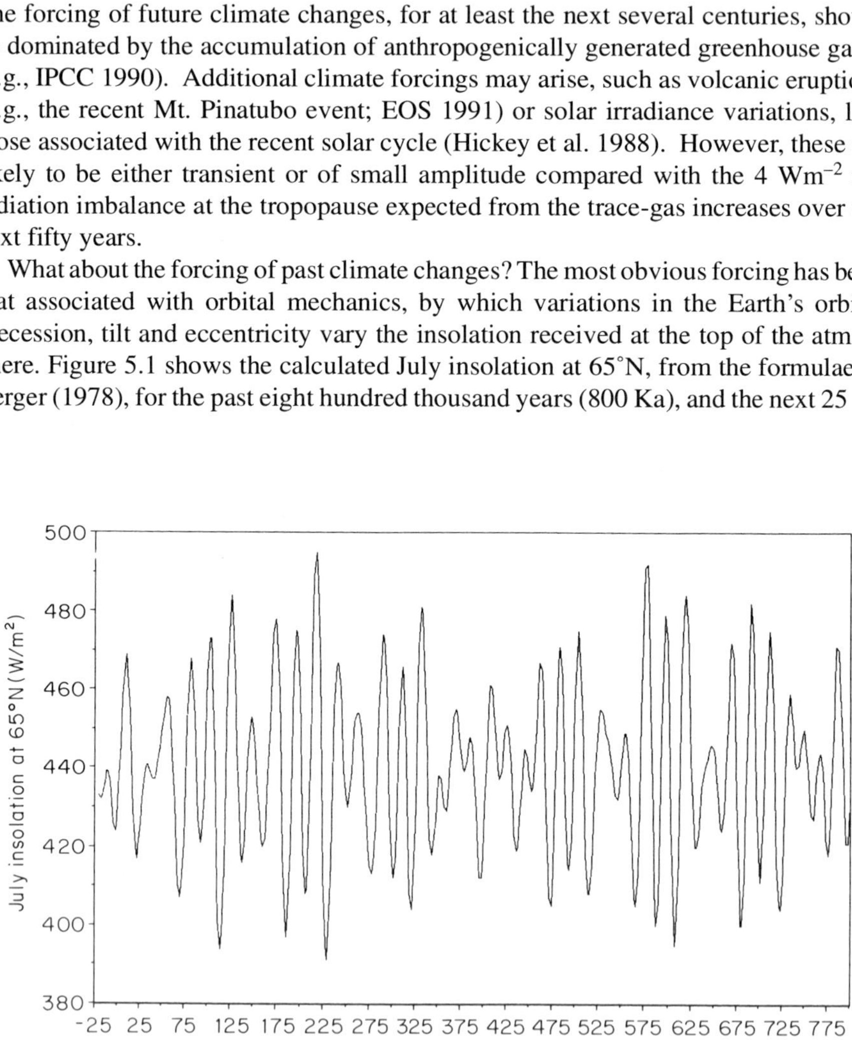

Figure 5.1 July insolation at 65°N from 800 Ka B.P. to 25 Ka in the future (from formulae of Berger 1978). Courtesy of R.Webb.

("–25"). Past ice ages have apparently been initiated by solar radiation decreases at high northern latitudes during summer (e.g., circa 115 Ka B.P.), although there is some question about the timing of such events (e.g., Winograd et al. 1988). The deglaciation was apparently occasioned by the relatively high solar radiation around 14 Ka B.P., in conjunction with increasing CO_2 (Barnola et al. 1987). The so-called "climatic optimum," a climate presumed to be the warmest over the past 100 Ka years, was contemporaneous with the relatively high Northern Hemisphere summer insolation around 6–9 Ka B.P. (although the warmth was not globally synchronous; Crowley 1990). In contrast, for the next 25 Ka, the insolation maximum and minimum are exceedingly modest. It would be surprising if they were able to force substantial changes in the climate, and this possibility is even less likely for the time scale of the next few centuries.

Even as an analog for warmer climates, orbital forcing is unacceptable. Its effect maximizes at high latitudes in summer, with the opposite effect often prevailing in winter, so that annual mean insolation changes are small. In contrast, greenhouse-gas forcing is more spatially uniform (and, if one includes the water vapor feedback, maximizes at low latitudes), with little seasonal variation and substantial net annual forcing. In their effects on the dynamics of the climate system, which integrates responses from different seasons, and on atmospheric dynamics, which is sensitive to changes in latitudinal temperature gradients, the two forcings should be very different.

During the "Little Ice Age" (ca. 1430–1850), the Earth's climate was apparently cooler than that of today. Eddy (1976) suggested that it may have been the result of solar insolation reductions, coinciding with reduced solar activity (the "Maunder Minimum"). The observed decrease in solar irradiance during the sunspot minimum of the mid-1980s was about 0.04% below the average value; Lean et al. (pers. comm.) estimated that, for the Maunder Minimum, the decrease may have been as large as 0.25%. In contrast, to provide forcing as large as that expected from the doubled CO_2-equivalent climate of the next century, it would require a solar reduction of some 2%, which is large enough to be unlikely.

Porter (1986) suggested that the Little Ice Age may have been caused by an increase in volcanic aerosols. To match the additional greenhouse-gas forcing of the next century would require an "El Chichon"-type eruption every few years. This cannot be ruled out; since there have been just three events of this magnitude (including Mt. Pinatubo) over the last century (all in the past 30 years).

On a longer-time scale, it is generally believed that the warm climates of the Mesozoic (some 100–200 million years ago [Ma]) and the Tertiary (from 60 to ′3 Ma) were forced by higher concentrations of atmospheric CO_2. If true, this would appear to have some relevance for future climate, at least in terms of forcing. The evidence for higher CO_2 levels comes from multiple sources, such as seafloor spreading rates (Owen and Rea 1985; Caldeira and Rampino 1991), geochemical cycles (Berner et al. 1983), and chemical ocean modeling (Berger and Spitzy 1988). Values as high as 16× current levels have been proposed for the late Cretaceous (~60 Ma), and 2 to 4× for some periods within the Tertiary, such as the Eocene (~50 Ma), although there have

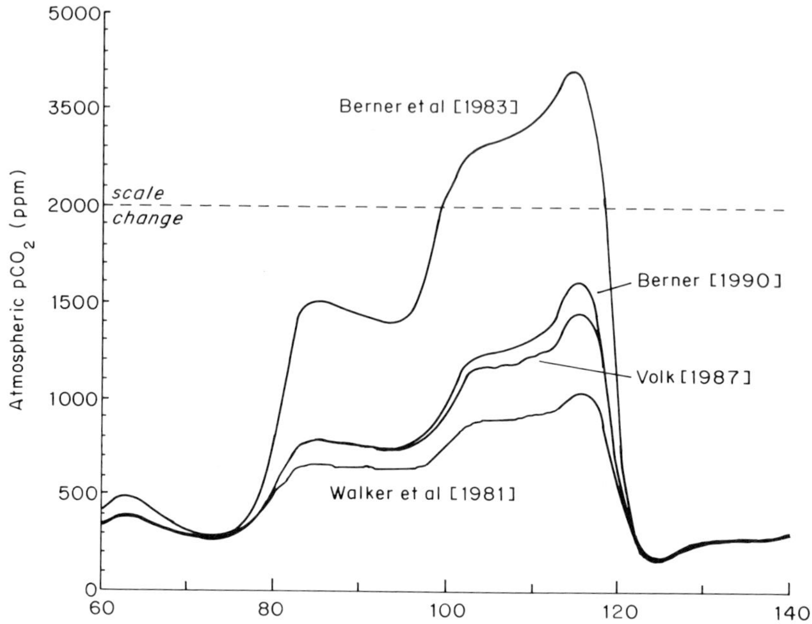

Figure 5.2 Estimates of paleo-CO_2 levels for 60–140 Ka B.P. From Caldeira and Rampino (1991).

been estimates as low as zero increase in the Cenozoic (see the discussion in Berger and Spitzy 1988). Some of these estimates are shown in Figure 5.2.

Higher CO_2 levels may have prevailed at these times; nevertheless, none of the evidence is foolproof, and quantification is extremely difficult (as indicated by the general lack of agreement among the different methods). Rind and Chandler (1991) have proposed that at least part of the warming of these paleoclimates may have been the result of increased poleward ocean-heat transports, associated with different ocean structures and reduced topography. Current general circulation models (GCMs) show that as CO_2 is increased, high latitudes warm more than low latitudes, but low latitudes *do* warm considerably. The paleo-isotope evidence for low-latitude oceans fails to show significant tropical warming for these times periods (e.g., Horrell 1990), while high latitudes are unambiguously warmer (e.g., Frakes 1979). This reconstruction of latitudinal temperature change is more successfully achieved in the GISS model with increased ocean-heat transports than with higher CO_2 levels (see the comparison shown in Figure 5.3).

This analysis emphasizes that ocean-heat transport changes are an additional climate-change forcing mechanism, or at least a climate-change feedback, which may be relevant for the future. Broecker et al. (1985) suggested that the North Atlantic may have more than one mode of oscillation, and that a reduced North Atlantic Deep

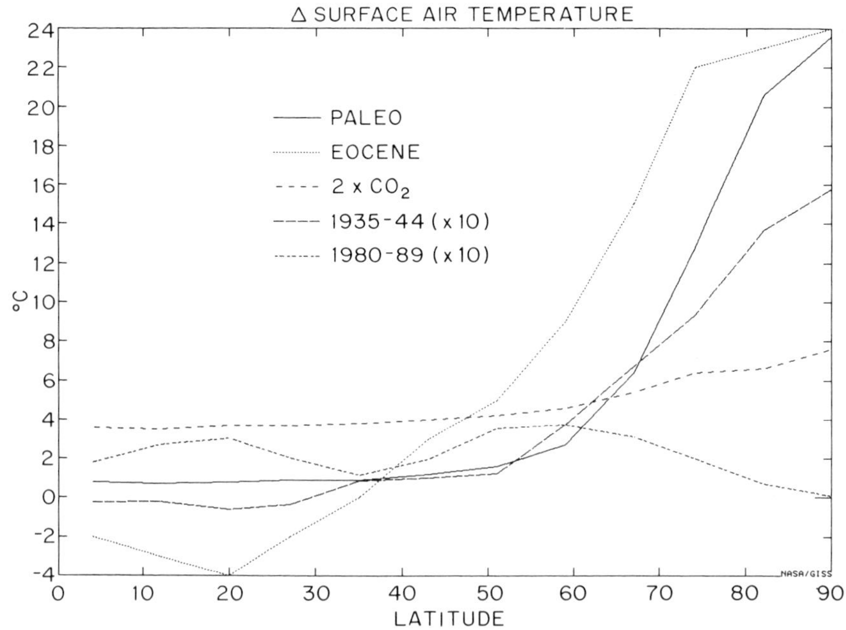

Figure 5.3 Change in Northern Hemisphere latitudinal average temperature for the Eocene and for general circulation model (GCM) experiments with increased ocean-heat transports (PALEO) and doubled CO_2. Also shown are the observed changes for the time periods 1935–1944 and 1980–1989 (both times ten). The 1930s and the Eocene resemble the ocean heat transport increase experiment, while the 1980s look more like the doubled CO_2 simulation (from Rind and Chandler 1991).

Water (NADW) mode existed during the last ice age. Accompanying deep-water reductions would be reduced poleward heat transport, cooling the climate at high latitudes. Broecker et al. (1985) give evidence to support the existence of this effect. It may even have been triggered during the Little Ice Age, and it may occur in the future. Altered ocean-heat transports have probably affected past climates, but our understanding of ocean processes and the development of current ocean models is not sufficient to allow us to estimate the likelihood or magnitude of future changes.

To summarize, future climate changes for the next several centuries should be forced primarily by increased concentrations of greenhouse gases, while the most recent large paleoclimate changes were likely to have been forced by orbital variations in conjunction with altered CO_2. We do not know the extent to which paleoclimates were influenced by higher CO_2 levels, changes in solar irradiance, increased volcanic aerosols, or altered ocean circulations. Outside of the projected greenhouse-gas increases, we do not know the likelihood of occurrence of these other forcing mechanisms in the future. They will probably not be of a magnitude commensurate with expected greenhouse trace-gas forcing, at least for the next several decades.

TIMING OF THE CHANGE

The projected increase in trace-gas concentrations provides the equivalent of a doubled CO_2 forcing within the next 50 years (IPCC 1990). The climate response time is still in question, with estimates ranging from 30 years to 100 years or more (Hansen et al. 1985). Regardless of this exact value, the climate system should be changing at a much faster rate than that associated with the predominant climate forcings of the past (e.g., orbital variations, with periodicities of tens of thousands of years).

A possible exception to this disparity in timing is related to the rapid oscillations seen in the ice-core and terrestrial record associated with events such as the Younger Dryas Oscillation (~11 Ka B.P.; e.g., Rind et al. 1986). Within perhaps a few decades, climate in certain regions changed from deglacial conditions to almost a full glacial regime, and then approximately 1000 years later, warmed back up just as quickly (Dansgaard et al. 1989). There is evidence that this may have happened at other times in the past (Figure 5.4) (see the discussion in Broecker et al. 1985). We do not know the cause of such events, although ocean circulation variations and perhaps glacial melting could have been involved (Broecker et al. 1990). Therefore, we do not know whether they are likely to occur in the future, but their paleoclimate signature could potentially be used to assess how the system responds to a rapid climate change if the data have high enough temporal precision.

Since greenhouse-gas releases will presumably continue during the time the climate itself is changing, the climate system may well be seriously out of equilibrium with the forcing. This is presumably already true, as the surface warming experienced so far is not in balance with the net radiative forcing already provided by the additional greenhouse gases (Hansen et al. 1985). However, the effect will become worse as the ocean, with its greater heat capacity, lags behind the warming land. An increase of this nature in land/ocean contrast is unlikely to have occurred in the past (in the Younger Dryas, the extratropical land conditions were much different, for the land still had significant ice cover). Warmer temperatures over the land will encourage increased evaporation, while cooler ocean temperatures will inhibit the evaporation which provides most of the rainfall. Thus an increase in this contrast is likely to intensify terrestrial droughts (Rind et al. 1990). A possible analog for this effect, although in a seasonal rather than an annual framework, is the 6–9 Ka B.P. time period for the summer season, when increased insolation warmed the land with apparent mid-continental drying.

BACKGROUND CONDITIONS

The climate warming forecast for the next century is occurring against a background of an interglacial, but not overly warm, climate. This is apparently a unique situation. Even if Mesozoic and Tertiary warmth was associated with higher greenhouse-gas levels, there was little or no land ice available at that time (a point emphasized by Crowley 1990). During the deglaciation, the warming occurred slowly by comparison, and in a climate with extensive land ice. The last interglacial (120 Ka B.P.) is thought to have been warmer than the current one (CLIMAP 1984), but again this warming presumably was associated with orbital effects (Figure 5.1) and generated slowly. We therefore have no appropriate paleoclimate situation to gauge how ice sheets in their current configuration will respond to rapid warming.

We also cannot estimate, from paleoclimate studies, how ocean circulation will respond. Without high-latitude land ice in the pre-Pleistocene climates, it is unlikely that significant deep-water production was forced through high-latitude cooling. The land configuration changes of those times, including the connections of ocean basins, should also have affected ocean-heat transports (Maier-Reimer et al. 1990). Later, during the deglaciation warming, NADW production intensified from its diminished glacial state, in response, or in spite of, to the extensive ice-sheet melting.

The ice-sheet/deep-water problems are linked, for the possibility of melting of the West Antarctic ice sheet (producing a sea-level rise of 5 m) appears to hinge on ocean temperature changes at 50 m depth (DOE 1984). Several modeling studies have noted the possibility of cooling of the high-latitude Southern Ocean due to circulation changes as the climate warms (e.g., Manabe et al. 1990). We are therefore in a situation in which the ice-sheet response would be affected by ocean circulation changes, while the oceanic response would undoubtedly be influenced by ice-sheet melting. The reaction of the ice sheets and ocean circulation are crucial for an accurate determination of the future climate, and no appropriate paleoclimate analog exists.

One other component of the climate system that needs to be forecast is the biosphere. Biological processes control much of the hydrologic state of the land surface during the growing season. We need to know how the biosphere will respond to a rapidly changing climate with more CO_2. The higher CO_2 levels that may have prevailed in the pre-Pleistocene era suggest those times as possible analogs. Yet, although increased CO_2 could have augmented photosynthesis, for this effect to have been efficient, other factors (soil moisture, nutrients, etc.) must not have been limiting. Furthermore, they would have been affected by the background state of those climates (much warmer in winter, with possible non-freezing conditions). If CO_2 were not the dominant cause of those warmer climates, and if their latitudinal temperature gradient differed from that which has been forecast for the future (Figure 5.3), the precipitation patterns in the two cases would be dissimilar. Land configuration and vegetation distribution would also differ significantly (e.g., Frakes 1979).

The projected rapid change in the climate system also raises the question as to how ecological systems will be able to respond. Without suitable analogs for rapid warming, it is hard to estimate the response time of natural vegetation. Furthermore, the vegetation and landscape are far from natural. Land use has altered the vegetation distribution and, with the building of roads and industrial complexes, pathways previously available for plant migration have been interrupted.

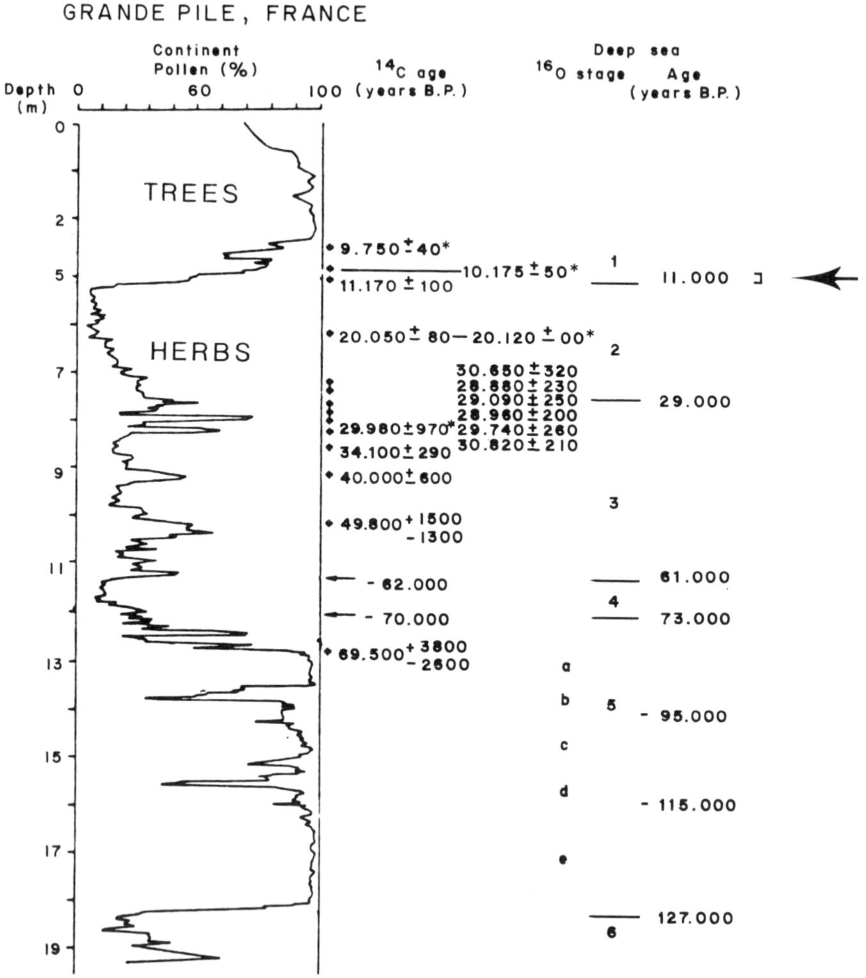

Figure 5.4a Pollen percentage diagram for Grand Pile, France (Woillard and Mook 1982). Records show fluctuations possibly indicative of rapid climate changes, with arrows marking the time of the Younger Dryas.

We therefore do not know how vegetation will be affected by the forecast climate change, and without knowing the vegetative response, we cannot properly estimate what will happen to transpiration and summer soil moisture. Ecosystem response models predict losses of soil moisture when using the forecast temperature and precipitation from GCMs, while the same GCMs (which ignore vegetation) show soil moisture gains (Rind et al. 1990).

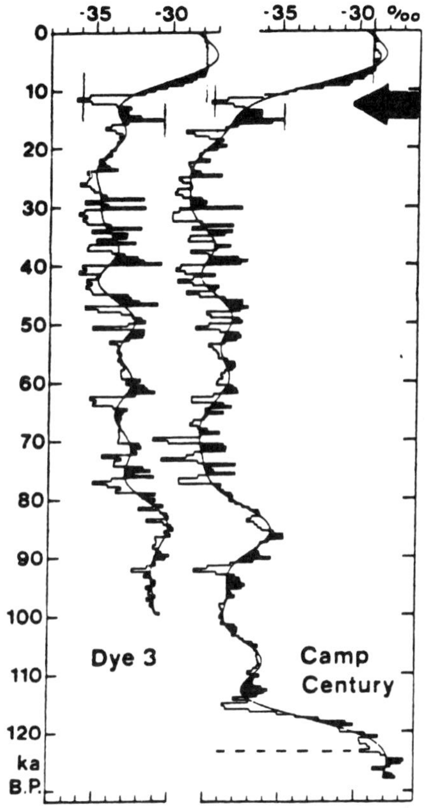

Figure 5.4b Pollen percentage diagram for $^{18}O/^{16}O$ record from Greenland (Dansgaard et al 1984). Records show fluctuations possibly indicative of rapid climate changes, with arrows marking the time of the Younger Dryas.

PALEOCLIMATE PERSPECTIVE

The lack of strict analogs for future warming does not, however, mean that the paleoclimate perspective is irrelevant. There have apparently been warmer climates in the past, as well as times of rapid warming and probably higher CO_2 levels. Paleoclimate data can help us understand how different components of the system (i.e., the ocean, cryosphere, biosphere, and climate sensitivity in general) reacted to these situations, allowing us to understand the processes by which the climate system operates. It is this basic understanding that can be incorporated into models, which may then be capable of assessing the unique situation with which the coming century presents us. An awareness of the sensitivity of fundamental processes is the most valuable knowledge that the study of past climate changes can offer.

CONCLUSION

To answer the question posed in the title, future climate changes will be forced in a manner different from most of the climate changes in the past. They will likely occur more rapidly and they will take place against an ocean/cryosphere/ biosphere background, which has never experienced rapid warming. Given this situation, it is doubtful whether paleoclimate analogs can be found to enable us to specify the characteristics of the future climate. However, knowledge of paleoclimates should enable us to study both the sensitivity of the climate system and the processes that operate within it, through its responses to forcings of different magnitudes and rates. Ideally, we may then compare our model simulations of these situations to what actually happened, thereby helping us to validate or improve the models we are using to forecast the future climate. It is in this indirect manner that the "past may be the key to the future."

ACKNOWLEDGEMENTS

Climate modeling at NASA/GISS is supported by the NASA Climate Program Office, while increased greenhouse-gas studies are supported by the U.S. Environmental Protection Agency.

REFERENCES

Barnola, J.-M., D. Raynaud, Y.S. Korotkevich, and C. Lorius. 1987. Vostok ice core provides 160,000-year record of atmospheric CO_2. *Nature* **329**:408–414.

Berger, A.L. 1978. Long-term variations of daily insolation and quaternary climatic changes. *J. Atmos. Sci.* **35**:2362–2367.

Berger, W.H., and A. Spitzy. 1988. History of atmospheric CO_2: Constraints from the deep-sea record. *Paleoceanog.* **3**:401–411.

Berner, R.A. 1990. Atmospheric carbon dioxide levels of Phanerozoic time. *Science* **249:**1382–1386.

Berner, R.A., A.C. Lasaga, and R.M. Garrels. 1983. The carbonate-silicate geochemical cycle and its effect on atmospheric carbon dioxide over the last 100 million years. *Amer. J. Sci.* **283**:641–683.

Broecker, W.S., G. Bond, M. Klas, G. Bonani, and W. Wolfli. 1990. A salt oscillator in the glacial Atlantic? 1. The concept. *Paleoceanog.* **5**:469–478.

Broecker, W.S., D.M. Peteet, and D. Rind. 1985. Does the ocean–atmosphere system have more than one stable mode of operation? *Nature* **315**:21–26.

Caldeira, K., and M.R. Rampino. 1991. The mid-Cretaceous super plume, carbon dioxide and global warming. *Geophys. Res. Lett.* **18**:987–990.

CLIMAP. 1984. The last interglacial ocean. *Quat. Res.* **21**:123–224.

Crowley, T.J. 1990. Are there any satisfactory geologic analogs for a future greenhouse warming? *J. Climate.* **3**:1282–1292.

Dansgaard, W., S.J. Johnsen, H.B. Clausen, D. Dahl-Jensen, N. Gundestrup, C.U. Hammer, and H. Oeschger. 1984. North Atlantic climatic oscillations revealed by deep Greenland ice cores. In: Climate Processes and Climate Sensitivity, ed. J.E. Hansen and T. Takahashi. *Geophys. Monog.* **29**:288–298. Washington, D.C.: American Geophysical Union.

Dansgaard, W., J.W.C. White, and S.J. Johnsen. 1989. The abrupt termination of the Younger Dryas climate event. *Nature* **339**:532–536.

DOE. 1984. Glaciers, Ice Sheets, and Sea Level: Effect of a CO_2-Induced Climatic Change. Rep. DOE/ER/60235–1. Washington, D.C.: U.S. Dept. of Energy.

Eddy, J.A. 1976. The Maunder Minimum. *Science* **192**:1189–1202.

EOS. 1991. Pinatubo cloud measured. *EOS Trans.* **72**:306–307.

Frakes, L.A. 1979. Climates Throughout Geologic Time. New York: Elsevier.

Hansen, J., G. Russell, A. Lacis, I. Fung, D. Rind, and P. Stone. 1985. Climate response times: Dependence on climate sensitivity and ocean mixing. *Science* **229**:857–859.

Hickey, J.R., B.M. Alton, H.L. Kyle, and D. Hoyt. 1988. Total solar irradiance measurements by ERB/NIMBUS–7. A review of nine years. *Space Sci. Rev.* **48**:321–342.

Horrell, M.A. 1990. Energy balance constraints on ^{18}O-based paleo-sea surface temperature estimates. *Paleoceanog.* **5**:339–348.

IPCC. 1990. Climate Change. Cambridge: Cambridge Univ. Press.

Maier-Reimer, E., U. Mikolajewicz, and T. Crowley 1990. Ocean general circulation model sensitivity experiment with an open Central American isthmus. *Paleoceanog.* **5**:349–366.

Manabe, S., K. Bryan, and M.J. Spelman. 1990. Transient response of a global ocean–atmosphere model to a doubling of atmospheric carbon dioxide. *J. Phys. Ocean.* **20**:722–749.

Owen, R.M., and D.K. Rea. 1985. Seafloor hydrothermal activity links climate to tectonics: The Eocene carbon dioxide greenhouse. *Science* **227**:166–169.

Porter, S.C. 1986 . Pattern and forcing of northern hemisphere glacier variations during the last millennium. *Quat. Res.* **26**:27–48.

Rind, D., and M. Chandler. 1991. Increased ocean-heat transports and warmer climate. *J. Geophys. Res.* **96**:7437–7461.

Rind, D., R. Goldberg, J. Hansen, C. Rosenzweig, and R. Ruedy. 1990. Potential evapotranspiration and the likelihood of future drought. *J. Geophys. Res.* **95**:9983–10004.

Rind, D., D. Peteet, W. Broecker, A. McIntyre, and W. Ruddiman. 1986. The impact of cold North Atlantic sea surface temperatures on climate: Implications for the Younger Dryas cooling (11–10 Ka). *Clim. Dynam.* **1**:3–33.

Volk, T. 1987. Feedbacks between weathering and atmospheric CO_2 over the last 100 million years. *Am. J. Sci.* **287:**763–779.

Walker, J.C.G., P.B. Hays, and J.F. Rasting. 1981. A negative feedback mechanism for the long-term stabilization of the Earth's surface temperature. *J. Geophys. Res.* **86:**9776–9782.

Winograd, I.J., B.J. Szabo, T.B. Coplen, and A.C. Riggs. 1988. A 250,000-year climatic record from Great Basin vein calcite: Implications for Milankovitch Theory. *Science* **242**:1275–1280.

Woillard, G.M., and W.G. Mook. 1982. Carbon-14 dates at Grand Pile: Correlation of land and sea chronologies. *Science* **215**:159–161.

Standing, left to right:
T. Webb, B. Frenzel, D. Rind, T.J. Crowley, A.-K. Gliemeroth, A. Zwick

Seated, left to right:
C. Prentice, W.F. Ruddiman, M. Sarnthein, J. Jouzel, L. Labeyrie

6

Group Report: Use of Paleoclimatic Data as Analogs for Understanding Future Global Changes

T. WEBB III, Rapporteur

T.J. CROWLEY, B. FRENZEL, A.-K. GLIEMEROTH, J. JOUZEL, L. LABEYRIE, I.C. PRENTICE, D. RIND, W.F. RUDDIMAN, M. SARNTHEIN, A. ZWICK

INTRODUCTION

Increasing awareness of the potential rate and magnitude of future climate changes has stimulated the study of how the climate system works. Much interest is currently being focused on the mechanisms and regional patterns of recent and ancient climate changes associated with (a) increased concentrations of CO_2, (b) warmer conditions than today, (c) periods of large and/or rapid warming, and (d) times of higher-than-present sea levels or less-than-present sea ice. The paleoclimatic record is critical here because it provides information about most of these conditions for which standard meteorological instruments and satellite images contain no record. In addition, studies of past climates have contributed much to climate theory through data compilation, time series analysis, and systematic comparisons between paleoclimatic data and climate model results (Berger et al. 1984; COHMAP 1988; Rind et al. 1986). Despite this and other climatic research, however, we still lack an adequate theory and understanding of the climate system to make accurate predictions of future climate changes. The scenarios such predictions could provide would be useful aids in establishing economic and political policy.

In the absence of a complete theory on the basis of which predictions could be made, climate modeling experiments with increased levels of CO_2 have been a leading

Global Changes in the Perspective of the Past
Edited by J.A. Eddy and H. Oeschger © 1993 John Wiley & Sons Ltd.

source of scenarios of future climate change (Houghton et al. 1990; chapter 5 in MacCracken et al. 1990; Heinze et al. 1991). However, doubts have been expressed about the accuracy and realism of such models and about their ability to predict correctly regional patterns of future change (Schlesinger and Mitchell 1985). Uncertainty about modeling results has led to the use of data about past climates as an alternative source of scenarios of potential future climates (Kellogg 1978; Budyko et al. 1985; Budyko and Izrael 1987; chapter 6 in MacCracken et al. 1990). Key times suggested have included the mid-Holocene (5 to 6 Ka, thousand of years ago), the last interglacial period (~125 Ka), and the early Pliocene (4.3 to 3.3. Ma, million of years ago). The hope has been that the estimates for climates at these and other selected intervals can serve as *analogs* for future climates. Our group unanimously rejects such a simplistic application of paleoclimatic data but strongly favors the use of such data to check the predictions of climate models, to learn about key climatic processes, and to characterize the behavior of the climate system and its major components during periods of rapid global change. Paleoclimatic data are also essential for characterizing the natural variability of the climate system. We subscribe to a reversed version of the uniformitarian principle, i.e., that the past is *a*, but not necessarily *the*, key to understanding the present and the future.

TYPES OF ANALOGS FROM PAST CLIMATES

In discussing the use of paleoclimatic data to help understand the climate system and to aid in the prediction of future changes, we found it useful to distinguish between climatic homologs and analogs (Schumm 1991). Homologs show similarities in structure, function, origin, and cause (e.g., much like homologous organs in animals, such as the limbs among tetrapods), whereas analogs are similar only in appearance or in certain characteristics but not necessarily in genesis or cause. Schumm (1991) notes that whereas "analogy is like a metaphor," homology may be "like a synonym." One problem with this distinction for climates is that the nonlinear nature of the climate system may make it intransitive and, therefore, similar causes or initial states may produce different patterns or final states. Hence, although homologous climates may share similar causes, they may differ in form or pattern just as a bat's forelimb differs from that of a cat.

A central criticism of the use of past climates in predicting the effects of rapid greenhouse warming (Crowley, this volume) has been that past climates may neither be analogous nor homologous to future ones. Many past climates qualify as partial homologs because they occurred at times when CO_2 concentrations were higher than today, or as partial analogs because they occurred when temperatures were warmer than today or increasing, when global mean temperatures (GMTs) increased rapidly, and/or when sea levels were higher or sea ice less than they are today. However, none of these climates match in all aspects or in origin the predicted greenhouse-gas-induced warming of the next 50 to 100 years. These past climates, therefore, are imperfect as either homologs or analogs.

In judging the use of past climates as analogs, we distinguished among three types of analogs: (a) time-slice analogs, (b) component or process analogs, and (c) rate-of-change analogs. *Time-slice analogs* are past climates that were warmer than today, either globally or regionally, or had higher atmospheric concentrations of CO_2. Certain of these climates, e.g., those of the early to mid-Holocene (9000 to 6000 radiocarbon years ago or 9 to 6 kra), have been used to generate scenarios for the future (Kellogg 1978). *Component or process analogs* are climates from times or periods chosen to show how components or processes within the climate system operate and interact. Therefore, while they may be critical in illustrating how the climate system works, such analogs should not be used to provide scenarios. *Rate-of-change* analogs are past periods chosen to illustrate the behavior of the climate system during times of rapid change, with times of rapid temperature increase being of particular interest. These analogs are most useful if the changes to which they refer occurred on a global scale; however, times of rapid regional or local changes can also be helpful in understanding impacts and interactions. The end of the Younger Dryas period (ca. 10 kra), when regional temperatures increased rapidly, is an example of a possible rate-of-change analog (Rind et al. 1986).

The main body of this report contains a discussion of each of these types of analogs. Recommendations for future research on as yet understudied topics are listed at the end.

TIME-SLICE ANALOGS

Because of the prediction that increased concentrations of CO_2 will warm the Earth's climate, all climates chosen as time-slice analogs are purported to have been warmer than present. These include the following periods: (1) the Medieval "Warm Epoch" (A.D. 100 to 1300), (2) the early to mid-Holocene (9 to 6 kra), (3) the last interglacial period (Eemian, ~125 Ka, $\delta^{18}O$ Stage 5e), (4) other interglacials during the Quaternary (e.g., Holstein interglacial, ca. 400 Ka, $\delta^{18}O$ Stage 11), (5) the early Pliocene (4 to 3 Ma), (6) the early Eocene (55 to 50 Ma), and (7) the mid-Cretaceous (120 to 90 Ma). We emphasize that global conditions at these times "may" have been warmer than present because definitive evidence for some of them is not yet available. To obtain such evidence is a key objective for future research.

The use of time-slice analogs as scenarios has evoked much critical comment (Webb and Wigley 1985; Crowley 1990, and this volume). Much of the recent criticism was set off by the use, in Budyko et al. (1985) (see also Budyko and Izrael 1987; Folland et al. 1990; chapter 6 in MacCracken et al. 1990), of reconstructed climate patterns for 5.5 ± 0.5 kra, 125 ± 5 Ka, and 3.8 ± 0.5 Ma to provide scenarios for climates with GMTs 1°C, 2°C, and 3°C higher than today.

We identified six concerns that hamper use of these past warm periods as analogs for potential future changes (Crowley 1990, and this volume). Some of the criticisms are listed as caveats in Kellogg (1978), Folland et al. (1990), and chapter 6 in MacCracken et al. (1990).

1. *Mismatch of forcing.* Elevated concentrations of CO_2 are neither the cause nor the sole or major cause of most of the past warmer climates. Other factors, such as orbital forcing, retreating ice sheets, changes in ocean circulation patterns, or tectonic activity, often appear to be the dominant cause of past warming. This holds true for all periods except the early Pliocene, in which increased concentrations of CO_2 may have been the main cause of its being warmer than today, despite a slightly different continental arrangement.
2. *Inadequate temporal resolution.* The danger exists that data of significantly different climatic states may be lumped together as synchronous even when they are not. Important leads and lags among variables, processes, and regions in the climate system may not be resolved, and time-transgressive phenomena may be mapped as having occurred simultaneously. Here, the early to mid-Holocene period ranks better than the earlier time periods in terms of showing reasonably correct geographic patterns, but even its temporal resolution is most accurate in terms of centuries.
3. *Uneven or incomplete spatial coverage.* This can make it difficult to estimate the GMT or other variables, such as global biomass, ocean ventilation, productivity, etc. Evidence for warmer-than-present conditions in certain areas may lead to overestimates of the past GMT. The Medieval warm epoch suffers here as do all periods; however, the early to mid-Holocene time has better coverage than the other periods.
4. *Slow rate of past global change relative to predicted future change.* Processes and responses important over long time scales (e.g., deep-ocean circulation) may not vary significantly over the short term. Certain equilibrium responses that take 1000 or more years to develop may not occur in the immediate future, when transient responses may be the rule. These are classic problems with scale. All but the Medieval warm period suffer from this restriction, but even it may only offer limited information, as its GMT may not have been higher than today's.
5. *Different boundary conditions from present.* Certain key processes or contemporary estimates of time constants may not apply in the future. With different continental or ice-sheet configurations, bottom-water formation may occur at different rates. This problem plagues the Eocene and mid-Cretaceous and is potentially true for the early Pliocene (see later discussion).
6. *Uncertainties in interpreting the data.* These can lead to inaccurate and imprecise estimates of past climate variables and thus to inaccurate estimates of the differences between past climates and today's. This problem exists to some extent for all periods, but is greatest for the most ancient times.

The climates for each of the warm periods are imperfect analogs or homologs and should not be used to construct specific regional climatic scenarios. When, however, each period is studied for understanding the climate system, it can provide information helpful in predicting certain aspects of future global changes.

Medieval Warm Epoch (ca. A.D. 1000 to 1300)

Historical records from Eurasia demonstrate that the Medieval period (A.D. 1000 to 1300) was a period of relative warmth there (LeRoy Ladurie 1967; Lamb 1965, 1988; Alexandre 1987; Bradley 1992). The exact dates are uncertain, but A.D. 1120 to 1280 bracket part of the warm time. The dates listed in the subheading come from Lamb (1965), who defined the period as "a prevailing warm epoch with dry anticyclonic character, especially in summer...between A.D. 1000 and 1300" (for Europe). Summer temperatures in some regions may have been 1° to 1.5°C higher than they are at present. Tree-ring records from northern Scandinavia (Briffa et al. 1990) and the northern Ural Mountains (Graybill and Shiyatov 1992) show evidence of warmer summer temperatures. The global extent of the higher-than-present temperatures is still unknown (Webb 1991), however, and Bradley (1992) concludes that "the available evidence suggests that there was not a long, continuous warm interval which was synchronous throughout the world, or even throughout the Northern Hemisphere." Further study of historical and other records is strongly encouraged in order to broaden the data base available for this period.

Early to Mid-Holocene (9 to 6 kra)

Global data bases for this period have already been published (Webb 1985; Street-Perrott et al. 1989; Wright et al. 1993). Data in the oceans, in the tropical rain forests, and in southern latitudes are sparse, but estimated isotherm maps exist for Europe and eastern North America (Huntley and Prentice 1988, 1993; Webb and Wigley 1985; Webb et al. 1993). In addition, less well-documented maps for several climatic variables exist for the Northern Hemisphere (Budyko and Izreal 1987; chapter 3 in MacCracken et al. 1990). In comparison to 18 kra, relatively little is known about sea-surface temperatures (SSTs), but what is known indicates that temperatures differed little from today (Ruddiman and Mix 1992). No spatially averaged estimate of the GMT has been attempted; however, Webb and Wigley (1985) estimated the GMT to be at most 1° ± 1°C higher than today for 6 kra. Summer temperatures are estimated to have been around 1°–2°C warmer than present in the mid- to high latitudes of North America and Eurasia. Winters were also warm in high-latitude Eurasia (e.g., Velichko et al. 1991; Klimanov 1984; Huntley and Prentice 1993), but colder than present in the mid latitudes of North America (Webb et al. 1993). The increased summer-season insolation led to enhanced monsoons in Eurasia and central Africa. Annual precipitation was much higher than today in southern Asia and much of Africa north of the equator (COHMAP 1988; Petit-Maire 1989). Annual precipitation in central North America, however, was less than today at 6 kra (Bartlein et al. 1984). During this time, permafrost in Siberia retreated for the last time. In contrast, in the Southern Ocean (Indian sector), cores dated by accelerator mass spectrometry indicate that the warmest temperatures were about 11 kra, and that the period from 9 to 6 kra was significantly colder than today (Labracherie et al. 1989).

Concerning the factors creating the warmer conditions, we know that the average summer-season insolation in the Northern Hemisphere was 5% to 8% higher than today, but that insolation during winter was less than today (COHMAP 1988). Annual mean insolation was substantially greater than today in the high latitudes of both the Northern and Southern Hemispheres. This, however, was balanced by annual mean insolation slightly less than present over the much larger area of the Subtropics and Tropics. Given the much greater amount of terrestrial data that have been collected in the high latitudes and the stronger temperature signal over the high latitudes, it is easy to see how this spatial bias might have created a false impression of overall global warmth.

From 9 to 6 kra, CO_2 levels were 270 ± 30 ppm (i.e., at preindustrial levels), and ocean productivity, a major factor for atmospheric CO_2 concentrations, was as high as today (Sarnthein and Winn 1990). In eastern North America and Europe, the slow disappearance of the Laurentide ice sheet delayed the period of highest summer temperatures until 6 kra or later, which was after the time of peak summer insolation at 11 to 9 Ka (Webb et al. 1993; Huntley and Prentice 1993). These facts indicate that the cause of the warmer-than-present temperatures was not enhanced concentrations of CO_2. Knowledge of the factors controlling the climates from 9 to 6 kra and the fact that the climate differences from present are large in certain regions, e.g., the northern Tropics, make these climates ideal for testing the results of general circulation models (COHMAP 1988).

Last Interglacial: The Eemian/Mikulino/Sangamon (125 ± 5 Ka)

Evidence exists that sea levels during the last interglacial at 125 ± 5 Ka (SPECMAP time scale; Martinson et al. 1987) were 6 to 7 m higher than during the Holocene and that glacial ice volume was less (CLIMAP 1984; Bard et al. 1990). This evidence suggests that the GMT might have been higher than during the Holocene, perhaps by 1° to 2°C. Pollen data from Eurasia and North America also indicate that the ranges of many plant taxa extended farther north at that time than during the Holocene (Watts 1988). Difficulties exist in dating, however, with resolution no better than ± 5 Ka for absolute age estimates. This imprecision restricts accurate mapping of the past climatic pattern at a submillenial scale, but within individual cores relative ages allow for improved resolution. At the beginning of this interglacial, the atmospheric concentrations of CO_2 were somewhat higher than preindustrial levels during the Holocene (Barnola et al. 1987), but the global maps of estimated SSTs show average temperatures no higher than those of today (CLIMAP 1984). The exact location and extent of ice sheets is unknown; however, evidence exists that the area of Eurasian permafrost was much smaller than it is today (Frenzel 1967; Velichko et al. 1984). Valuable information about fossil soils in China is found in Kukla (1987) and Liu and Yuan (1987). These soils seem to resemble those of present-day natural areas but also indicate more moisture and biological activity than are found today. Hemispheric maps for selected climatic variables exist (Velichko et al. 1984, 1991; Folland et al. 1990; chapter 3 in MacCracken et al. 1990), but their contours often represent large extrapolations from sparse or no data. A cooperative effort to redo these maps with indications of dating uncertainties at each site is a top priority project.

Other Interglacials during the Quaternary

The Holstein interglacial in Europe has been tentatively correlated with $\delta^{18}O$ Stage 11 in the deep-sea record (~400 Ka) (Sarnthein et al. 1986), although Watts (1988) discusses its correlation with $\delta^{18}O$ Stage 7. Marine records show that $\delta^{18}O$ Stage 11 was characterized by an almost complete lack of precessional forcing. Its onset lasted about 3 to 4 times longer than the subsequent glacial terminations. This time interval may have been at least as warm as or even warmer than the last interglacial (~125 Ka), with polar sea ice retreating in both hemispheres. Pollen studies of this and other interglacials in Europe suggest that the sequence and magnitude of vegetation changes differ for the various interglacials (Watts 1988; Frenzel 1967). Comparisons among these time periods may provide some insight into the potential differential response of the biosphere to slightly differing climatic forcing and to the different climate conditions they produce (Watts 1988). Comparisons may also show the degree of independence of the biological response when climatic forcing is weak in selected interglacials; however, poor time resolution may make such conclusions uncertain.

Early Pliocene (3 to 4 Ma)

During the early Pliocene, when the GMT may have been 1° to 3°C higher than today, Northern Hemisphere ice sheets were not present, and levels may have been a significant causative force. Budyko et al. (1985) and Crowley (1991) give an estimate of 600 ppm for CO_2 concentrations; however, this estimate needs verification. Because close time control is difficult to obtain during this interval, orbital-scale changes are less easy to resolve, although many orbital-scale climate changes were probably smaller because of reduced feedbacks from Northern Hemisphere snow and ice sheets (Tiedemann 1991). In our judgment, this period may be the "least bad" homolog for the predicted future climates after global warming. Tectonic boundary conditions were very similar to the present, and high-latitude warmth seems to have been synchronous (within dating uncertainties of $>10^4$ yr) in both hemispheres. Planktonic foraminiferal species are sufficiently similar to those at present to justify the use of transfer function techniques for SST estimates. A fairly large number of terrestrial sites have moderately good stratigraphy. Earlier studies on the Pliocene averaged stratigraphies over several Milankovitch-scale cycles (Budyko and Izrael 1987); more recent studies (e.g., Cronin and Dowsett 1991) indicate that good temporal resolution (<10 Ka) can be achieved within many Pliocene sites. However, the temporal resolution for maps of these data is much greater than 10 Ka.

Like the mid-Holocene (~6 kra) and the last interglacial (~125 Ka), hemispheric maps of selected climatic variables for this period have been published (Budyko and Izreal 1987); however, these maps lump together data which range in age by 1 million years (a span equivalent to half the Quaternary) and require full documentation in terms of site locations, dating accuracy, and paleoclimatic uncertainties. Changes in climate are evident during the Pliocene. For example, Sarnthein and Fenner (1988) published global maps of ocean productivity, which was lower during the latter part of the Pliocene warm phase (3.2 to 3.4 Ma) than during the earliest glacials (approximately 2.6 to 2.4 Ma). Caution, therefore, is required in interpreting the maps of Pliocene data.

Early Eocene (55 to 50 Ma)

This period was the warmest of the Cenozoic, with GMTs perhaps 4° to 6°C higher than today (Crowley and North 1991). Tropical vegetation grew in relatively high latitudes (Tallis 1991), and fossils of tropical plants were at least 10° to 15° poleward of their location today (Crowley and North 1991). Although this peak warmth may have been associated with the highest CO_2 concentrations of the Cenozoic, there is relatively little concrete evidence of this association at this stage of analysis. The chief attraction of this time period is the extreme warmth and fairly good geographic coverage of samples.

Mid-Cretaceous (120 to 90 Ma)

GMTs of this period are estimated to have been 6° to 14°C higher than today, and coral reefs and floral provinces were 15° poleward of their present locations (Crowley and North 1991). CO_2 levels in the Cretaceous were likely considerably higher than at present. One piece of circumstantial evidence for this estimate is the inability of any factor other than CO_2 to explain the absence of permanent ice on the globe at this time, particularly on Antarctica, which was already in a polar position 100 million years ago. Recently, several geochemical techniques have begun to give promise of resolving past CO_2 levels, which are thought to have been in the range of those reached in later centuries. Berger and Spitzy (1988) describe evidence for CO_2 concentrations 4 to 7 times those in the atmosphere today.

The mid-Cretaceous interval, however, does not represent a good analog for future climate change. Continental positions were substantially different from today, including a much smaller Atlantic Ocean without a deep connection to the northern seas, an open subtropical Tethys Sea, and a grouping of Southern Hemisphere continents closer to (or connected with) Antarctica. In addition, global topography was lower than today, with no Tibetan Plateau, or lower elevations in East Africa, the Mediterranean

Alps, and the Cordilleran chain through the Americas. These differences in continental position and elevation would have had added climatic impacts on atmospheric and oceanic circulation beyond the effects of CO_2, thus compromising the degree of "analogy" with future CO_2 change.

COMPONENTS/PROCESS ANALOGS

An empirical view of interactions among different components of the climate system is critical to an understanding of how the climate system works. Because the instrumental record of climate change affords only limited examples of many of these interactions, the paleoclimatic record is essential for illustrating the patterns and interactions during climates with higher temperatures than present or with higher concentrations of atmospheric CO_2. Information is needed about how warmer climates or more CO_2-rich climates affect (or are affected by) ice sheets, sea levels, sea ice, ocean circulation and temperatures, the biosphere, terrestrial systems, and climatic sensitivity (lambda) to changes in atmospheric composition.

Ice Sheets

The projected future warming raises concerns about the vulnerability of the polar ice sheets (Warrick and Wigley 1991). Previous warm climates provide the possibility of assessing the equilibrium reaction of these ice sheets to climates associated with higher than present CO_2 concentrations. This assessment, however, will not provide a perfect analog for the transient response to increased greenhouse gases. In particular, sea level was estimated to be about 6 m higher during the last interglacial (~125 Ka, $\delta^{18}O$ stage 5e, CLIMAP 1984; Bard et al. 1990). Where did this additional water come from?

One possibility is that the Greenland ice sheet melted during this time (Koerner 1989), which may have been somewhat warmer than the current interglacial and did experience greater solar radiation in summer. Within the next two years, ice-core drilling programs in Greenland should establish whether ice was present at ~125 Ka, at least in its central part.

Another possibility for the rise in sea level comes from melting of the West Antarctic ice sheet, which would provide about 5 m of sea level. Its configuration apparently makes it vulnerable to melting during warming, because it is held in place via grounding below sea level. There are no immediate plans to drill for a new core deep enough to investigate its state during $\delta^{18}O$ stage 5e (~125 Ka).

The East Antarctic ice sheet covers a much greater area and has the potential for producing more than 60 m of sea-level rise. For the most part, this ice sheet appears to be less susceptible to disintegration; however, changes in thickness of 200 to 300 m (i.e., about 10%) could supply the necessary water for the 6 m sea-level rise. Estimates of past ice-sheet elevation from the air pressure of entrapped bubbles, in combination with dynamic ice-flow modeling, may lead to estimates for the volume of this ice sheet for ~125 Ka, but these estimates will have large uncertainties.

The equilibrium conditions associated with whatever relative warming existed at the previous interglacial, however, may prove an inaccurate guide to the transient response of the next 100 years. A relatively small amount of warming, insufficient to raise temperatures above freezing, may lead to added accumulation on ice sheets, as seems to be occurring today in some parts of Antarctica (Morgan et al. 1991), as warmer air can hold more moisture for precipitation. Substantially greater warming, which may require longer to achieve because of the slow response time of the oceans, could set in motion instability mechanisms whose end result could be recorded in the equilibrium record. Nevertheless, if the equilibrium response does imply ice-sheet vulnerability, then this condition could eventually be approached in coming centuries, a consequence which should be known as soon as possible.

Sea-level Changes

A major concern about global warming is that ice sheets may melt and sea level may rise and flood coastal cities. Earlier periods, like the last interglacial (the Eemian, ~125 Ka), can provide information about changes when sea levels were higher than today. The fast eustatic increase of sea level at the beginning of the Holocene and the Eemian (at a rate of between 2.0 and approximately 6.0 m/century) caused a rapid retreat of the coastline in low lying regions of Europe. This caused Scandinavia to become an island during the Eemian, and transgressions penetrated far to the south in the West Siberian lowlands during the Holstein and Eemian interglacials (~400 and ~125 Ka).

Sea-ice Changes

Sea ice is a fast-response part of the climate system that is particularly sensitive to high-latitude temperature and thus to changes in atmospheric CO_2 levels and ocean-heat transport. Sea ice, in turn, has a large impact on regional albedo feedback, on heat and moisture transfer between land and sea, and on atmospheric circulation.

In regions that were warm, some processes (e.g., sea-ice feedbacks) may have operated in a way similar to that which is anticipated under greenhouse warming. For example, all climate models predict a sea-ice feedback that will amplify greenhouse warming, especially in winter, at high latitudes. At 6 kra, the high latitudes experienced greater-than-present summer and annual insolation, and the climate models again predict amplification of the resulting warming by sea-ice feedback. This is even more true of ~125 Ka. These predictions are borne out by land evidence for exceptionally warm winters in northernmost Eurasia during the early/mid-Holocene (Huntley and Prentice 1993) and, even more so, during the last interglacial (~125 Ka).

Other sea-ice changes, such as reduced sea-ice limits in the early to mid-Holocene and previous interglacials, may be less closely linked to ice-sheet dynamics; however, even in these cases, orbitally induced insolation changes probably outweigh CO_2 in strength of forcing, and melting ice sheets may defer the maximum effect of insolation-induced heating until the maximum in summer insolation.

Important coring efforts are underway or planned in the Arctic. The exact limits of reduced sea ice in the Northern Atlantic and Arctic are not known for either the early Holocene, the last interglacial, or pre- and early Pliocene. Efforts to extract these past limits of sea ice in both the Arctic and Antarctic from gravity and piston cores, as well as from planned ocean drilling work, are vital.

Ocean Circulation and Temperatures

North Atlantic Deep Water (NADW)

Concern that ocean circulation may suddenly change during global warming has prompted interest in past changes in circulation. Evidence exists for changes (sometimes rapid) among different circulation states in the oceans. The best examples are from the late Pleistocene (Rind et al. 1986), and these appear to be causally linked to forcing from the ice sheets, either via changes in wind stress curl or changes in surface-ocean salinity (or both). The windstress effects tend to be slower (orbital-scale) changes with less relevance to possible future climate. The salinity changes, however, may well be relevant to future climate because portions of existing ice sheets may undergo rapid melting in a warmer world. This would directly affect mixed-layer salinity in the North Atlantic or Antarctic, and deep-water formation in both areas is strongly influenced by salinity.

Tropical Sea-surface Temperatures

The response of tropical (20°N–20°S) SSTs to the projected increase in greenhouse gases remains one of the greatest uncertainties in predicting future climates. Different models predict different magnitudes of tropical warming (Rind 1987); nevertheless, all models predict significantly more warming than that which has been reconstructed for times of supposed increased levels of CO_2, such as during the Pliocene and Eocene. For models to provide the substantial high-latitude warming apparent in the fossil flora and fauna of these time periods (Tallis 1991), in conjunction with isotopic reconstructions of little tropical temperature change, they must incorporate increased ocean-heat transports of about 20% globally, with the largest changes at high latitude.

Various warm periods (e.g., ~125 Ka, and previous interglacials) can be used to investigate the tropical response; however, the most appropriate time periods may be in the Tertiary because the warming in the past may be closely associated with particular forcing (e.g., insolation maxima at high latitudes induced by orbital variations are less likely to provide tropical warming than are elevated CO_2 values that are globally distributed). To evaluate the equatorial response, and the appropriateness of a past warm period serving as an analog, we need to know: (a) what were CO_2 levels during the Tertiary, and (b) what were tropical land and SSTs. To evaluate the ultimate relevance and perhaps provide future projections, we also need to know how much heat the ocean transported at these times.

Various methods are being developed that involve $\delta^{13}C$ from carbonates and organic carbon in an attempt to reconstruct CO_2 levels for the Tertiary. Reevaluation of isotopic reconstruction for tropical SSTs is necessary, along with assessments of the temperature indications from land biota. The various methods used to estimate paleothermohaline circulations should be applied to data from these time periods, although the different climate/ocean structures and the possibility of different water masses such as North Pacific Deep Water must be considered.

Terrestrial Systems

Vegetation

From pollen and plant macrofossil data, we know that major tree taxa respond independently of one another and migrate at relatively fast rates (Davis 1989; Huntley 1991; Johnson and Webb 1989; Prentice et al. 1991). Plant communities and ecosystems do not migrate intact but form new assemblages and perform new functions under new climates. Over time scales of decades to centuries, lag and hysteresis effects are evident when taxa migrate and local extinctions occur (Prentice 1986; Grimm 1983). Such transient effects occur when the details of local-to-regional vegetation change are resolved (Davis 1991). Gene frequencies also change with migrating populations of a species (Cwynar and MacDonald 1987); however, these changes may not affect their climatic tolerance (Huntley et al. 1989). Opportunities exist to test the degree of genetic change between interglacials. These tests should be performed on (a) records from Europe that contain data from one or more interglacials and (b) records from deep cores containing pollen records of 150,000 or more years (Watts 1988). Such studies can help us sort out what role evolution plays in the response of species to climate change. Other questions that need answers include:

1. What will happen to the tropical rain forest if and when the climate in the Tropics becomes warmer than it is today?
2. What happens to moisture balance in the middle of continents under warmer conditions?
3. How critical is the feedback between vegetation and climate for estimating the rate and magnitude of changes in the moisture balance?

Global vegetation models that can be coupled with global climate models are needed; the vegetation models should incorporate the necessary vegetation processes to show the correct vegetational response to warming (Prentice and Solomon 1991).

Soils

The former distribution patterns of fossil soils can be mapped if broad time windows, with a duration of about 2000 to 3000 years, are used. This can be accomplished for the warmest periods of the Holsteinian and the Eemian interglacials (~400 and ~125 Ka), and the more important interstadials of the last glaciation, its stadials, and the Holocene up to the time of the formation of bronze-age burial mounds (~3.5 Ka). These maps for the last two interglacials and for the Brørup-Odderade/St. Pierre interstadials (70 to 30 Ka) already exist (Frenzel 1967). These maps can be extended to the east as far as Northern China and to North America, south of the last glacial ice margin, as well. This extension should be made in order to understand the contemporary terrestrial CO_2 budgets.

Permafrost

Dates on the former extent of permafrost can be obtained from various fossil indicators of permafrost as well as from frozen cadavers of well-dated animals in the present-day permafrost. These dates and other geomorphic evidence for permafrost make it possible to map the former full glacial extent of permafrost, at least in northern Eurasia (Velichko 1984). The extent of permafrost can also be mapped for the Holocene and the last interglacial (~125 Ka). The patterns on these maps may help us understand the changing methane production from the melting of peat bogs and fens.

Climate Sensitivity to Changes in Atmospheric Composition (lambda)

Although the paleoclimate record supports the general relationship between atmospheric CO_2 levels and temperature, the precise sensitivity (or lambda, given in units of w/m^2 per °C) of the climate system to boundary condition changes is uncertain to within a factor of 2 or 3. The sensitivity also depends on the time scale. Additional work on this topic requires more accurate determination of surface temperatures (especially in the Tropics) for various times in the past and a reconstruction of oceanic plankton production and carbon transfer, atmospheric aerosol loading, and atmospheric CO_2 levels. For any pre-Pleistocene simulations with climate models, accurate estimates are required of atmospheric CO_2 concentrations. To improve the accuracy of current estimates, we recommend that the CO_2 estimates be compared for different data sources, with the Pliocene being an attractive interval for these comparisons.

RATE-OF-CHANGE ANALOGS

Current predictions suggest that greenhouse-gas-induced changes in climate of 2°C or more may occur in 50 to 200 years (Houghton et al. 1990). Intervals in the past are therefore being studied that contain rapid changes in either regional or global climate, especially rapid rates of warming. Ideally such a period would contain rates of GMT increases of 1°C/100 years or faster. The hope is that these intervals will reveal patterns of change and interactions among variables, which will help us predict future climate patterns. We certainly need to learn more about how the climate system behaves when changing rapidly. Proposed periods and events warranting study include early in the Medieval warm epoch (A.D. 1100 to 1150), rapid changes in African lake levels in the early to mid-Holocene (9 to 6 kra), the period of rapid warming as the Younger Dryas ended (~10.4 to 9.1 kra), rapid temperature changes in Greenland ice cores in $\delta^{18}O$ stage 3 (~40 to 25 Ka), and rapid cooling in the Southern Ocean in the middle of $\delta^{18}O$ stage 5e (~126 to 124 Ka).

Examples

1. *Early in the Medieval warm epoch (A.D. 1120 to 1150).* The early part of the Medieval warm epoch may serve as an example for rapid climatic change during the recent past. At that time, summer temperatures across Europe rose by about 2° to 3°C in comparison to the preceding phase of colder climates (Lamb 1988). We lack evidence, however, to show that this temperature increase was global or all that abrupt (Webb 1991; Bradley 1992).
2. *Early to mid-Holocene (9 to 6 kra): African Lake Levels.* Many records from lakes show abrupt, large changes in water levels over the scale of centuries (Street-Perrott and Perrott 1990). These fluctuations appear to be correlative among lakes over broad areas, but highly precise dating is required. This potential synchroneity indicates large amplitude, broad-scale responses during a time when ice volume was not much larger than today. These rapid changes are therefore less complicated by ice-sheet dynamics than those just after the Younger Dryas (~10.4 kra). Determining the cause of such large-scale responses is a key problem for paleoclimatic research. Nonlinear systems like the climate and hydrological systems may well produce rapid responses to gradual changes in forcing. If so, documentation of such changes is important to help avoid climatic surprises in the future.
3. *Post-Younger Dryas warming/Termination IB (10.4 to 9.1 kra).* One of the most striking candidates for a rate-of-change analog occurred during the last deglaciation, when regional climates warmed just after the cool climatic episode during the Younger Dryas interval (from ~11 to 10.4 kra). Because large biospheric changes are evident from eastern North America to Europe, this period can illustrate a high-resolution biospheric response to rapid warming. Observed changes for the post-Younger Dryas interval include:
   - A fast rise in sea level of about 40 m over about 2000 cal (calender) years as documented in coral reefs (Fairbanks 1989) and perhaps up to more than 50 m over less than 2000 years, linked to preceding meltwater events, recorded in various $\delta^{18}O$ records from the deep sea. The fast shift of sea level implies that in some adjacent seas, such as the Persian Gulf, the coastal marine biosphere had not yet reached an ecological equilibrium, as documented by the lack of coral reefs during the Holocene.

— A fast horizontal migration of the coastline in the region of the Persian Gulf, where the coastline shifted by up to several hundred meters per year (Sarnthein 1972). Reports of deluges in Sumerian records may refer to these changes.
— A rapid warming of high-latitude seas, such as the Norwegian–Greenland Sea, without a major shift in the circulation patterns or the location of fronts in the surface water. Sarnthein et al. (1992) present evidence for an increase of 7°C in 2000 to 2500 cal. yrs just west of Norway. This response, however, may be induced by nearby changes in the Scandinavian ice sheet.
— An even more marked rise in atmospheric methane of ~60% of the total glacial-to-interglacial change (Chappellaz 1990).
— A short-term meltwater-induced breakdown in NADW circulation, lasting for less than 50 (radiocarbon) years (about 10 to 9.5 kra) as resolved in a few high-resolution cores from the deep sea (Boyle and Keigwin 1987; Fairbanks 1989; Sarnthein and Tiedemann 1990), though other studies (Labeyrie et al. 1992) show an active NADW circulation with reduced flow or flow intermediate between that at 18 Ka and today.
— An abrupt increase in continental humidity in (northern) subtropical latitudes, linked to both an expanded monsoon regime and enhanced westerly rains along the Mediterranean belt (Sirocko et al. 1991).
— A striking shift from pine to oak forest in European riverine forests within less than 150 years (Becker et al. 1991), associated with a rise of the Alpine treeline by 500 to 600 m (Burga 1988). Becker et al. (1991) also recorded a unique unexplained shift in the $\delta^{13}C$ composition of tree rings in Europe over 300 cal yrs during the Preboreal (9.5 to 9.2 kra).

Caution is required in any use of these rate-of-change analogs because of:

— Potential imprecise dating of the observed events due to (a) lack of sediment resolution and (b) a "10 Ka ^{14}C plateau," i.e., the constancy of radiocarbon ages over more than 400–600 cal yrs from 11.9 to 11 cal Ka (Becker et al. 1991).
— Imprecise chronostratigraphic correlation of various events, observed on land and in the ocean, although ash layers (e.g., the Vedde ash in the northern North Atlantic at 10.6 ± kra) provide unique tools for correlation.
— An antecedent regime of glaciation that differed markedly from present day. Large ice sheets still existed in North America and Europe. Also, the African monsoon was already enhanced in the northern Tropics.
— The geographic extent of the colder conditions during the Younger Dryas is still unknown, and the global pattern of the post-Younger Dryas warming requires mapping.

Stage 3 (40 to 25 Ka): Rapid Changes in $\delta^{18}O$ in Greenland Ice-core Records

In ice cores from Greenland, the period between 40 and 30 Ka is characterized by a series of rapid changes in $\delta^{18}O$. These changes imply that large temperature increases

(~5°C) occurred at the century time scale (Stauffer et al. 1985). The same events may be recorded in North Atlantic deep-sea cores (Broecker et al. 1990), suggesting that there may be changes of ocean circulation of a regional character (North Atlantic and adjacent regions) to these temperature changes. In these intervals, temperature changes are in phase with changes in CO_2 (~60 ppm) recorded in the Greenland ice core (Stauffer et al. 1985); however, these CO_2 changes are not recorded in data from the Byrd Antarctic ice core for the same time interval nor are the temperature changes of the same character. This discrepancy has to be resolved for consensus to develop that the rapid changes in CO_2 observed in the air occluded in Greenland ice do indeed reflect changes in atmospheric CO_2 concentrations and in climate. New measurements of CO_2 and $\delta^{18}O$ are planned for the GRIP and GISP2 ice cores from central Greenland.

Middle of $\delta^{18}O$ Stage 5e (126 to 124 Ka): Rapid Cooling in the Southern Ocean

The Southern Ocean provides a record of rapid change that involves changes in temperature, surface albedo, and ocean circulation. Cores with $\delta^{18}O$ measurements from benthic foraminifera indicate that the first initiation of the last glaciation occurred as early as the middle of $\delta^{18}O$ stage 5e (~126 to 124 Ka, SPECMAP time scale), when a ~5°C cooling occurred within about 2000 years and diatoms associated with sea ice invaded lower latitudes (to ~50°S). Both the surface albedo and deep-water formation were affected (Pichon et al. 1992; Labeyrie et al. 1992).

RECOMMENDATIONS

With an aim towards a better understanding of the climate system and towards making better predictions of climate patterns for any climate changes induced by increased concentrations of greenhouse gases, we support the continued study of past climates. Much relevant research is already underway. Therefore, our recommendations focus on areas that have been underemphasized so far, yet which offer good opportunities for significant progress:

1. Early Pliocene (4.5 to 3.0 Ma). This period may prove to be the "least bad" homolog, in view of the facts that CO_2 concentrations at that time were probably higher than today, continental and ocean configurations were very much as they are today, and ice-volume variations were considerably less than today. Estimates are needed of surface temperatures and of past concentrations of CO_2 from multiple data sources. Records of deep-water circulation are also needed.
2. Better estimates of GMTs are needed for all periods, and better temperature estimates are needed from the Tropics.

3. Better estimates of cryosphere and sea-level variations during warm intervals. New research should address (a) the position of sea ice during the early Holocene and the last interglacial (~125 Ka) and (b) the causes of past sea levels being higher than today during the last interglacial (~125 Ka).
4. Global and regional descriptions of the terrestrial biosphere: an effort is needed to gather together existing and reliably dated paleovegetation records at a global scale, both for key time periods (e.g., 6 kra, 125 Ka) and from long records spanning the past 150 Ka (1 Ma or more) for analysis and comparison with climate and biome model results. High-resolution records also need to be analyzed for periods of rapid climate change.
5. Deep-water/surface-water connections and response times in the oceans: high-resolution records are needed for the past 1000 years to cover the Medieval warm epoch, and we need to know the rate of deep-water formation during previous interglacials.
6. Better estimates of the sensitivity of the climate system to changes in atmospheric composition are needed from paleoclimatic data. We need to estimate the response of lambda to changes in boundary conditions, tropical SSTs, and other factors.
7. More studies and improved dating of data from tropical lakes are needed to test for the synchroneity and spatial patterns of short-term (100 yr) changes in water levels during the last 12,000 years.

These studies, in addition to active paleoclimatic research, will contribute to our understanding of the climate system and to our ability to predict future patterns and rates of climate change.

REFERENCES

Alexandre, P. 1987. Le Climat en Europe au Moyen Age. *Rech. Hist. Sci. Soc.* **24**:1–827.

Bard, E., B. Hamelin, and R.G. Fairbanks. 1990. U–Th ages obtained by mass spectrometry in corals from Barbados: Sea level during the past 130,000 years. *Nature* **346**:456–458.

Barnola, J.M., D. Raynaud, Y.S. Korotkevich, and C. Lorius. 1987. Vostok ice core provides 160,000-yr record of atmospheric CO_2. *Nature* **329**:408–414.

Bartlein, P.J., T. Webb III, and E. Fleri. 1984. Holocene climatic change in the northern Midwest: Pollen-derived estimates. *Quat. Res.* **22**:361–374.

Becker, B., B. Kromer, and P. Trimborn. 1991. A stable-isotope tree ring timescale of the late Glacial/Holocene Boundary. *Nature* **353**:647–649.

Berger, A., J. Imbrie, J. Hays, G. Kukla, and B. Saltzman, eds. 1984. Milankovitch and Climate. Dordrecht: Kluwer.

Berger, W.H., and A. Spitzy. 1988. History of atmospheric CO_2: Constraints from the deep-sea record. *Paleoceanog.* **3**:401–411.

Boyle, E.A., and L. Keigwin. 1987. North Atlantic thermohaline circulation during the past 20,000 years linked to high-latitude surface temperature. *Nature* **330**:35–40.

Bradley, R.S. 1992. Reconstructions of climate from A.D. 1000 to the present. In: Climate Variability and Global Change, ed. S. Tibaldi. Bologna: Univ. of Bologna Press, in press.

Briffa, K.R., T.S. Bartholin, D. Eckstein, P.D. Jones, W. Karlen, F.H. Schweingruber, and P. Zetterberg. 1990. A 1400 yr tree-ring record of summer temperatures in Fennoscandia. *Nature* **346**:434–439.

Broecker, W.S., G. Bond, M. Klas, G. Bonani, and W. Wolfli. 1990. A salt oscillator in the glacial Atlantic? I. The concept. *Paleoceanog.* **5**:469–477.

Budyko, M.I., and Y.A. Izrael. 1987. Anthropogenic Climatic Changes. Leningrad: Gidrometeoizdat. (in Russian).

Budyko, M.I., A.B. Ronov, and A.L. Yanshin. 1985. The History of the Earth's Atmosphere. Leningrad: Gidrometeoizdat. (English trans. Springer-Verlag, 1987).

Burga, C.A. 1988. Swiss vegetation history during the last 18,000 years. *New Phytol.* **110**:581–620.

Chappellaz, J. 1990. Etude du methane atmospherique au cours du denier cycle climatique. Thesis, Univ. of Grenoble.

CLIMAP. 1984. The last interglacial ocean. *Quat. Res.* **21**:123–224.

COHMAP. 1988. Climatic changes of the last 18,000 years: Observations and model simulations. *Science* **241**:1043–52.

Cronin, T.M., and H.J. Dowsett, eds. 1991. Pliocene Climates. *Quat. Sci. Rev.* **10**:115–296.

Crowley, T.J. 1990. Are there any satisfactory geological analogs for a future greenhouse warming? *J. Climate* **3**:1282–1292.

Crowley, T.J. 1991. Modeling Pliocene warmth. *Quat. Sci. Rev.* **10**:275–282.

Crowley, T.J., and G.R. North. 1991. Paleoclimatology. New York: Oxford Univ. Press.

Cwynar, L., and G. MacDonald. 1987. Geographical variation of lodgepole pine in relation to population history. *Amer. Nat.* **129**:463–469.

Davis, M.B. 1989. Lags in vegetation response to greenhouse warming. *Clim. Change* **15**:75–82.

Davis, M.B. 1991. Research questions posed by the paleoecological record of global change. In: Global Changes of the Past, ed. R.S. Bradley, pp. 385–395. Boulder, CO: UCAR/Office for Interdisciplinary Earth Studies.

Fairbanks, R.G. 1989. A 17,000-yr glacio-eustatic sea-level record: Influence of glacial melting rates on the Younger Dryas event and deep-ocean circulation. *Nature* **342**:637–642.

Folland, C.K., T.R. Karl, and K.Y. Vinnikov. 1990. Observed climate variations and change. In: Climate Change: The IPCC Scientific Assessment, ed. J.T. Houghton, G.J. Jenkins, and J.J. Ephraums, pp. 195–255. Cambridge: Cambridge Univ. Press.

Frenzel, B. 1967. Die Klimaschwankungen des Eiszeitalters. Braunschweig: Vieweg. (English trans. by A.E.M. Nairn, Cleveland: Press of Case Western Reserve University, 1973).

Graybill, D.A., and S.G. Shiyatov. 1992. Dendroclimate evidence from the northern Soviet Union. In: Climate Since A.D. 1500, ed. R.S. Bradley and P.D. Jones, pp. 393–414. London: Routledge.

Grimm, E.C. 1983. Chronology and dynamics of vegetation change in the prairie-woodland region of southern Minnesota. *New Phytol.* **93**:311–350.

Heinze, C., E. Maier-Reimer, and K. Winn. 1991. Glacial pCO_2 reduction by the world ocean: Experiments with the Hamburg carbon cycle model. *Paleoceanog.* **6**:395–430.

Houghton, J.T., G.J. Jenkins, and J.J. Ephraums, eds. 1990. Climate Change: The IPCC Scientific Assessment. Cambridge: Cambridge Univ. Press.

Huntley, B. 1988. Europe. In: Vegetation History, ed. B. Huntley and T. Webb III, pp. 341–383. Amsterdam: Kluwer.

Huntley, B. 1991. How plants respond to climate change: Migration rates, individualism and the consequences for plant communities. *Ann. Botany* **67**:15–22.

Huntley, B., P.J. Bartlein, and I.C. Prentice. 1989. Climatic control of the distribution and abundance of beech (*Fagus* L.) in Europe and North America. *J. Biogeog.* **16**:551–560.

Huntley, B., and I.C. Prentice. 1988. July temperatures in Europe from pollen data, 6000 years before present. *Science* **241**:687–690.

Huntley, B., and I.C. Prentice. 1993. Holocene vegetation and climates of Europe. In: Global Climates since the Last Glacial Maximum, ed. H.E. Wright Jr., T. Webb III, W.F. Ruddiman, F.A. Street-Perrott, J.E. Kutzbach, and P.J. Bartlein. Minneapolis: Univ. of Minnesota Press, in press.

Johnson, W.C., and T. Webb III. 1989. The role of blue jays in the postglacial dispersal of Fagaceous trees in eastern North America. *J. Biogeog.* **16**:561–571.

Kellogg, W.W. 1978. Global influences of mankind on climate. In: Climatic Change, ed. J. Gribben, pp. 205–227. Cambridge: Cambridge Univ. Press.

Klimanov, V.A. 1984. Paleoclimatic reconstructions based on the information statistical method. In: Late Quaternary Environments of the Soviet Union, ed. A.A. Velichko, pp. 297–303. Minneapolis: Univ. of Minnesota Press.

Koerner, R.M. 1989. Ice core evidence for extensive melting of the Greenland ice sheet in the Last Interglacial. *Science* **244**:964–969.

Kukla, G. 1987. Loess stratigraphy in central China. *Quat. Sci. Rev.* **6**:191–219.

Labeyrie, L.D., J.C. Duplessy, J. Dupat, A.J. Juillet-Leclerc, J. Moyes, E. Michel, N. Kallel, and N.J. Shackleton. 1992. Changes in the vertical structure of the North Atlantic Ocean between glacial and modern times. *Quat. Sci. Rev.*, in press.

Labracherie, M., J.C. Duplessy, and P.L. Blanc. 1989. The last deglaciation in the Southern Ocean. *Paleoceanog.* **4**:629–638.

Lamb, H.H. 1965. The early Medieval warm epoch and its sequel. *Palaeogeog. Palaeoclim. Palaeoecol.* **1**:13–37.

Lamb, H.H. 1988. Climate and life during the Middle Ages, studied especially in the mountains of Europe. In: Weather, Climate and Human Affairs, pp. 40–74. London: Routledge.

LeRoy Ladurie, E. 1967. Histoire du Climat depuis l'An Mille, Paris: Flammarion.

Liu, T., and B. Yuan. 1987. Paleoclimatic cycles in northern China: Luochuan loess section and its environmental implications. In: Aspects of Loess Research, ed. T. Liu, pp. 3–26. Beijing: China Ocean.

MacCracken, M.C., A.D. Hecht, M.I. Budyko, Y.A. Izrael. 1990. Prospects for Future Climates. Chelsea, MI: Lewis.

Martinson, D.G., N.G. Pisias, J.D. Hays, J. Imbrie, T.C. Moore, and N.J. Shackleton. 1987. Age dating and the orbital theory of the ice ages: Development of a high-resolution 0 to 300,000-yr chronology. *Quat Res. 21*:1–29.

Morgan, V.I., I.D. Goodwin, D.M. Etheridge, C.W. Wookey. 1991. Evidence from Antarctic ice cores for recent increases in snow accumulation. *Nature* **354**:58–60.

Petit-Maire, N. 1989. Interglacial environments in presently hyperarid Sahara: Paleoclimatic implications. In: Palaeoclimatology and Palaeometeorology: Modern and Past Patterns of Global Atmospheric Transport, ed. M. Leinen and M. Sarnthein, pp. 637–661. Dordrecht: Kluwer.

Pichon, S.S., L.D. Labeyrie, G. Bareille, M. Labracherie, J. Duprat, J. Jouzel. 1992. Surface water temperature changes in the high latitudes of the Southern Hemisphere over the last glacial-interglacial cycle. *Paleoceanog.* **7**:289–318.

Prentice, I.C. 1986. Vegetation responses to past climatic changes. *Vegetatio* **67**:131–141.

Prentice, I.C., P.J. Bartlein, and T. Webb III. 1991. Vegetation and climate change in eastern North America since the Last Glacial Maximum. *Ecology* **72**:2038–2056.

Prentice, I.C., and A.M. Solomon. 1991. Vegetation models and global change. In: Records of Past Global Change, ed R. Bradley, pp. 365–383. Boulder, C0: UCAR/Office of Interdisciplinary Earth Studies.

Rind, D. 1987. Components of the ice-age circulation. *J. Geophys. Res.* **92**:4241–4281.

Rind, D., D. Peteet, W.S. Broecker, A. Mclntyre, and W.F. Ruddiman. 1986. The impact of cold North Atlantic sea-surface temperatures on climate: Implications for the Younger Dryas cooling (11–10 Ka). *Clim. Dynam.* **1**:3–33.

Ruddiman, W.F., and A.C. Mix. 1993. Paleoceanography of the North and equatoriai Atlantic at 9000 and 6000 yr B.P. In: Global Climates Since the Last Glacial Maximum, ed. H.E. Wright Jr., J.E. Kutzbach, T. Webb III, W.F. Ruddiman, F.A. Street-Perrott, and P.J. Bartlein. Minneapolis: Univ. of Minnesota Press, in press.

Sarnthein, M. 1972. Sediments and history of the Postglacial transgression in the Persian Gulf and NW Gulf of Oman. *Marine Geol.* **12**:245–266.

Sarnthein, M., and J. Fenner. 1988. Global wind-induced change of deep-sea sediment budgets, new ocean production and CO_2 reservoirs about 3.3–2.35 Ma. *Phil. Trans. R. Soc. Lon. B* **318**:487–504.

Sarnthein, M., E. Jansen, M. Arnold, J.C. Duplessy, H. Erlenkeuser, A. Flatoy, T. Veum, E. Vogelsang, and M.S. Weinelt. 1992. $\delta^{18}O$ time-slice reconstruction of meltwater anomalies at Termination I in the North Atlantic between 50 and 80°N. In: The Last Deglaciation: Absolute and Radiocarbon Chronologies, ed. E. Bard and W.S. Broecker. Berlin: Springer.

Sarnthein, M., H. Stremme, and A. Mangini. 1986. The Holstein interglaciation: Time-stratigraphic position and correlation to stable-isotope stratigraphy of deep-sea sediments. *Quat. Res.* **26**:283–298.

Sarnthein, M., and R. Tiedemann. 1990. Younger Dryas-style cooling events at glacial terminations I–IV at ODP site 658: Associated benthic $\delta^{13}C$ anomalies constrain melt water hypothesis. *Paleoceanog.* **56**:1041–1055.

Sarnthein, M., and K. Winn. 1990. Reconstruction of low and mid-latitude export productivity, 30,000 yr B.P. to Present: Implications for global carbon and nutrient reservoirs. In: Climate and Ocean Interaction, ed. M. Schlesinger, pp. 319–342. Dordrecht: Kluwer.

Schlesinger, M.E., and J.F.B. Mitchell. 1985. Model projections of the equilibrium climatic response to increased carbon dioxide. In: Projecting the Climatic Effects of Increasing Carbon Dioxide, ed. M.C. MacCracken and F.M. Luther, pp. 81–147. Washington, D.C.: U.S. Dept. of Energy, Carbon Dioxide Research Division.

Schumm, S.A. 1991. To Interpret the Earth. Cambridge: Cambridge Univ. Press.

Sirocko, F., M. Sarnthein, H. Lange, and H. Erlenkeuser. 1991. Glacial to interglacial changes in Arabian Sea sediment accumulation rates: History of atmospheric summer circulation and coastal upwelling over the last 30,000 years. *Quat. Res.* **36**:72–93.

Stauffer, B., A. Neftel, H. Oeschger, J. Schwander, and U. Siegenthaler. 1985. Atmospheric CO_2 concentrations during the last glaciation. *Ann. Glaciol.* **5**:106–164.

Street-Perrott, F.A., D.S. Marchand, N. Roberts, and S.P. Harrison. 1989. Global Lake-level Variations from 18,000 to 0 Years Ago: A Palaeoclimatic Analysis. U.S. DOE Technical Report TR046. Washington, D.C.: U.S. Dept. of Energy, Carbon Dioxide Research Division.

Street-Perrott, F.A., and R.A. Perrott. 1990. Abrupt climatic changes in the Tropics: An Atlantic Ocean feedback mechanism. *Nature* 343:607–612.

Tallis, J.H. 1991. Plant Community History: Long-term Changes in Plant Distribution and Diversity. New York: Chapman and Hall.

Tiedemann, R. 1991. Acht Millionen Jahre Klimageschichte von Nordwest Afrika und Paläo-Ozeanographie des angrenzenden Atlantiks: Hochauflösende Zeitreihen von ODP-Sites 658–661. Ph.D. Thesis, University of Kiel.

Velichko, A.A., ed. 1984. Late Quaternary Environments of the Soviet Union. Minneapolis: Univ. of Minnesota Press.

Velichko, A.A., M.P. Barash, M.P. Grichuk, E.E. Gurtovaya, and E.M. Zelikson. 1984. Climate of the Northern Hemisphere in the last intergiacial (Mikulino). *Izv. Acad. Sci. USSR Ser. Geog.* **N1**:5–18.

Velichko, A.A., B. Frenzel, and M. Pecsi. 1991. Atlas of paleoclimates and paleoenvironments of the Northern Hemisphere: Late Pleistocene-Holocene. Budapest: Geographic Research Institute, Hungarian Academy of Sciences.

Warrick, R.A., and T.M.L. Wigley, eds. 1991. Climate and Sea Level Change. Cambridge: Cambridge Univ. Press.

Watts, W.A. 1988. Europe. In: Vegetation History, ed B. Huntley and T. Webb III, pp. 155–192. Dordrecht: Kluwer.

Webb, T. III. 1985. A Global Paleoclimatic Data Base for 6000 yr B.P. U.S. DOE Technical Report TR018. Washington, D.C.: U.S. Dept. of Energy, Carbon Dioxide Research Division.

Webb, T. III. 1991. The spectrum of temporal climatic variability: Current estimates and the need for global and regional time series. In: Records of Past Global Change, ed R. Bradley, pp. 61–81. Boulder, CO: UCAR/Office of Interdisciplinary Earth Studies.

Webb, T. III, P.J. Bartlein, S.P. Harrison, and K.H. Anderson. 1993. Vegetation, lake levels, and climate in eastern North America. In: Global Climates since the Last Glacial Maximum, ed. H.E. Wright Jr., J.E. Kutzbach, T. Webb III, W.F. Ruddiman, F.A. Street-Perrott, and P.J. Bartlein. Minneapolis: Univ. of Minnesota Press, in press.

Wright, H.E., Jr., J.E. Kutzbach, T. Webb III, W.F. Ruddiman, F.A. Street-Perrott, and P.J. Bartlein, eds. 1993. Global Climates since the Last Glacial Maximum. Minneapolis: Univ. of Minnesota Press, in press.

7

How Reliable and Consistent Are Paleodata from Continents, Oceans, and Ice?

J. WHITE[1], B. MOLFINO[2], L. LABEYRIE[3], B. STAUFFER[4], and G. FARQUHAR[5]

[1]Institute of Arctic and Alpine Research, Campus Box 450, and Dept. of Geological Sciences
University of Colorado, Boulder, CO 80309, U.S.A.
[2]Lamont-Doherty Earth Observatory, P.O. Box 1000,
Palisades, NY 10964, U.S.A.
[3]Centre des Faibles Radioactivités, CNRS, Avenue de la Terrasse,
Gif-sur-Yvette Cedex 91198, France
[4]Physikalisches Institut, Universität Bern, Sidlerstr. 5, 3012 Bern, Switzerland
[5]Research School of Biological Sciences, Institute of Advanced Studies,
Australian National University, GPO Box 475,
Canberra City, ACT 2601, Australia

ABSTRACT

In this chapter, paleodata are evaluated in terms of intrinsic and extrinsic uncertainties. The data examined are from continents, oceans, and ice. Completeness in terms of types of paleodata is neither achieved nor attempted, rather the data presented are meant to demonstrate the kinds of uncertainties associated with paleodata. The continental data are primarily isotopic data from trees. A more complete table of paleodata, along with an assessment of their reliability, is given in Appendix 11.1 of our group report (see Bradley et al., this volume).

INTRODUCTION

One of the central issues common to all reconstructions of paleoenvironmental conditions is the reliability of the data collected. Each researcher, when paleodata are collected, does his/her best to place limits of confidence on this data, in the hope that subsequent users will responsibly adhere to those limitations. Since each study is in many ways unique, and since the number of paleoenvironmental indicators is large and growing larger, it is not within the scope of this review to treat the subject of

Global Changes in the Perspective of the Past
Edited by J.A. Eddy and H. Oeschger

paleodata reliability on a property-by-property basis. Nonetheless, the different types of paleodata, whether collected from the land, ocean, or ice, do share some common characteristics that impact on the reliability of the data. For example, paleodata collected from material that is derived from biomass, or influenced by biological processes, will share similar problems in reliability. These problems stem from the fact that plants and animals, while they may be sensitive indicators of past environmental conditions, are not necessarily passive recorders, rather they may alter their behavior in a changing climate to optimize their chances of survival. This, in turn, impacts the reliability of the transfer functions used to relate paleodata to climate. The difficulties involved in designing reliable transfer functions for paleodata based on biological remains are thus one area of common problems in reliability.

Perhaps the most difficult issue to address in assessing the reliability of paleodata is determining which of several possible environmental influences are recorded in, and can be extracted from, the paleodata. For each paleorecorder, one must determine what forcing factor(s) is (are) primarily responsible for the variability in the data. Recorders that are responsive to too many factors may lead to more confusion than progress in a field. In this regard, inorganic recorders (e.g., ice cores and carbonate precipitation) are perhaps simpler than organic recorders (such as tree rings), as the number of variables which must be directly considered is generally smaller.

Scale, both spatial and temporal, also plays a role in limiting the reliability of paleodata. The time period and space scale that data represent must somehow be determined. Space scales can be assessed by assembling a network of sample sites, although it is worth noting that, for a number of reasons, good and bad, this is not always done. Time scales represented by paleodata are often harder to assess, although biologically based data sometimes have the advantage here because of the time limits naturally placed on organisms for their growth. Once time and space scales have been assigned to specific paleodata, there must often be a rather large scaling-up in both time and space in order to make the data useful for modern models. This extrapolation process can often decrease the reliability of the data. Scaling-up will, however, continue to be a necessity in using paleodata until models can make use of site-specific and regional data.

Finally, an ever-present class of problems stems from the actual measurements. Vigilant and careful lab intercalibrations must be commonplace. Nevertheless, experience has taught us that even in fields where intercalibration is rigorously practiced, significant differences between laboratories can and do occur. As our understanding of what paleodata can reveal increases and thus our demands on the data increase, we must step up efforts to ensure data quality, from the initial steps of field collecting to the final laboratory analyses.

In this chapter, we address the reliability and consistency of paleodata from ice cores, tree rings, and ocean sediments. This, of course, does not cover all of the sources of paleodata, or even all of the major ones (pollen is notably absent, for example). We will attempt to use these types of paleodata, however, as examples of the problems encountered in producing and using paleodata. Appendix 11.1 (Bradley et al., this volume) was a joint effort with our working group (notably Barbara Molfino and rapporteur Ray Bradley) to give a broad overview of paleodata types and their reliability. The Appendix is certainly not complete, but we have attempted to list the major data types. As a final note, the reader should recognize that reliability is often difficult to quantify, and compilations such as Appendix 11.1 are bound to offend somebody. In reading this review and using the table, please keep in mind that this review is intended not only for the experts who collect the data, but also for those who use the data in models of environmental dynamics and who want an accessible and brief treatment of paleodata reliability.

PALEODATA FROM ICE CORES

Glaciers and ice sheets are sediments of precipitation. They consist of layers of snow compacted under their own weight to ice. A core taken from an ice sheet or glacier provides a detailed and, in the central parts of polar ice sheets, complete sequence of precipitation samples from the past. By analyzing the isotopic composition and chemistry of the ice and the composition of the air enclosed in the bubbles, information about the climate and environment of the past is obtained. In discussing the reliability and consistency of this information, we will concentrate on (a) the reconstruction of the temperature record by measuring the isotopic composition, (b) the estimation of volcanic activity by measuring the aerosol concentration, and (c) the reconstruction of fluctuations of the composition of the atmosphere by analyzing the air in the bubbles of the ice.

One ice core provides local climatic information. However, since polar ice sheets are remote from all source regions of pollutants and water vapor, we expect a rather uniform distribution of the deposition of precipitation and pollutants and therefore a regional significance of the results obtained from one ice core. Discussions of the consistency are based on this assumption, but they have to take into account the local noise.

Isotopic Composition and Surface Temperature

A relationship between the isotopic ratio in precipitation and temperature is due to the tendency of heavier water molecules (${}^{1}H_2{}^{18}O$ or ${}^{1}H_2H^{16}O$) to evaporate less rapidly from water surfaces but to condense more readily in clouds than normal water molecules (${}^{1}H_2{}^{16}O$). This tendency explains the lower concentration of heavy isotopes in precipitation condensed at lower temperatures. Very good linear correlations are found empirically, both on the Greenland and Antarctic ice sheet. A linear relationship between the isotopic concentration (or its δ-value) of an annual snow layer and the mean annual surface temperature at the deposition site is not fully explained by simple equilibrium considerations. More complex models involving kinetic isotope effects can, however, reproduce the observed relationships (Jouzel and Merlivat 1984). The coefficient of the observed linear relation for the $\delta^{18}O$-value is 0.62‰ per °C for

Greenland (Dansgaard et al. 1973) and about 0.75‰ per °C for East Antarctica (Lorius et al. 1989). These empirical relations are valid for the present climatic conditions. For earlier climatic regimes, these relations are valid only if the isotopic composition of the surface waters in the source region and the cooling mechanisms of the water vapor along its trajectory to the ice sheet have not changed significantly and if the ratio between summer and winter precipitation have remained the same. Based on detailed analyses of the $\delta^{18}O$ and $\delta^{2}H$ values, Johnsen et al. (1989) conclude that this was approximately the case for Greenland during the last glaciation. For East Antarctica, Lorius suggests, based on meteorological considerations and a simple model, that the same coefficient be used as at present (Lorius et al. 1989).

The isotopic records of three ice cores from Antarctica (Lorius 1989) and of two ice cores from Greenland (Dansgaard and Oeschger 1989) are so similar in the general trend, and also in most of the details, that there can be no doubt that they reflect not only the *general* global trend but also detailed regional climatic changes. Short-term variations are more affected by local noise, depending on parameters such as accumulation rate and wind speed. For East Antarctica, the noise of samples representing a ten-year average is estimated to be about 1‰ (Lorius 1983) for drilling sites 10 km apart. A decade with a mean temperature 1°C warmer than the long time average can therefore not be reconstructed, at least not based on one ice core.

Volcanic Activity and Aerosols

In violent volcanic eruptions, large amounts of SO_2 are injected into the stratosphere and are transformed there into sulfate aerosols. Sulfuric acid becomes the dominant stratospheric aerosol. It cools the lower troposphere due to backscattering of solar radiation, spreads all over the hemisphere, and continues to reenter the troposphere for up to three years after the eruption. Some of it is deposited on the ice sheet by washout or dry deposition processes, thereby increasing the acidity of the ice. Hammer (1980) established a simple method to measure the acidity of ice samples by measuring a kind of electrical conductivity. By comparing the deposition patterns of fission products from bomb tests with patterns of volcanic eruptions, it is possible to estimate the magnitude of the injected amount of sulfuric acid into the stratosphere. If the location of an eruption is known, it is possible to estimate the injected amount within a factor of two. For example, the magnitude for the Tambora eruption (1815) was estimated to be:

- 150×10^6 tons by Hammer et al. (1980), based on one ice core from Greenland, and by Legrand and Delmas (1987) based on one ice core from Dome C (East Antarctica);
- 360×10^6 tons by Langway et al. (1988), based on data from a core from South Pole; and
- 222×10^6 tons by Clausen and Hammer (1988) based on ice cores from 11 different sites in Greenland.

It should be possible to estimate approximately the latitude of unknown eruptions based on comparisons of ice core data from Greenland and Antarctica. The estimated amount of sulfuric acid injected into the stratosphere for such eruptions should be accurate within a factor of three, at least for large eruptions.

Composition of the Atmosphere

In regions with a mean annual surface temperature below –25°C, ice is formed through a dry sintering process. In this case, air in the bubbles is expected to have an atmospheric composition, and therefore it is possible to reconstruct the composition of the atmosphere at the time of the ice formation. We will discuss the reliability of CO_2 analyses, as it is the component that attracts the greatest interest and which has been investigated most intensively. These investigations have determined that:

- Atmospheric preindustrial CO_2 concentration was 280 ppm (Neftel et al. 1985).
- During the transition from the last glaciation to the Holocene, the atmospheric CO_2 concentration level rose from about 180–200 ppm to the preindustrial level of 280 ppm (Neftel et al. 1988).
- During the last glaciation, fast and drastic variations in the atmospheric CO_2 concentration occurred (Stauffer et al. 1984).

The main arguments for the reliability of the first of these results came from evidence of atmospheric composition of the main air components (N_2, O_2, and Ar) in the bubbles and the overlapping of results from ice cores with direct measurements on atmospheric air (e.g., Mauna Loa). Due to a steady improvement of the analytical techniques, there are rather important differences between data published ten years ago and at present. At present, the reproducibility of the measurement is about ± 3 ppm. Accuracy, however, has to take some additional systematic errors into account: uncertainty in dating the air, enrichment of CO_2 in the firn due to gravitation separation, and enrichment with CO_2 entrapped already in snow crystals. The reliability of the preindustrial CO_2 concentration is estimated to be ± 7 ppm.

The CO_2 increase at the end of the last glaciation would not be very convincing if it had only been measured in one ice core. Interactions of CO_2 with impurities in the ice or a depletion of CO_2 in the extracted gases due to air hydrates replacing the bubbles at great depth could be responsible for the lower concentrations in the ice (Neftel et al. 1983). These two effects have been investigated with negative results. The most convincing argument for an interpretation of these results, as atmospheric CO_2 increases, is the fact that the increase is the same in cores from different sites with different temperatures, different crystal size, and different impurity concentrations. The fast variations during the last glaciation have been observed in two ice cores from Greenland but not yet in Antarctic cores. The interpretation of the atmospheric CO_2 concentration as fast variations is certainly less reliable than the first two results and needs to be further investigated. Studies being conducted on new ice cores from Central Greenland (GISP II and GRIP) will allow these investigations.

For every other air component, all questions discussed above for CO_2 have to be answered separately. Methane has already been tested intensively (Stauffer et al. 1985). Tests show that methane results can be interpreted as atmospheric methane concentrations with the same reliability as CO_2 results.

Time Scale and Site Selection

Results from ice cores are only valuable if a time scale can be established. Different methods for dating ice cores are available, but most of them are based on certain assumptions. Time scales are therefore still a matter of debate. For the Dye 3 record, Hammer et al. (1986) gives an age of the end of the Younger Dryas of 10,720 ± 150 years. Such precise error limits are an exception in ice-core dating, but generally dating uncertainties for ice from about 10,000 to 20,000 B.P. are estimated to be in the range of 500 years. Assuming that time scales are accurate within this error limit, the last deglaciation started about 20,000 B.P. at Byrd Station, 16,000 B.P. ago at Vostok, and 13,000 B.P. ago at Dye 3 (end of older Dryas). A time lag of the onset of the deglaciation at different locations cannot be excluded. However, the isotopic changes parallel the increase of the CO_2 concentrations. If their interpretation of the increase of atmospheric CO_2 concentration is correct, it would have to have been synchronous over the globe. Therefore, some inconsistencies remain, both in the dating of cores and in the interpretation of some of their results. We urgently need to increase the reliability of the dating of ice cores. If future drilling sites are selected, the dating aspect (ice divide, flat bedrock, high accumulation) should be given high priority.

PALEODATA FROM TERRESTRIAL PLANTS

The most commonly used source of paleodata from terrestrial plants is that preserved in tree rings. Such rings occur where there is a seasonality of growth, e.g., at higher latitudes, and under tropical conditions where strong seasonality of wet and dry conditions occur. However, most analyses performed to date have featured cold temperature-induced limitations of growth.

An attractive feature of tree rings is that temporal sequences are usually sufficiently dependable, especially when a number of adjacent logs are examined, for reliable time scales to be placed on the parameters of interest.

The parameters most studied are the widths of the rings themselves, which take into account the drift that occurs naturally with the diameter of the ring, and the isotopic composition of C, O, and H in the rings. With respect to the widths, inferences can only be drawn about conditions occurring during the season of photosynthesis which formed them, if the factors controlling growth are known. In cool environments, one can therefore find correlations with temperature, and in dry sites one can find correlations with rainfall. There are obviously confounding effects of changes in light intensity, nutrient availability, and so on. However, in this section we concentrate on the isotopic composition of the rings.

Factors Affecting Carbon Isotopic Composition

Carbon isotope discrimination, Δ_p, of photosynthate with respect to source air is defined as (R_a/R_p-1), where the subscripts a and p denote source air and photosynthate, respectively, and R is the $^{13}C/^{12}C$ molar ratio (Farquhar et al. 1982). To a good approximation, Δ_p is given by

$$\Delta_p = a + (b - a)\, c_i/c_a - d\,, \tag{7.1}$$

where a is the fractionation during diffusion in air (4.4‰), b is the fractionation during primary carboxylation (29 or 30‰ , for ribulose biphosphate carboxylase [Rubisco]), c_i/c_a is the ratio of intercellular to atmospheric concentration (mole fraction) of CO_2, and d is a term which corrects for a number of effects, notably diffusion from the substomatal cavity to the site of carboxylation in the chloroplasts. A more precise formulation of the above processes is given by Farquhar et al. (1989).

Further changes in isotopic composition may occur in the conversion from carbohydrate to total plant carbon and, in particular, in the conversion to specific organic compounds. These are less well understood. Lignin is usually ~3‰ lighter than cellulose, and the lignin-to-cellulose ratio varies (Narayanamurti and Das 1955; Wilson and Grinsted 1977). For this reason, cellulose is the component normally analysed for greater consistency. By modifying the parameter d to d' the discrimination occurring from source CO_2 to cellulose, Δ_c, is then given by

$$\Delta_c = a + (b - a)\, c_i/c_a - d'\,. \tag{7.2}$$

Most studies examine the isotopic composition of cellulose (in relation to a standard), δ_c, because the precise composition, δ_a, of the source air is unknown. An approximate expression for δ_c is then

$$\delta_c = \delta_a - a - (b - a)\, c_i/c_a + d'\,. \tag{7.3}$$

Strictly,

$$\Delta_c = (\delta_a - \delta_c)/(1 + \delta_c)\,, \tag{7.4}$$

and since δ_c is typically -27×10^{-3}, δ_a *is* about 3% proportionally greater than $\delta_a - \delta_c$ (i.e., about 0.5‰ greater). However, the differences in δ_c are usually closely matched by differences in $\delta_a - \delta_c$.

From the above, it is obvious that past climates may have affected δ_c in many ways. The variation in δ_a relates to changes in atmospheric CO_2 concentration and will be discussed later. The parameter associated with diffusion, *a*, should, in theory, be independent of temperature and other effects.

The fractionation associated with the primary carboxylase, *b*, may well have a slight temperature dependence. Estimates of the temperature coefficient for *b* are (‰ per °C) 1.2 for Rubisco from sorghum (Whelan et al. 1973) and –0.22 from soybean (Christeller et al. 1976). These and earlier estimates of the value of *b* showed considerable experimental variability. More recent, reliable estimates have only been carried out at 25°C (Roeske and O'Leary 1984; Guy et al. 1987). There may also be genetic variation in *b*. *Anacystis nidulans* and *Rhodospirillum rubrum* are reported to have Rubisco fractionation factors of 21‰ and 18‰, respectively (Roeske and O'Leary 1984). Clearly, further investigations are needed here.

The instantaneous value of c_i is given by

$$c_i = c_a - A/g, \tag{7.5}$$

where A (mol CO_2 m^{-2} s^{-1}) is the rate of CO_2 assimilation by the leaves and g (mol CO_2 m^{-2} s^{-1}) is the stomatal conductance to the diffusion of CO_2. Therefore, factors which reduce *A* through effects primarily on the mesophyll capacity for photosynthesis (e.g., low light intensities [Ehleringer et al. 1987; von Caemmerer and Evans 1991]) and deficiencies of certain mineral nutrients) will increase c_i/c_a and reduce δ_c (Farquhar 1980; Francey and Farquhar 1982). Factors which reduce *A* primarily through large atmospheric vapor pressure deficits (Winter et al. 1982) or soil water deficits (Winter 1981; Farquhar and Richards 1984; Hubick et al. 1986) will decrease c_i and increase δ_c. Supporting evidence is reviewed by Farquhar et al. (1989).

The parameter *d′* is less well defined. The component *d* is positively related to the assimilation rate (Evans et al. 1986; Farquhar et al. 1989; von Caemmerer and Evans 1991). Typical values are estimated to be about 4‰, decreasing at low light intensities (von Caemmerer and Evans 1991; Lloyd and Syversten 1992). Whole plant carbon appears to be depleted in ^{13}C by about 2‰ compared to photoassimilate (von Caemmerer and Evans 1991; Henderson et al. 1992). This may, in part, be explained by processes leading to lipid production, since lipids are depleted in ^{13}C (Park and Epstein 1961). The latter effect has been ascribed to fractionation by pyruvate dehydrogenase during the oxidation of pyruvate to acetyl coenzyme *A* (De Niro and Epstein 1977). On the other hand, cellulose in leaf tissue is enriched in ^{13}C compared to the whole leaf by about 1.5‰ (Schleser 1990), which would make it similar in isotopic composition to photosynthate. Leavitt and Long (1982) reported that cellulose of wood was enriched in ^{13}C by, on average, 2‰ compared to cellulose of leaves in juniper trees (*Juniperus monospenna*), but Schleser found no such difference in beech trees (*Fagus sylvatica L.*).

Temperature Effects on δ_c

There is a long history of attempts to retrieve paleotemperatures from $^{13}C/^{12}C$ ratios in tree rings. Here, we largely restrict discussion to the examination of cellulose in order to avoid complications described above.

Pearman et al. (1976) reported a temperature effect on δ_c of whole wood of 0.24 to 0.48‰/°C in Tasmanian *Arthrotaxis selaginoides* trees. Wilson and Grinsted (1977) also reported 0.2‰/°C in cellulose of New Zealand *Pinus radiata*. However, Grinsted et al. (1979) found that the temperature effect in bristlecone pine (*Pinus longeava*) cellulose from North America, over the past 1,000 years, roughly follows the English high summer temperature variations for the same period, but with a reversed (and much larger) coefficient (ca. –1.5‰ /°C). Farmer (1979) observed a relationship in American elm trees of –0.7‰ per °C, Tans and Mook (1980) reported 0.4‰ for *Quercus rubra* trees in Holland, and Freyer and Belacy (1983) found 0.18‰ per °C for *Pinus sylvestris*, while Leavitt and Long (1983) reported –0.27‰ per °C for *Juniperus* species in Arizona. Bender and Berge (1982) found a small, but not statistically significant, positive coefficient relating δ_c to average summer temperatures in *Quercus macrocarpa* growing in Wisconsin. Stuiver and Braziunas (1987) reported a coefficient of 0.32‰ per °C for a set of North American conifers.

Another way of examining temperature effects on carbon isotope composition is to examine latitudinal variation. Stuiver and Braziunas (1987) deduced from their data that d_c increased with temperature, whereas Lancaster (1990), who compared δ of CO_2 emitted from soil with that of ambient air, found a 5‰ decrease in moving from Arctic tundra to tropical forest. Körner et al. (1991) examined leaves of fully sunlit, nonwater-stressed herbaceous C_3 plants and found a decrease of 2.7‰, over a similar range, which yields a coefficient of –0.1‰ per °C when related to the mean temperature of the growing period. The latter two studies also showed that elevation alone is not a good way of comparing temperature effects, as increases in elevation (at constant, low temperature) also cause an increase in δ. Körner et al. ascribed these results to the effects of decreasing oxygen partial pressure and supported their conclusion with physiological studies. In the present context, however, latitudinal and elevational studies suffer from the potential problem of genetic adaptation to the environment.

How then might such diverse results be explained? Farquhar (1980) suggested that at very low and very high temperatures, assimilation rate, A, is suppressed and intercellular CO_2 concentration, c_i, increases. He suggested that on this basis, and ignoring possible effects of temperature on Rubisco fractionation, etc., δ_c should be maximal at, or slightly above, the temperature optimum for photosynthesis, and that the temperature optimum should increase with increasing light intensity and also with increasing CO_2 concentration. This would suggest a positive relation between δ_c at cool temperatures and a negative one at hot temperatures. Thus it could be that the Tasmanian, New Zealand, Dutch, and North American conifer records reflect cool growth conditions. The negative coefficients related to high summer temperatures

(Grinsted et al. 1979) would also be consistent. The results for *Juniperus* species were obtained by correlating δ_c with the December temperature in Arizona, as this month gave the greatest values of r^2 and since the winter months were regarded as important preconditioning periods. However, it is unlikely that the proportion of carbon fixed in this period is very great. Possibly, the correlations were smaller in warmer months because of the confounding effects of vapor pressure deficits and soil water deficits with higher temperatures. Stuiver and Braziunas (1987) pointed out that their data could equally well be interpreted as a decrease in δ_c, with increasing relative humidity.

Precipitation Effects on δ_c

From the earlier discussion we would expect that c_i should increase and d_c decrease with increasing available soil water. Tans and Mook (1980) found a coefficient of –0.2‰ mm^{-1} yearly total precipitation. Bender and Berge (1982) observed a negative but nonsignificant coefficient between δ_c and spring or summer rainfall. Leavitt and Long (1983) reported a coefficient of –0.04‰ mm^{-1} for the correlation between δ_c and December precipitation. However, quantitative estimates of rainfall from δ_c are problematic, given that soil moisture depends upon evaporation as well. Leavitt and Long (1988) reported that in high-frequency fluctuations in δ_c, from A.D. 1700 to the present, *Pinus edulis* and *P. monophylla* from the American southwest correlated negatively with ring width and positively to drought index, as they expected for that environment. The trends are in accord with a much larger range of physiological studies on herbaceous plants (reviewed by Farquhar et al. 1989).

Reconstructing Past Changes in δ_a: The Isotopic Composition of Atmospheric CO_2

A number of studies have attempted to deduce changes in d_a from observations of time series of δ_c. Leavitt and Long (1988) summarized previous findings based on tree-ring studies, many of which appeared to be ambiguous. They then reported their measurements on δ_c of *P. edulis* and *P. monophylla* after removing climatic influences. From 1952–1982, δ_c decreased by 0.7‰ over the period, matching observations in δ_a made by Keeling. However, between 1957 and 1977 there was a 0.4‰ disparity. The results show the difficulty of reconstructng δ_c using C_3 terrestrial plant material.

A promising development is the analysis of the carbon isotope composition of ancient cobs of *Zea mays* (Marino and McElroy 1991), which has the C_4 pathway of photosynthesis. The latter differs from the C_3 pathway (common to all trees used for tree-ring studies) in that CO_2 is first fixed by phosphoenolpyruvate (PEP) carhoxylase in the mesophyll cells before being decarboxylated (at a high concentration) in the bundle sheath cells where Rubisco is found. The majority of the CO_2 is then fixed, but a proportion, f, leaks back to the mesophyll.

A simple expression for discrimination during photosynthesis is then (Farquhar 1983):

$$\Delta = a + (b_4 + (b_3 - s) f - a)\, c_i / c_a\,, \tag{7.6}$$

where b_4 denotes the fractionation, with respect to gaseous CO_2, in equilibrium with bicarbonate, by PEP carboxylase (–5.7‰ at 25°C), b_3 is the Rubisco fractionation (30‰), and s is the fractionation during leakage (1.8‰).

Henderson et al. (1992) recently showed that this form describes C_4 discrimination well if the leakage, f, is taken as 0.231 at a temperature of 29°C for which b_4= –5.3‰ , over a range of other environmental conditions. With this particular coordination of the C_4 system, discrimination is then only slightly (negatively) dependent on c_i/c_a:

$$\Delta = 4.4 - 3.8\, c_i / c_a \ (‰)\,. \tag{7.7}$$

Thus the isotopic composition of recent photosynthate, close to that of cellulose, should be given by:

$$\delta = \delta_a - 4.4 + 3.8\, c_i / c_a\,. \tag{7.8}$$

For *Z. mays*, c_i/c_a is typically ca. 0.4 ± 0.05 (Wong et al. 1979) so that the equation becomes:

$$\delta_c = \delta_a - 2.9 + 0.2\,, \tag{7.9}$$

that is to say,

$$\delta_a = \delta_c + 2.9 + 0.2‰\,. \tag{7.10}$$

Without the benefit of the results from Henderson et al. (1992), Marino and McElroy (1991) made the above analysis, and although they neglected the term s, they derived above relationship, but with 3.3 instead of 2.9. However, their observations on trends in δ_c from kernels of maize grown in Iowa, from 1948 to 1976, actually yielded the value 2.9 as predicted above. The perfect agreement is fortuitous, given the slight temperature dependence of Δ, but reinforces their optimism about the value of information, which they hope to obtain from samples of maize dating back to A.D. 1500.

Factors Affecting the Oxygen and Hydrogen Isotopic Compositions

In plants, particularly in tree rings, δD and $\delta^{18}O$ values have been used to reconstruct air temperatures and humidities (e.g., Epstein and Yapp 1976; Gray and Thompson 1976; Yapp and Epstein 1982a; Edwards and Fritz 1986). The air temperature reconstruction is via the isotopic composition of precipitation, which itself varies as a function of temperature (Dansgaard 1964). The relative humidity reconstruction is via the effect of evaporation of water from plant leaves on isotopes, the degree of fractionation being partly a function of the relative humidity of the air. The reliability of these isotopic tools for reconstructions of paleoenvironmental parameters is still the subject of some debate.

In general, the utility of reconstructing a single variable from δD and $\delta^{18}O$ values in plants is hampered by the potentially large number of environmental factors which can cause variations in the δD and $\delta^{18}O$ values. These include variations in the isotopic composition of the source water for plants, which can range from isotopically heavy summer rains to lighter time-averaged groundwaters, or any combination in between (White et al. 1985). How plants tap, in a relative sense, surface-fed rain versus deeper groundwaters is not fully understood. Concentrating on trees in very dry sites, which only tap summer rains, or in very wet sites dominated by groundwaters seems to yield isotopic simplicity, however, and thus an increased measure of reliability.

Once water gets to the leaves, almost all of it is evaporated, a process which greatly increases the heavy isotope relative to the light isotope. The factors which influence this process are the temperature of the leaves, the humidity of the air, and the isotopic composition of atmospheric moisture. The dominant variable appears to be the humidity of the air (Burk and Stuiver 1981; Yapp and Epstein 1982b), which in a strict sense is calculated relative to the leaf, and not air, temperature. Variability in the isotopic composition of air vapor is often handled by assuming that the moisture is in isotopic equilibrium with local precipitation. This is a risky assumption and has been shown to be wrong at least for the northeastern United States (White and Gedzelman 1984). The impact of changes in the isotopic composition of atmospheric moisture on leaf water isotopic compositions is probably small for $\delta^{18}O$ values, but potentially much larger for δD values.

This distinction between δD and $\delta^{18}O$ values also holds for sensitivities to humidity changes. The relative impacts of the phase change (equilibrium) and diffusive (kinetic) fractionations during transpiration are different for the hydrogen and oxygen isotopes. It is the diffusive fractionation which is sensitive to changes in the humidity of the air. The sensitivity of hydrogen isotopes to the diffusive fractionation is considerably less than that for the oxygen isotopes. Hence, $\delta^{18}O$ values are more reliable recorders of humidity changes, whereas δD values are more reliable indicators of source water (temperature) changes.

The manufacture of plant material from leaf water also carries an isotopic shift that is not completely understood. Leaf water is heterogeneous, and how evaporation affects the water actually used in photosynthesis is the subject of intense study, not only among the paleoclimate community but also within the plant physiological community. Finally, recent evidence by De Niro and Cooper (1989) suggests that products of photosynthesis may not retain their isotopic composition acquired in the leaves, but rather have their δD and $\delta^{18}O$ values reset via exchange with unevaporated water in the trunk and stems of plants. This would greatly simplify the interpretation of isotopes in plants, as leaf water effects would be eliminated; however, it is still not clear whether this reexchange actually occurs. It is worth noting that humidity reconstructions would not be possible if this was true. Alternative explanations for the observed relationship between isotopes and humidity would thus have to be found.

Despite the potential for confusion in the interpretation of δD and $\delta^{18}O$ values in plants, several studies have clearly shown strong relationships between isotopes in plants and either the water used to grow the plant and/or climatic parameters (e.g., Epstein and Yapp 1976; and Gray and Thompson 1976; Yapp and Epstein 1982a; Edwards and Fritz 1986). Clearly at some level of scrutiny, one variable—most often temperature—becomes dominant. Hence, δD and $\delta^{18}O$ values in plants can certainly yield qualitative climatic information and, if conditions are well controlled, even quantitative information. The assignment of error bars must vary from situation to situation, depending primarily on the degree to which other possible influences can be ignored or controlled.

PALEODATA FROM THE OCEANS

Paleodata from the oceans can be classified as biological (organic) and nonbiological (inorganic) in origin (see Appendix 11.1, this volume). Both tell us about the chemical and physical processes of past oceans. The biological data represent responses of an active system, since organic matter interacts with its environment. These data are potentially more sensitive recorders of ocean processes than nonbiological ones, which are passive environmental recorders. It is worth noting that the oceanic biological data sets, used for global change interpretations, are based upon a small percentage of all life in the oceans. The question thus arises, are ocean data biased recorders?

Appendix 11.1 outlines the state-of-the-art in ocean paleodata, listing the proxy measures with their limitations and reliability. The biological data is subdivided into biotal census, biogenous sediment, and trace element data; nonbiological data is subdivided into eolian, volcanic ash, and runoff data. We stress that no single proxy measure is perfect: Some are spatially restrictive, some rely on unproven assumptions, some are calibration-specific, and some are qualitative. Most proxy measures would benefit from a greater understanding of dissolution and its effect. However, it is the joint use of these proxy data which strengthens our belief in our interpretations and reconstructions of the past. Much of the impetus in paleo-oceanography is in finding independent proxy measures. This section reviews such biological paleo-ocean data, addressing some of the physical and chemical processes recorded by proxy signals. This is equivalent to reviewing the statistical, biological, and geochemical approaches in paleo-oceanography.

Before proceeding, we briefly review some basic principles with respect to the pelagic sediments of the ocean basins and their biotal components (the tests, or fossils, of once-living organisms). Sediment samples represent long-term averages (> 500 years) of the benthos which lived on the seafloor and of the microplankton which lived in the upper layer of the ocean; though some zooplankton species have been found to live at deeper depths, the phytoplankton are restricted to the photic zone (0–200 meters). The preservation (or dissolution, if one is a pessimist) of calcareous and siliceous organisms on the seafloor is a function of the chemistry of the bottom waters. What is preserved on the seafloor may only be a portion of what actually lived within the water column. Surface sediment samples constitute the paleo-ontologic modern-world data base. By calibrating variations in surface sediments (core tops) with the observed physics and chemistry of the modern world, paleo-oceanographers seek to interpret downcore variations of ocean sediments. These variations are the depth and time series which define rate of change. Alternatively, a datum horizon like the last glacial maximum (LGM) can be identified in a suite of cores, and the variation within that horizon can be compared to the paleo-ontologic modern world. This method introduces new sources of uncertainty, principally stratigraphic and chronologic, which we shall not discuss.

As the mixing time of the oceans is on the order of 200–1000 years, sediment samples record ocean processes whose time constants exceed these values. However, sediment samples (as recorders of ocean processes) are further restricted by sedimentation rate (the continuous rain of material to the seafloor) and bioturbation (the mechanical mixing action of bottom-burrowing animals): the lower the sedimentation rate, the lower the probability that short-term events are recorded, while higher benthic activity causes short-term (Ka) variability to be smoothed out. The latter is particularly a problem when sedimentation rates are less then 3 cm Ka^{-1}. Bioturbation varies in both time and space (Ruddiman et al. 1980), thus producing nonhomogeneous smoothing. Deconvolution shows that this may bias sediment samples (Bard et al. 1987).

Biological Census Counts

There are four major microfossil groups preserved in ocean sediment: foraminifera (calcareous animal), radiolara (siliceous animal), coccolithophorids (calcareous plants), and diatoms (siliceous plant). The singular most difficult problem is the assessment of the dissolution overprint since it preferentially alters the species composition of sediment samples and hence introduces paleoestimation bias. This is particularly a problem for siliceous organisms since more than 50% of the exported production is dissolved before burial in the sediment, and the distribution of warm versus cold water diatoms could be affected (Shemesh 1989). Certain ocean regions are best studied by particular organisms. In regions where we are fortunate to have good preservation of both carbonates and silicates, our confidence in our inferences is increased, though discordances may exist.

Planktic census data of each biotic group provide independent evidence of the surface paleo-ocean. In statistical paleo-oceanography, multivariate data are calibrated to sea-surface temperature (SST) using transfer functions or nonparametric techniques. In biological paleo-oceanography, single-species responses are shown to be indicative of ocean circulation phenomenon. We discuss each to determine whether the different ocean measures available are concordant or discordant.

Sea-Surface Temperature

Paleoestimates of SST derived from planktic census data (relative abundances) are statistical constructs: they effectively summarize the information from a large number of species distributions into a single measure of climate change. Every summary has its associated error, the cost of simplification. For paleotemperature estimates developed using standard regression techniques, the error (standard error of estimate) is on the order of ±1° to ±2°C. There are a number of paleoestimation techniques currently available, but that of Imbrie and Kipp (1971) is the one most commonly used. This method first resolves the relative abundances of species in a suite of core-tops into a few assemblages (latent variables) using factor analysis and then relates the assemblages to modern SST at the core-top site using multivariate linear regression. While sediment samples represent long-term averages with a mean value of ~1000 years, calibration data are short-term averages based on observations of less than 100 years. This is an unavoidable dilemma. The use of assemblages carries with it the prophetic tenet of the constancy of biotic relationships over time. Yet we know that no-analog conditions existed in the past and that evolution occurs; the latter makes the coccolith SST equations unreliable before 70 Ka and the foram SST equations questionable before 1.8 Ma.

Given that there are different depth preferences between biotic groups, preferential preservation of species, counting procedure differences, and a degree of arbitrariness in the selection of which species to use in the analysis, it is (from the skeptics' point of view) remarkable that paleoestimates of SST from different biotic groups are so similar. Molfino et al. (1982) showed 87% concordancy among Atlantic paleotemperatures from forams, radiolaria, and coccoliths in the modern world and during the LGM; discordancy among LGM SST estimates was observed to be spatially restricted to the equatorial divergence and subtropical convergence regions. Howard and Prell (1984) reported a similar discordancy along the subtropical convergence in the Indian Ocean; foram and radiolara SST estimates disagreed along the deglacial transition at 127 Ka in two cores. However, at the LGM, 18 Ka, CLIMAP (1981) successfully utilized estimates from all four biotic groups to reconstruct the global oceans. The Pacific presented the most difficult problems due to the use of regional equations and the varying degrees of microfossil preservation (Moore et al. 1980).

Alternate methods for computing SST include nonparametrical statistical techniques ("modern analog technique" of Prell [1985]; "response surface technique" of Molfino [1993]) and geochemical (oxygen isotope, alkenones) techniques. Concordancy among paleotemperatures from the three independent statistical approaches is shown to be dependent on choice of data treatment, with discordancy spatially restricted to the subtropical convergence region (Molfino 1993). Oxygen isotopes can also provide paleoestimates of temperature, presuming that the isotopic composition of seawater is a known quantity in the past. Broecker (1986), however, applied this method to the LGM data base and found that the results of CLIMAP (1981) could not be refuted. Thus, there appears to be consistency, within the limits stated, between the statistical and isotopic approaches.

Foraminiferal assemblages have also been interpreted as recorders of productivity (Mix 1989), hydrographic structure (Ravelo et al. 1990), and salinity (Kipp 1976; Cullen 1981). Each of these interpretations is based upon different regional data sets. The unresolved questions are: what information does the integrated foram assemblage record, and, are foram assemblages spatially biased? Molfino (1993) has shown, via statistical and ecological argument with an expanded and revised global foram data base, that temperature is the dominant signature on Atlantic foram species, with salinity being the secondary signature. This remains to be studied in other oceans.

We conclude this section with some cautionary statements:

1. Each equation is calibration-specific, meaning that it should only be used within the region in which it was calibrated. Prell (1985) nicely showed the effect of using different ocean-calibrated equations for the foraminifera on the sediment samples for other oceans, noting that each equation has its own bias. Further work by Molfino (1993) confirms these results by establishing the inherent ocean-bias effect on the foram species distributions.
2. When regional equations are developed for an ocean, a paleoceanographic dilemma is introduced, that is, which equation is optimal for reconstruction of a given core site. As in numerical modeling, boundary constraints can induce paleoestimation bias. What is accurately calibrated for the modern world may lack the flexibility to account for variation in the past. Fine-tuning regression equations are not necessarily desirable.
3. Multi-biotic equations offer a test of the statistical methodology used to develop paleotemperatures, since it is logical to assume that erroneous paleoestimates will not be simultaneously generated by each planktonic group (Molfino et al. 1982).
4. The correlation between environmental gradients and biotic assemblages (or individual species) may be spurious. The consistency of response in downcore records, however, would argue that the relationship is valid, thus supporting the spectral interpretation of climatic indicators.

Single-Species Indices

Biotic single-species indices provide evidence of consistency within ocean paleodata. The polar front movements of the North Atlantic over the past 300 Ka are indeed highly correlated with SST estimates (Ruddiman and McIntyre 1977) because both estimates are strongly dependent on *Neogloboquadrina pachyderma* left-coiling. Within the

Indian Ocean, the movement of the subtropical convergence (STC) is linked to SST estimates over the last 30 Ka (Prell et al. 1979) because STC movement is defined by the dominant assemblages which also control SST estimates. A more challenging test is when the indicators involved are from different biotic groups.

In the low-latitude Atlantic, variations in the coccolithophorid species *Florisphaera profunda* have been shown to be in-phase with the foraminiferal SST estimates over the last 200 Ka (Molfino and McIntyre 1990). *F. profunda* is interpreted as being a response to nutricline dynamics and a relative measure of coccolith productivity: When the nutricline is deep, nutrient depletion in the upper waters results in higher values of *F. profunda* relative to all other coccoliths in the sediments; and vice versa, hence the signal is a measure of relative coccolith productivity. *In situ* studies support this interpretation. The signal has been linked to precessional forcing, with abundance minima of *F. profunda* coinciding with times of December perihelion and abundance maxima coinciding with June perihelion. In a different tropical Atlantic core, but spanning the same time interval, the advective index of *Melosira* (freshwater diatom), interpreted as an aridity indicator, is also linked to precessional control and the monsoon circulation (Pokras and Mix 1987). *Melosira* is coherent ($k = 0.89$) and lags foraminiferal SST by nearly 7 Ka in the 23-Ka band (unpublished analyses, B. Molfino).

Trace Element Data

Geochemical paleoceanography has dominated the field of paleoceanography during the 1980s. The major tools are the isotopic measurements of carbon and oxygen in the shells of foraminifera.

$\delta^{13}C$ in Foraminifera

Photosynthesis fractionates against ^{13}C. This fractionation has been recently shown to be especially important when dissolved CO_2 is high (Rau et al. 1991). When organic matter is oxidized, the derived low $\delta^{13}C$ CO_2 is reincorporated into the dissolved CO_2 reservoir. If only deep waters are considered (and benthic foraminifera), the older the water (the higher the amount of dissolved CO_2), the lower the $\delta^{13}C$ value. The gradient in benthic foraminifera $\delta^{13}C$ between communicating water masses is therefore a measurement of the dissolved oxygen loss during water mass transport (Duplessy et al. 1984). Globally, the mean $\delta^{13}C$ changes of the ocean follow the transfer of carbon between the ocean and continental biosphere (Shackleton 1977). Recent estimates of that effect have been made by Curry et al. (1988), who calculate a mean glacial–interglacial $\delta^{13}C$ shift of 0.45‰ , and Duplessy et al. (1988), who propose a value of +0.32‰, taking into account the fact that during the last glacial, intermediate waters of the Atlantic and Indian oceans were relatively more ventilated than the modern ocean.

Some species of benthic foraminifera, in particular *Cibicides wuellerstorfi*, appear to be better recorders of the deep-water total dissolved CO_2 $\delta^{13}C$ (Duplessy et al. 1984). *Uvigerina* sp. gives the same information, with a larger uncertainty, in the

Pacific and Indian oceans (± 0.2‰; Duplessy et al. 1984). In the Atlantic Ocean, Zahn et al. (1986) have shown that the amplitude of the *Uvigerina* $\delta^{13}C$ changes is always larger than that for *Cibicides*. Significantly lower $\delta^{13}C$ values are obtained when the sediment is rich in organic matter (the difference may reach 0.5‰). This phenomenon has been attributed to the habitat of *Uvigerina*, within the upper few millimeters of sediment (Lutze and Thiel 1987), as opposed to *Cibicides*, which develops at the true surface.

Records of planktonic foraminifera $\delta^{13}C$ are more difficult to interpret quantitatively, as the $\delta^{13}C$ of dissolved CO_2 exhibits a high variability in surface waters (over 1‰, both with depth and time). The dissolved CO_2 $\delta^{13}C$ is generally higher by 1‰ to 2‰ within the upper few meters of the water column, when compared to values below the mixed layer. Two basic reasons can help explain this difference:

1. Because of the fractionation during photosynthesis and the fall of a large part of the organic remains below the mixed layer, the upper water $\delta^{13}C$ increases during phytoplankton blooms by about 1%. The isotopic exchange rate of the surface water dissolved CO_2 is slow (several months of e-folding time). Therefore, during the summer, surface water $\delta^{13}C$ integrates the effect of productivity. In middle to high latitudes, winter mixing rehomogenizes in part the $\delta^{13}C$ of the upper waters.
2. The $\delta^{13}C$ value of dissolved CO_2 in isotopic equilibrium with preindustrial atmosphere is about +2 to +3. (This value is now lowered by ~1‰, because of the input to the atmosphere of low $\delta^{13}C$ fossil-fuel derived CO_2). The equilibrium value is temperature dependent, with a slope of ~0.1‰ /°C.

Below the mixed layer, the $\delta^{13}C$ of the dissolved CO_2, (enriched in CO_2 derived from organic matter oxidation) is in the –0.5‰ to + 0.5‰ range everywhere except in the well-ventilated North Atlantic. The effect of the first fractionation process is significantly larger and faster. The relative shifts of $\delta^{13}C$ in surface-dwelling planktic foraminifera (as referred to representative benthics) may therefore be taken as a proxy for productivity ("the biological pump" record of Shackleton and Pisias 1985). There is, however, not enough information on the surface water $\delta^{13}C$ distribution to attach uncertainties onto these estimates.

$\delta^{18}O$ in Foraminifera

The oxygen isotopic fractionation is temperature dependent. The experimental relationship by O'Neil et al. (1969) has been extended at low temperatures by Shackleton (1974):

$$T(°C) = 16.9 - 4.38\,(\delta^{18}O_{foraminifera} - \delta^{18}O_{water}) + 0.1 + (\delta^{18}O_{foraminifera} - \delta^{18}O_{water})^2\ ‰\,. \tag{7.11}$$

Erez and Luz (1983) obtained a slightly different relationship from living warm temperature planktic foraminifera:

$$T = 17 - 4.52\ (\delta^{18}O_{foraminifera} - \delta O_{water}) + 0.03$$
$$(\delta^{18}P_{foraminifera} - \delta^{18}O_{water})^2\ ‰\ . \qquad (7.12)$$

A species/genus-dependent correction factor has to be applied for benthic foraminifera to follow the "equilibrium relationship" of Shackleton (1974): *Uvigerina*, 0.0; *C. wuellerstorfi*, + 0.64‰ ; *Oridorsalis tener*, + 0.36‰ ; *Melonis barleanum*, + 0.4‰ ; *Globobulimina pacifica*, –0.4‰ (Shackleton and Opdyke 1973; Duplessy et al. 1984; CFR/Cambridge unpublished). These values are now widely accepted, as significantly different values occur only in untypical environments or during climatic transitions, where faunas from different levels are mixed by bioturbation.

The ocean water $\delta^{18}O$ varies in space and time. Spatial water $\delta^{18}O$ changes are well correlated with salinity, as evaporation and precipitation are the causes for these changes. The mean ocean $\delta^{18}O$ varies with waxing and waning of continental ice with low $\delta^{18}O$. Recent estimates of the glacial–interglacial amplitude, deriving from different hypotheses, are + 1.1‰ (Labeyrie et al. 1987), + 1.2‰ (Shackleton 1987), and + 1.3‰ (Fairbanks 1989). A mean glacial–interglacial shift of + 1.2‰ is now generally accepted.

Deep-water temperature changes may be derived from the benthic $\delta^{18}O$ records once the effect of the changes in mean ocean $\delta^{18}O$ have been taken into account. Labeyrie et al. (1987) have estimated that the different deep ocean basins cooled to approximately 0°C during the LGM. Dodge et al. (1983) obtained the same result from a comparison of benthic $\delta^{18}O$ and sea-level records from raised terraces. Zahn et al. (1991) and Labeyrie et al. (1992) went a step further, discussing the possibilities of using benthic foraminifera $\delta^{18}O$ to estimate deep-water density distribution in the past; however, more work is necessary to reach quantitative reconstructions.

These interpretations are limited by the following factors: (a) the lack of calibration between the different isotopic laboratories (several tenths of per mil often separate measurements from the same area and period) and (b) the uncertainties in the distribution of salt and on the deep water $\delta^{18}O$ versus salinity relationship in the past oceans (see Zahn et al. 1991 and Labeyrie et al. 1992 for a discussion of the problems).

Duplessy et al. (1991) have shown that the Shackleton's relationship also holds for planktic foraminifera if the individuals lived within their optimum temperature range, and if corrections are introduced to take into account seasonality and the water depth of development. For *Neogloboquadrina -pachyderma senestra*, the optimum summer temperature range is + 3° to + 10°C, and growth temperature is mean summer temperature – 2.2°C. For *Globigerina bulloides*, optimum temperature range is + 8° to + 20°C, and growth temperature is mean summer temperature minus 1°C. Duplessy et al. (1992) proposed using these relationships to estimate past sea-surface salinity gradients. The isotopic fractionation of $\delta^{18}O$ is derived from the summer SST estimated with transfer functions. Then $\delta^{18}O$ of the water is estimated using the relationship:

$$\delta^{18}O = \delta^{18}O_{foraminifera} - \delta^{18}O_{water} . \tag{7.13}$$

The surface water salinity may be derived from this value because, in most surface waters, $\delta^{18}O$ water is proportional to salinity (with a slope of 0.55‰ in areas and periods where reliable summer SST estimates exist). The statistical uncertainty is ± 0.5‰ in salinity for ± 1°C errors in SST.

$\delta^{18}O$ in Diatoms

Diatoms, an algae with siliceous skeleton, need light to develop and are therefore limited to the upper mixed layer. They constitute most of the fossil deposits in the Southern Ocean. Measurement of their $\delta^{18}O$ values requires special precautions since part of the oxygen atoms bound to silicon are in hydroxyl radicals which can easily exchange (Labeyrie and Juillet 1982). A calibration of the fractionation has been proposed by Juillet-Leclerc and Labeyrie (1987):

$$T°C = 17.2 - 2.4\,(\delta^{18}O_{diatoms} - \delta^{18}O_{water} - 40) - 0.2\,(\delta^{18}O_{diatoms} - \delta^{18}O_{water} - 40)^2\ ‰ . \tag{7.14}$$

This relationship has been applied to high-resolution temperature reconstructions over the last few hundred to thousand years in the Gulf of California (Juillet-Leclerc et al. 1991; Juillet-Leclerc and Schrader 1987) In that time range, uncertainties are mostly limited to the analytical errors, about ± 0.4°C. There is apparently no species dependency in the diatom-silica fractionation factor (Juillet-Leclerc and Labeyrie 1987).

$\delta^{18}O$ in Coccoliths

These planktonic algae present the same potential interest that the diatoms do, but for warm waters. However, strong species-dependent isotopic fractionations exist, which limit the interest of the method until the different species are well-calibrated isotopically (Dudley and Nelson 1989). In most of the areas where coccoliths are found, planktic foraminifera constitute an easier system on which to work, yet results to date indicate significantly larger $\delta^{18}O$ changes for nannofossils than for planktic foraminifera in the same cores. This could point to a more surficial habitat of coccoliths or perhaps to selective dissolution of the warmer-living foraminifera (Anderson and Steinmetz 1981).

Cd/Ca and Ba/Ca Ratios in Foraminifera

The dissolved Cd/Ca and Ba/Ca ratios are not uniform in ocean water because both Cd and Ba are, in part, associated with organic matter. The Cd/Ca ratio appears especially suited to reconstruct the dissolved nutrient distribution in deep waters. Uncertainties in the calibration of the Cd/Ca ratio versus dissolved Cd content in the ocean are of the order of ± 20%. Paleochemical reconstructions of the glacial ocean using a combination of benthic foraminifera $\delta^{13}C$ and Cd/Ca are quite complimentary. However, large unexplained discrepancies exist, such as the large increase in nutrient content of the Southern Ocean deep waters, as evidenced by a 1‰ to 1.5‰ decrease in $\delta^{13}C$, while the Cd/Ca ratio is constant. Boyle (1992) gives a thorough comparison of the Cd/Ca versus $\delta^{13}C$ distribution for modern and glacial oceans.

The Ba/Ca ratio in benthic foraminifera may be considered as a proxy for "refractory" nutrients, as the concentration increases more slowly with depth that the Ca/Cd ratio (Lea and Boyle 1990a, b). The paleochemical data base is still rather limited, and it is too early to appreciate fully the possibilities of the method for paleocirculation reconstructions.

^{14}C in Foraminifera

With the advent of Accelerator Mass Spectrometry (AMS), ^{14}C ages may be derived from milligram quantities of foraminifera. Because it allows the estimation of age differences between surface waters (obtained from dated planktic foraminifera) and deep waters (obtained from the benthic foraminifera at the same levels in the cores), ^{14}C is a splendid tracer of deep-water circulation for the past 30 Ka (Duplessy et al. 1992; Broecker et al. 1984, 1988, 1990). Uncertainties involved in these reconstructions derive mostly from the bioturbation and other perturbations in the regularity at which sediment accumulates. This can, for example, create inversions in the age succession exceeding the analytical errors.

Oxygen ($\delta^{18}O$) and Carbon ($\delta^{13}O$ and $\Delta^{14}C$) Isotopes in Corals

It is rather easy to study interannual variability of tropical surface waters from the isotopic analysis of coral annual bands (Weber and Woodhead 1970, 1972; McConnaughey 1989). As in foraminifera, $\delta^{18}O$ is a function of surface water temperature and $\delta^{18}O$ of the water (salinity), $\delta^{18}O$ and $\Delta^{14}C$ a function of the surface water ventilation, exchange with deeper waters, and productivity. Numerous studies have covered the last few hundred years. Typical applications are the evolution of the Southern Ocean Oscillation (ENSO) (Druffel 1985; Cole and Fairbanks 1990) or the estimation of the variability in the ventilation of surface ocean (Nozaki et al. 1978; Druffel 1989). The accuracy of the reconstructed changes in surface water properties is limited by the changes in the hydrology of these waters introduced by the coral reefs themselves. However, generalization of these studies to the whole intertropical area should allow a much better understanding of the interannual variability of its surface oceanography. Eventually, similar studies could be developed for older periods, from cores collected along the submarine slopes of the coral reefs.

Geochemistry of Organic Matter

Recent interest by organic chemists in the changes of ocean hydrology has triggered research into the applicability of biomarkers for SST reconstitutions. Alkenones (long chain ketones) are the better-defined components for these studies. They have been linked with prymniesiophytes algae (in particular the coccolithophoride *Emiliana huxleyi*). The overall degree of unsaturation in the long chain alkenone series (Uk37) biosynthesized by these plants varies inversely to water temperature (Brassell et al. 1986; Prahl and Wakeham 1987). Measurement is easy (gas chromatography and limited organic chemistry). Covariance with planktic foraminifera $\delta^{18}O$ in downcore studies is striking ($r = 0.93$; Brassell et al. 1986). The calibration of the method is still under progress, in particular for testing of possible species-specific sensitivity for Uk37. Present precision of the estimates obtained from controlled cultures of *E. Huxlei* are on the order of 0.2°C over a range in seawater temperatures of 8° to 25°C (Prahl and Wakeham 1987). Uncertainties also exist in the stability of this marker to different diagenic alterations, although the first results appear to indicate a sufficient stability over the last several 10^5 to 10^6 years in marine sediments (Brassell et al. 1986). These studies have been recently reviewed, with other applications of the lipid biomarkers, by Prahl and Muehlhausen (1989).

$\delta^{13}C$ in Organic Matter

Since Sackett (1964), numerous workers have analyzed $\delta^{13}C$ in organic matter from oceanic sediment, a convenient tracer of the input of continental organic matter (see, for example, the recent paper by Jasper and Gagozian 1990). Fontugne and Duplessy (1978, 1981) have demonstrated a dependance between $\delta^{13}C$ fractionation and phytoplankton characteristics. Renewed interest in this tracer has evolved from the work of Popp et al. (1989), who proposed to use it as a proxy for atmospheric pCO_2. This method derives from the hypothesis that marine phytoplankton build their organic matter primarily from dissolved CO_2. The higher the available dissolved CO_2 concentration in the euphotic layer, the more photosynthesis fractionates against ^{13}C. Descolas-Gros and Fontugne (1990) have shown that this property derives in part from the changes in photosynthetic pathways that marine phytoplankton utilize to adapt to lower dissolved CO_2 availability. Several workers have used these results to reconstruct changes in surface water pCO_2 and have compared it to the Vostok CO_2 records. Both total organic matter (Rau et al. 1991; Fontugne and Calvert 1991) and purified compounds (Jasper and Hayes 1990) have been used. Both methods have their limitations: (a) organic matter is not homogeneous in $\delta^{13}C$, with several per mil differences, for example, between lipids and sugars (Degens 1969), and (b) organic extraction by solvents and gas chromatography increases the analytical uncertainties by several tens of per mil. Jasper and Hayes have derived from Popp et al. (1989) the following relation:

$$e_p = a \log c + b \,, \tag{7.15}$$

with e_p being the photosynthetic isotope fractionation (‰), c the concentration of dissolved CO_2, and a and b constants which are functions of physiological, environmental and ecological factors. For the C37 alkadienones compound analyzed, Jasper and Hayes (1990) propose:

$$e_p = 1000[(\delta_p + 1000)/(\delta_d + 1000) - 1] \text{ with}$$
$$\delta_p(‰) = \delta_{37.2} + 3.8 \text{ and } \delta_d(‰) = \delta_{G.ruber} - 8.3\ , \qquad (7.16)$$

the inorganic dissolved CO_2 $\delta^{13}C$ being monitored by the $\delta^{13}C$ of the planktic foraminifera *Globigerinoides ruber*. The comparison of the changes in e_p along a record covering the last climatic cycle in Gulf of Mexico with the CO_2 Vostok record gives the calibration $a = -32.9$ and $b = 14.3$. Results from McCabe and Rau et al. (1989) would point more towards $a = -20$ (see discussion in Jasper and Hayes 1990). These authors consider a total analytical uncertainty in their prediction of surface water pCO_2 of the order of ± 15 ppm, with added uncertainties (of 20–40 ppm ?) deriving from the estimation of the constant a. Rau et al. (1991) have proposed the simpler relation mean plankton

$$\delta^{13}C_{org} = -0.08\ [CO_2(aq)]\ (\mu M) - 12.6\ ‰\ . \qquad (7.17)$$

The dissolved $[CO_2(aq)]$ is linked to pCO_2 via the solubility constant of CO_2, a function of temperature and salinity. It is too early to discuss errors on these estimates before the completion of the calibration, but predicted glacial pCO_2 concentrations are compatible with the Vostok record, taking into consideration that, in most of the surface water masses, pressure equilibrium is never achieved. An historical review of some of the most important papers on the subject may be found in Sackett (1991).

CONCLUSIONS

In this chapter and in Appendix 11.1 we have attempted to cover the subject of paleodata reliability. Except in the most general of terms, such a broad topic cannot be given justice in any single paper. Nonetheless, it is hoped that this treatment will benefit those who seek an broad overview of the subject. As a final comment, it is worth noting that reliability is a subject that must ultimately be considered in the context of what one wants from the data. Paleodata is collected to answer questions. As an initial measure of reliability, the data need only be reliable enough to answer the question asked. As our understanding grows and our questions become more sophisticated, our need for reliability increases. For example, if we wish to know whether the concentration of atmospheric CO_2 has changed over the last glacial–interglacial cycle, we ask this question of ice cores and ocean sediment cores and receive a positive reply. The reliability of these paleorecorders is very high for this simple question. If we wish to know exactly when the changes occurred, and exactly how large they were, the reliability of these recorders decreases, to the point where key questions such as the interhemispheric timing and the existence of rapid fluctuations remain unanswered. Yet the answers will come with improved reliability: perhaps through improvements in analytical techniques, perhaps via the addition of a new paleorecorder. In essence, then, the answer to the question which forms the title of this chapter must be given by the current state of our knowledge of global change. In short, the overall reliability is quite good and is constantly improving.

REFERENCES

Anderson, T.F., and J.C. Steinmetz. 1981. Isotopic and biostratigraphical records of calcareous nannofossils in a Pleistocene core. *Nature* **294**:741–744.

Bard, E., M. Arnold, J. Moyes, and J.C. Duplessy. 1987. Reconstruction of the last deglaciation: deconvolved records of 810 profiles, micropaleontological variations, and accelerator ^{14}C dating. *Clim. Dynam.* **1**:101–112.

Bender, M.M., and A.J. Berge. 1982. Carbon isotope records in wisconsin trees. *Tellus* **34**:500–504.

Boyle, E.A. 1992. Cadmium and $\delta^{13}C$ Paleochemical ocean distrbutions during the stage 2 Glacial maximum. *Ann. Rev. Geophysics*, in press.

Brassell, S.C., G. Eglington, I.T. Marlowe, U. Pflaumann, and M. Sarnthein. 1986. Molecular stratigraphy: a new tool for climatic assesment. *Nature* **320**:129–133.

Broecker, W.S. 1986. Oxygen isotope constraints on surface ocean temperatures. *Quat. Res.* **26**:121–134.

Broecker, W., A. Mix, M. Andree, and H. Oeschger. 1984. Radiocarbon measurements on coexisting benthic and planktic foraminifera shells: Potential for reconstructing ocean ventilation times over the past 20,000 years. *Nucl. Instr. Methods* **B5**:331–339.

Broecker, W.S., M. Andree, G. Bonani, W. Wolfli, H. Oeschger, M. Klas, A. Mix, and W. Curry. 1988. Preliminary estimates for the radiocarbon age of deep water in the glacial ocean. *Paleoceanogr.* **3**: 659–669.

Broecker, W.S., T.H. Peng, S. Trumbore, G. Bonani, and W. Wolfli. 1990. The distrbution of radiocarbon in the glacial ocean. *Global Biogeochem. Cycles* 4: 103–l 17.

Burk, R.L. and M. Stuiver. 1981. Oxygen isotope ratios in trees reflect mean annual temperature and relative humidity. *Science* **211**: 1417–1419.

Clausen H.B. and C.U. Hammer. 1988. The Laki and Tambora eruptions as revealed in Greenland ice cores from 11 locations. *Ann. Glaciology* **10**: 16–22.

CLIMAP Project Members (A. Mclntyre, Leader LGM Project). 1981. Seasonal reconstructions of the Earth's surface at the last glacial marimum. *Geological Soc. Am. Map and Chart Series 36.*

Christeller,J.T, W.A. Lang, and J.H.Troughton. 1976. Isotope discrimination by ribulose l ,5 diphosphate carhoxylase. *Plant Physiol.* **57**: 580–582

Cole, J.E., and R.G. Fairbanks. 1990. The Southern Oscillation recorded in the 8180 of corals from Tarawa Atoll. *Paleoceanography* **5**: 669–683.

Cullen, J.L. 1981. Microfossil evidence for changing salinity patterns in the Bay of Bengal over the last 20,000 years. *Palaeogeo. Palaeoclim. Palaeoecol.* **35:**315–356.

Curry, W.B., J.C. Duplessy, L.D. Labeyrie, and N.J. Shackleton. 1988. Quaternary deep-water circulation changes $\delta^{13}C$ of deep-water CO_2 between the last glaciation and the Holocene. *Palaeogeo* **3**: 317–342.

Dansgaard W., S.J. Johnsen, H.B. Clausen, and N. Gundestrup. 1973. Stable isotope glaciology. *Meddelelser om Groenland* **197(2):**1–53.

Dansgaard W., and H. Oeschger. 1989. Past environmental long-term records from the Arctic. In: The Environmental Record in Glaciers and Ice Sheets, ed. H. Oeschger and C.C. Langway, Dahlem Workshop Report PC 8, pp. 287–317. Chichester: Wiley.

Degens, E.T. 1969. Biogeochemistry of stable carbon isotopes, in Organic geochemistry, ed G. Eglington and M.T.J. Murphy, pp. 304–329. New York: Springer.

De Niro, M.J., and S. Epstein. 1977. Mechanism of carbon isotope fractionation associated with lipid synthesis. *Science* **197**:261–263.

DeNiro, M.J., and L.W. Cooper. 1989. Post-photosynthetic modification of oxygen isotope ratios in carbohydrates in the potato: Implications for paloeclimatic reconstruction based on isotopic analysis of wood cellulose. *Geochimica et Cosmochimica Acta* **53**:2573–2580.

Descolas-Gros, C., and M. Fontugne.1990. Stable carbonate isotope fractination by marne phytoplancton during photosynthesis. *Plant Cell Environment* **13**.

Dodge, E.D., R.G. Fairbanks, L.K. Benninger, and F. Maurasse. 1983. Pleistocene sea level from raised coral reefs of Haiti. *Science* **219**:1423–1425.

Dudley, W.C., and C.S. Nelson. 1989. Quaternary surface-water stable isotope signal from calcareous nannofossils at DSDP Site 593, Southern Tasman Sea. *Mar. Geol.* **13**:353–373.

Druffel, E.R.M. 1985. Detection of E1 Niño and decadal time scale variations of sea surface temperature from banded coral records: Implications for the carbon dioxide cycle. In: The Carhon Cycle and Atmospheric CO_2: Natural Varations Archean to Present. *Geophysical Monograph 32*, pp. 111–122. American Geophysical Union.

Druffel, E.R.M. 1989. Decadal time scale variability of ventilation in the north Atlantic: High-precision measurements of bomb radiocarbon in banded corals. *J. Geophys. Res.* **94**:3271–3285.

Duplessy, J.C., M. Arnold, E. Bard, A. Juillet-Leclerc, N. Kallel, and L. Labeyrie. 1992. AMS ^{14}C study of transient events and of the ventilation rate of the Pacific intermediate water during the last deglaciation. *Radiocarbon*, in press.

Duplessy, J.C., L. Labeyrie, A. Juillet-LeClerc, F. Maitre, J. Duprat, and M. Sarnthein. 1991. Surface salinity reconstruction of the North Atlantic Ocean during the last glacial maximum. *Oceanol. Acta* **14:**311–324.

Duplessy, J., N.J. Shackleton, R.G. Fairbanks, L.D. Labeyrie, D. Oppo, and N. Kallel. 1988. Deep-water source variations during the last climatic cycle and their impact on the global deep-water circulation. *Paleoceanography* **3**:343–360.

Duplessy, J.C., N.J. Shackleton, R.K. Matthews, W.L. Prell, W.F. Ruddiman, M. Caralp, and C. Hendy. 1984. ^{13}C record of benthic foraminifera in the last interglacial ocean: Implications for the carbon cycle and the global deep water circulation. *Quat. Res.* **21**:225–243.

Edwards, T.W.D., and P. Fritz. 1986. Assessing meteoric water composition and relative humidity from ^{18}O and ^{2}H in wood cellulose: Paleoclimatic implications for southern Ontario, Canada. *Appl. Geochem.* **1**:715–723.

Ehleringher, J.R., Z.F. Lin, C.B.Field, and C.Y. Kuo. 1987. Leaf carbon isotope ratios of plants from a subtropical monsoon forest. *Oecologia* **72**:109–114.

Epstein, S., and C.J. Yapp. 1976. Climatic implications of the D/H ratio of hydrogen in C–H groups in tree cellulose. *Earth Planet. Sci. Lett.* **30**:252–260.

Erez, J., and B. Luz. 1983. Experimental paleotemperature equation for planktonic foraminifera. *Geochim. Cosmochim. Acta* **47**:1025–1031.

Evans, J.R., T.D. Sharkey, J.A. Berry, G.D. Farquhar. 1986. Carbon isotope discrimination measured concurrently with gas exchange to investigate CO_2 diffusion in leaves of higher plants. *Aust. J. Plant Physiol.* **13**:281–292.

Fairbanks, R.G. 1989. A 17,000 year glacio-eustatic sea-level record: Influence of glacial melting rates on the Younger Dryas event and deep ocean circulation. *Nature* **342**:637–642.

Farmer, J.G. 1979. Problems in interpreting tree-ring $\delta^{13}C$ records. *Nature* **279**:229–231.

Farquhar, G.D. 1980. Carbon isotope discrimination by plants and the ratio of intercellular and atmospheric CO_2 concentrations. In: Carbon Dioxide and Climate: Australian research, ed. G.I. Pearman, pp. 105–110. Canberra: Australian Academy of Science.

Farquhar, G.D., M.H. O'Leary, and J.A. Berry. 1982. On the relationship between carbon isotope discrrmination and the intercellular carbon dioxide concentration in leaves. *Aust. J. Plant Physiol.* **9**:121–137

Farquhar, G.D. 1983. On the nature ofcarbon isotope discrimination in C_4 species. *Aust. J. Plant Physiol.* **10**:205–226.

Farquhar, G.D., and R.A. Richards. 1984. Isotopic composition ofplant carbon cornelates with water-use efficiency of wheat genotypes. *Aust. J. Plant Physiol.* **11**:539–552

Farquhar, G.D., S.C.Wong, J.R.Evans, and K.T. Hubick. 1989. Photosynthesis and gas exchange. In: Plants Under Stress: Physiology and Ecology and Their Application to Plant Improvement, ed H.G. Jones, T.J. Flowers, and M.B. Jones. Cambridge: Cambridge Univ. Press.

Francey, R.J., and G.D. Farquhar. 1982. An explanation of $^{13}C/^{12}C$ variations in tree rings. *Nature* **297**:28–31.

Freyer, H.D., and N.Belacy. 1983. $^{13}C/^{12}C$ records in northern hemispheric trees during the past 500 years: Anthropogenic impact and climate superpositions. *J. Geophys. Res.* **88**:6844–6852.

Fontugne, M., and S.E. Calvert. 1991. Late Pleistocene variability of the carbon isotopic composition of organic matter in the eastern Mediterranean: Monitor of changes in carbon sources and atinospheric CO_2 concentrations. *Paleoceanography* **7**:1–20.

Fontugne, M.R., and J.-C. Duplessy. 1978. Carbon isotope ratio of marine plankton related to surface water masses. *Earth Planet. Sci. Lett.* **41**:365–371.

Fontugne, M.R., and J.-C. Duplessy. 1981. Organic carbon isotopic fractionation by marine plankton in the temperature range –1° to 31°C. *Oceanol. Acta* **4**:85–90.

Grinsted, M.J., A.T. Wilson, and C.W. Ferguson. 1979. $^{13}C/^{12}C$ ratio variations in *Pinus longaeva* (Bristlecone pine) cellulose during the last millennium. *Earth Planet. Sci. Lett.* **42**:251–253.

Gray, J., and P. Thompson. 1976. Climatic information from $^{18}O/^{16}O$ ratios of cellulose in tree rings. *Nature* **262**:481–482.

Guy, R.D., M.F. Fogel, J.A. Berry, and T.C. Hoering. 1987. Isotope fractionation during oxygen production and consumption by plants. In: Progress in Photosynthetic Research III, ed. J.Biggins, pp. 597–560. Dordrecht: Martinus Nijhoff.

Hammer, C.U. 1980. Acidity in polar ice cores in relation to absolute dating, past volcanism, and radio-echoes. *J. Glaciology* **25(93)**:359–372.

Hammer, C.U., H.B. Clausen, and W. Dansgaard. 1980. Greenland ice sheet evidence of postglacial volcanism and its climatic impact. *Nature* **288(5788)**:230–235.

Hammer, C.U, H.B. Clausen, and H. Tauber. 1986. Ice-core dating of the Pleistocene/Holocene boundary applied to a calibration ofthe ^{14}C time scale. *Radiocarbon* **28**:284–291.

Henderson, S.A., S. von Caemmerer, and G.D. Farquhar. 1992. Short-term measurements of carbon isotope discrimination in several C4 species. *Aust. J. Plant Physiol.* **19**:263–286.

Howard, W.R., and W.L. Prell. 1984. A comparison of radiolarian and foraminiferal paleoecology in the Southern Indian Ocean: New evidence for the interhemispheric timing of climatic change. *Quat. Res.* **21**:244–263.

Hubick, K.T., G.D. Farquhar, and R. Shorter. 1986. Correlation between water-use efficiency and carbon isotope discrimination in diverse peanut (*Arachis*) germplasm. *Aust. J. Plant Physiol.* **13**:803–816.

Imbrie, J., and N. Kipp. 1971. A new micropaleontological method for quantitative paleoclimatology: Application to a late Pleistocene Caribbean core. In: The Late Cenozoic Glacial Ages, ed. K.K. Turekian, pp. 71–181. New Haven: Yale Univ. Press.

Jasper, J.P., and J.M. Hayes. 1990. A carbon isotope record of CO_2 levels during the late Quaternary. *Nature* **347**:462–464.

Jasper, J.P., and R.B. Gagosian. 1990. The sources and deposition of organic matter in the Late Quaternary Pigmy Basin, Gulf of Mexico. *Geochim. Cosmochim. Acta* **54**:1117–1132.

Johnsen, S.J., W. Dansgaard, W., and J.W.C. White. 1989. The origin of arctic precipitation under present and glacial conditions. *Tellus* **41B**:452–468.

Jouzel, J., and L. Merlivat. 1984. Deuterium and oxygen-18 in precipitation: Modelling the isotopic effect during snow fonnation. *J. Geophys. Res.* **89**:11,749–11,757.

Juillet-Leclerc, A., and L.D. Labeyrie. 1987. Temperature dependence of the oxygen isotopic hactionation between diatom silica and water. *Earth Planet. Sci. Lett.* **84**:69–74.

Juillet-Leclerc, A., L.D. Labeyrie, J.L. Reyss, and H. Schrader. 1991. Temperature variability in the Gulf of California during the last century: A record of the recent strong El Niño, *Geophys. Res. Lett.* **18:**1889–1893.

Juillet-Leclerc, A., and H. Schrader. 1987. Variations of upwelling intensity record of varved sediment from the Gulf of California during the past 300 years. *Nature* **329**:146–149.

Kipp, N.G. 1976. New transfer function for estimating past sea-surface conditions from sea-bed distrbution of planktonic foraminiferal assemblages in the North Atlantic. In: Investigation of Late Quaternary Paleoceanography and Paleoclimatology, ed. R.M. Cline and J.D. Hays., Memoir 145. Boulder: Geol. Soc. of Amer.

Körner, Ch., G.D. Farquhar, and S.C. Wong. 1991. Carbon isotope discrimination by plants follows latitudinal and altitudinal trends. *Oecologia* **88:**30–60.

Labeyrie, L.D., and A. Juillet. 1982. Oxygen isotopic exchangeability of diatom valve silica. *Geochim. Cosmochim. Acta* **46**:967–975.

Labeyrie, L.D., J.C. Duplessy, and P.L. Blanc. 1987. Variations in mode of fonnation and temperature of oceanic deep waters over the past 125,000 years. *Nature* **327**:477–482.

Labeyrie, L.D., J.C. Duplessy, J. Duprat, A.J. Juillet-Leclerc, J. Moyes, E. Michel, N. Kallel, and N.J. Shackleton. 1992. Changes in the vertical structure of the N. Atlantic Ocean between glacial and modern times. *Quat. Sci. Rev.* **11:**401–413.

Lancaster, J. 1990. Carbon-13 fractionation in carbon dioxide emitted diurnally from soils and vegetation at ten sites on the North American continent. Ph.D. Thesis, Univ. of San Diego. Ann Arbor: Microfilms Intemat. No. 902408.

Langway C.C., H.B. Clausen, and C.U. Hammer. 1988. An inter–hemispheric time-marker in ice cores from Greenland and Antarctica. *Ann. Glaciology* **10**:102–108.

Lea, D.W., and E.A. Boyle. 1990. Foraminiferal reconstruction of barium distribution in water masses of the glacial oceans. *Paleoceanography* **5**:719–742.

Lea, D.W., and E.A. Boyle. 1990. A 210,000-year record of barium variability in the deep northwest Atlantic Ocean. *Nature* **347**:269–272.

Leavitt, S.W., and A. Long. 1982. Evidence for $^{13}C/^{12}C$ fractionation between tree leaves and wood. *Nature* **298**:742–743.

Leavitt, S.W., and A. Long. 1983. An atmospheric $^{13}C/^{12}C$ reconstruction generated through removal of climate effects from tree-ring $^{13}C/^{12}C$ measurements. *Tellus* **35B**:92–102.

Leavitt, S.W. and Long, A. 1988. Stable carbon isotope chronologies from trees in the southwestern United States. *Global Bioceochem. Cycles* **2**:189–198.

Legrand, M. and R.J. Delmas. 1987. A 220-year continuous record of volcanis H_2SO_4 in the Antarctic ice sheet. *Nature* **327**:671–676.

Lloyd, J., and J.P.Syversten. 1992. Low conductance for CO_2 diffusion from stomata to the sites of carboxylation in leaves of woody species. *Plant Cell and Environ.* **15:**873–901.

Lorius, C. 1983. Antarctica: Survey of near-surface mean isotopic values. In: The Climatic Record in Polar Ice Sheets, ed. G. de Q. Robin, pp. 52–56. Cambridge: Cambridge Univ. Press.

Lorius, C. 1989. Polar ice cores and climate. In: Climate and Geo-Sciences, ed. A. Berger et al., pp. 77–103. Nordrecht: Kluwer.

Lorius C., G. Raisbeck, J. Jouzel, and D. Raynaud. 1989. Long-term environmental records from antarctic ice cores. In: The Environmental Record in Glaciers and Ice Sheets, ed. H. Oeschger and C.C. Langway, Dahlem Workshop Report PC 8, pp. 343–361. Chichester: Wiley.

Lutze, G.F, and H. Thiel. 1987. *Cibicidoides wuellerstorfi* and *Planulina ariminensis*, elevated epibenthic foraminifera. *Ber. Sonderforschungsbereich* **313**:17–30.

McConnaughey, T. 1989. ^{13}C and ^{18}O isotopic disequilibrium in biological carbonates: I. Patterns. *Geochim. Cosmochim. Acta* **53**:151–162.

Marino, B.D. and McElroy, M.B. 1991. Isotopic composition ofatmospheric CO_2 inferred from carbon in C4 plant cellulose. *Nature* **349**:127–131.

Mix, A. C. 1989. Pleistocene paleoproductivity: Evidence from organic carbon and foraminiferal species. In: Productivity of the Ocean: Present and Past, ed. W.H. Berger, V.S. Smetacek, and G. Wefer, Dahlem Workshop Report LS 44, pp. 313–340. Chichester: Wiley.

Molfino, B. 1993. Statistical modeling in paleoceanography: Paleoestimarion of upper water-mass dynamics using marine microfossils. Ph.D. diss. New York: Columbia Univ., 411pp.

Molfino, B., N. Kipp, and J. Morley. 1982. Comparison of foraminiferal, coccolithophorid, and radiolarian paleotemperatiure equations: Assemblage coherency and estimate concordancy. *Quat. Res.* **17**:279–313.

Molfino, B., and A. Mclntyre. 1990. Precessional forcing of nutrcline dynamics in the equatorial Atlantic. *Science* **249**:766–769.

Moore, T.C. Jr., L.H. Burckle, K. Geitzenauer, B. Luz, A. Molina-Cruz, J.H. Robertson, H. Sachs, C. Sancetta, J. Thiede, P. Thompson, and C. Wenkam. 1980. The reconstriction of sea surface temperatures in the Pacific Ocean of 18,00 B.P. *Mar. Micropaleontology* **5**:215–247.

Narayanamurti, D., N.R. Das. 1955. Holz Rohstoff und Werkstoff **13**:52.

Neftel, A., Oeschger, H., Schwander, J., and Stauffer, B. 1983. Carbon dioxide concentration in natural cold ice. *J. Phys. Chem.* **87**:4110–4120.

Neftel A., E. Moore, H. Oeschger, and B. Stauffer. 1985. Evidence from polar ice cores for the increase in atmospheric CO_2 in the past two centuries. *Nature* **315**:45–47.

Neftel A., H. Oeschger, T. Staffelbach, and B. Stauffer. 1988. CO_2 record in Byrd ice core 20,000–5,000 years B.P. *Nature* **331**:609–611.

Nozaki, Y., D.M. Rye, K.K. Turekian, and R.E. Dodge. 1978. A 200-year record of carbon-13 and carbon-14 variations in a Bermuda coral. *Geophys. Res. Lett.* **5**:825–828.

O'Neil, J.R., R.N. Clayton, and T.K. Mayeda. 1969. Oxygen isotope fractionation in divalent metal carhonates. *J. Chem. Phys.* **51**:5547–5558.

Ostlund, H.G., C. Craig, W.S. Broecker, and D. Spencer, eds. 1987. GEOSECS Atlantic, Pacific, and Indian Ocean Expeditions. Shorebased Data and Graphics, GEOSECS Atlas Series , vol. 7, 200 pp. Washington D.C.: U.S. Govt. Printing Office.

Park, R., and S. Epstein. 1961. Metaholic fractionation of ^{13}C and ^{12}C in plants. *Plant Physiology* **36**:133–138.

Pearman, G.I., R.J. Francey, and P.J.B. Fraser. 1976. Climatic implications of stable isotopes in tree rings. *Nature* **260**:771–772.

Pichon, J.J., G. Bareille, M. Labracherie, L.D. Labeyrie, A. Baudrimont, and J.L. Turon. 1992. Quantification of the biogenic silica dissolution in Southern Ocean sediments. *Quat. Res.* **37:**361–378.

Pokras, E.M., and A.C. Mix. 1987. Earth's precession cycle and Quaternary climatic change in tropical Africa. *Nature* **326**:486–487.

Popp, B.N., R. Takigiku, J.M. Hayes, J.W. Louda, and E.W. Baker. 1989. The post-Paleozoic chronology and mechanism of ^{13}C depletion in primary marine organic matter. *Am. J. Sci.* **289**:436–454.

Prahl, F.G. and L.A. Muehlhause 1989. Lipid biomarkers as geochemical tools. In: Productivity of the Oceans: Present and Past, ed. W.H. Berger, V.S. Smetacek, and G. Wefer, pp. 271–289. Dahlem Workshop Report LS 44. Chichester: Wiley.

Prahl, F.G., and S.G. Wakeham. 1987. Calibration of unsaturated patterns in long chain ketone compostions for paleotemperature assessment. *Nature* **330**:367–369.

Prell, W.L., W.H. Hutson, and D.F. Williams. 1979. The subtropical convergence and late Quatemary circulation in the southern Indian Ocean. *Mar. Micropaleontology* **4**:225–234.

Prell, W. 1985. The stability of low-latitude SSTs: An evaluation of the CLIMAP reconstruction with emphasis on the positive SST anomalies. Washington, D.C.: Dept. of Energy Publ. 60176–1:1–60.

Rau, G.H., P.N. Froelich, T. Takahashi, and D.J. Des Marais. 1991. Does sedimentary organic $\delta^{13}C$ record variations in Quaternary ocean [CO_2 (aq)]. *Paleoceanography* **6**:335–347.

Ravelo, A.C., R.G. Fairbanks, and S.G.H. Philander. 1990. Reconstructing tropical Atlantic hydrography using planktonic foraminifera and an ocean model. *Paleoceanography* **5**:409–431.

Roeske, C.A., O'Leary,M.H. 1984. Carbon isotope effects on the enzyme catalyzed carboxylation of ribulose bisphosphate. *Biochemistry* **23**:6275–6284.

Ruddiman, W.F., and A. Mclntyre. 1977. Late Quaternary surface ocean kinematics and climatic change in the high-latitiude North Atlantic. *J. Geophys. Res.* **82**:3877–3887.

Ruddiman, W.F., G.A. Jones, T.-H. Peng, L.K. Glover, B.P. Glass, and P.J. Liebertz. 1980. Tests for size and shape dependency in deep-sea mixing. *Sed. Geology* **25**:257–276.

Sackett, W.M. 1964. The depositional history and isotopic organic carbon composition of marine sediments. *Mar. Geol.* **2**:173–185.

Sackett, W.M. 1991. A history of the $\delta^{13}C$ composition of oceanic plankton. *Mar. Chem.* **34**:153–156.

Schleser, G.H. 1990. Investigations of the $\delta^{13}C$ patterns in leaves of *Fagus syvatical. J. Exp. Bot.* **41**:565–572.

Shackleton, N.J., and N.D. Opdyke. 1973. Oxygen isotope and paleomagnetic stratigraphy of of equatorial Pacific core V28–238: Oxygen isotope temperatures and ice volumes on a 10^5 and l0-year scale. *Quat. Res.* **3**:39–55.

Shackleton, N.J. 1974. Attainment of isotopic equilibrium between ocean water and the benthonic foraminifera genus *Uvigerina*: Isotopic changes in the ocean during the last glacial. In: Les methodes quantitatives d'etude des variations du climat au cours du pleistocene, vol. 219, pp. 203–210. Colloque Intl. du CNRS.

Shackleton, N.J. 1977. Carbon-13 in *Uvigerina*: Tropical rainforest history and the equatorial Pacific carbonate dissolution cycles. In: The Fate of Fossil Fuel CO_2 in the Oceans, ed. N.R. Andersen and A. Malahoff, pp. 401–427, vol 6. New York: Plenum.

Shackleton, N.J. 1987. Oxygen isotopes, ice and sea level. *Quat. Sci. Rev.* **6**:183–190.

Shackleton, N.J., and N.G. Pisias. 1985. Atmospheric carbon dioxide, orbital forcing, and climate. In: The Carbon Cycle and Atmospheric CO_2: Natural Variations Archean to Present, ed. E. Sundquist and W.S. Broecker, pp. 303–317, 32 ed., vol. 32. Washington, D.C.: Am. Geophysical Union.

Shemesh, A., L.H. Burckle, and P.N. Froelich. 1989. Dissolution and preservation of Antarctic diatoms and the effect on sediment thanatocoenoses. *Quat. Res.* **31**:288–308.

Stuiver, M., and T.F. Braziunas. 1987. Tree cellulose $^{13}C/^{12}C$ isotope ratios and climate change. *Nature* **328**:58–60.

Stauffer B., H. Hofer, H. Oeschger, J. Schwander, U. Siegenthaler. 1984. Atmospheric CO_2 concentration during the last glaciation. *Ann. Glaciology* **5**:100–164.

Stauffer, B.G. Fischer, A. Neftel, and H. Oeschger. 1985. Increase of atmospheric methane recorded in Antarctic ice core. *Science* **229**:1389–1388.

Tans, P.P., and W.G. Mook. 1980. Past atmospheric CO_2 levels and the $^{13}C/^{12}C$ ratios in tree rings. *Tellus* **32**:268–283.

von Caemmerer, B., and J.R. Evans. 1991. Determination of the average partial pressure of CO_2 in chloroplasts from leaves of several C_3 plants. *Aust. J. Plant Physiol.* **18**:287–305.

Weber, I.N., and P.M.J. Woodhead. 1970. Carbon and oxygen isotopic fractionation in the skeletal carhonate of reef building corals. *Chem. Geology* **6**:93–117.

Weber, I.N., and P.M.J. Woodhead. 1972. Temperature dependance of oxygen-18 concentration in reef coral carbonates. *J. Geophys. Res.*

Whelan, T., W.M. Sackett, C.R. Benedict. 1973. Enzymatic fractionation of carbon isotopes by phosphenolpyruvate carhoxylase from C_4 plants. *Plant Physiol.* **51**:1051–1054.

White, J.W.C. and S.D. Gedzelman. 1984. The isotopic composition of atmospheric water vapor at Palisades, New York and the concurrent meteorological conditions. *J. Geophys. Res.* **89**:4937–4939.

White, J.W.C., E.R. Cook, J.R. Lawrence, and W.S. Broecker. 1985. The D/H ratios in trees: Implications for water sources and tree-ring D/H ratios. *Geochim. Cosmochim. Acta* **49**:237–246.

Wilson, A.T., and M.J. Grinsted. 1977. $^{12}C/^{13}C$ in cellulose and lignin as paleothermometers. *Nature* **265**:133–135.

Winter, K. 1981. C_4 plants of high biomass in arid regions of Asia: Occurrence of C_4 photosynthesis in *Chenopodiaceae* and *Polygonaceae* from the middle east and USSR. *Oecologia* **48**:100–106.

Winter, K., J.A.M. Holtum, G.E. Edwards, and M.H. O'Leary. 1982. Effect of low relative humidity on $\delta^{13}C$ value in two C3 grasses and in *Panicum milloides*, a C3–C4 intermediate species. *J. Exp. Bot.* **33**:88–91.

Wong, S.C., I.R. Cowan, and G.D. Farquhar. 1979. Stomatal conductance correlates with photosynthetic capacity. *Nature* **282**:424–426.

Yapp, C.J., and S. Epstein. 1982a. Climatic significance of the hydrogen isotope rstios in tree cellulose. *Nature* **5868**:636–639.

Yapp, C.J. and S. Epstein. 1982b A reexamination of cellulose carbon-bound hydrogen δD measurements and some factors affecting plant-water D/H relationships. *Geochim. Cosmochim. Acta* **46**:955–965.

Zahn, R., K. Winn, and M. Sarthein. 1986. Benthic foraminiferal $\delta^{13}C$ and accumulation rates of organic carbon: *Uvigerina peregrina* group and *Cibicidoides wuellerstorfi*. *Paleoceanography* **1**:27–42.

8

How Accurate Are Our Chronologies of the Past?

E. BARD[1], M. STUIVER[2], and N. SHACKLETON[3]
[1]Centre des Faibles Radioactivités, CNRS-CEA, 91198 Gif sur-Yvette, France; Géosciences de l'Environnement, Université d'Aix-Marseille III, 13397 Marseille, France; and Lamont-Doherty Geological Observatory, Palisades, NY 10964, U.S.A.
[2]Quaternary Research Center, University of Washington, Seattle WA 98195, U.S.A.
[3]Subdepartment of Quaternary Research, The Godwin Laboratory, Free School Lane, Cambridge CB2 3RS, U.K.

ABSTRACT

Divided into three separate contributions, this chapter describes recent results obtained by means of dating techniques that allow us to place chronological constraints on the rate of climatic changes derived from natural archives such as trees, corals, and ocean sediments.

Part one (by M. Stuiver) deals with the issue of time warps caused by the transformation of ^{14}C ages. Radiocarbon dating provides "model" ages that can be converted into cal (calibrated) ages through the use of high-precision calibration curves. The price of conversion is, more often than not, a cal age range that spans more years than the standard deviation associated with the radiocarbon age. The transformation of ^{14}C ages into cal ages results in time warps that seriously impair spectral analysis and correlation efforts.

Part two (by E. Bard) covers the chronology of the sea-level variation during the last 300 thousand years (Ka) by means of U-Th mass spectrometry. Precise U-Th ages in corals from Barbados have been measured by means of thermal ionization mass spectrometric technique (TIMS). The Barbados reefs analyzed so far can be correlated with the classical Emiliani's stages based on the oxygen isotopic variations in benthic foraminifera (Emiliani 1955). The available coral samples are representative of the following $\delta^{18}O$ stages: 1, 2, 3, 4, 5.1, 5.3, 5.5, 7.3, and 7.5 It can be shown that the age results are in broad agreement with the Milankovitch theory. These data also enable us to refine the calculation of lags between the extraterrestrial forcing and the climatic response. Nevertheless, the actual limit of U-Th dating in old corals is no longer due to the precision of the measurements but rather to the presence of important isotopic anomalies of uranium-234. Indeed, it is shown that all samples older than 50 Ka have initial $^{234}U/^{238}U$ ratios higher than present-day seawater, in contrast to modern or recent corals (< 30 Ka). This uncertainty on the initial ratio raises some doubt about the real accuracy of the age determination, despite the increased precision obtained by this new technique.

Global Changes in the Perspective of the Past
Edited by J.A. Eddy and H. Oeschger

Part three (by N. Shackleton) discusses the ages of the last six geomagnetic reversaly by means of Milankovitch tuning. A time scale has been developed for the $\delta^{18}O$ records of ODP Site 677 on the assumption that there has been a constant phase relationship between the astronomical forcing and the climate response over the past 2.6 Ma. For this entire time period, we report good coherencies between modeled and observed variations. However, the new time scale requires ages for the last few geomagnetic reversals that are around 5–7% greater than the currently accepted values.

PART I: TIME WARPS CAUSED BY THE TRANSFORMATION OF ^{14}C AGES

Crucial parameters for the calculation of radiocarbon (^{14}C) ages are (a) measured $^{14}C/^{12}C$ and $^{13}C/^{12}C$ isotope ratios and (b) the assumption of a constant $^{14}C/^{12}C$ ratio in the atmosphere during the past. The age calculation, as such, does not differ much from the ubiquitous "computer modeling" term applied to most Earth science calculations. The ^{14}C age (model age) has the standard flaws attached to computer modeling in that the assumptions influence the model results. Fortunately we are able to improve substantially on the model age by applying methods that provide dual information, such as a model age and a cal (calibrated) age (in solar or calendar years) associated with a single sample.

Samples formed at the same time may differ in their model ages (conventional ^{14}C ages as defined by Stuiver and Polach 1977). For instance, marine organisms are formed in media with $^{14}C/^{12}C$ ratios lower than that of the atmosphere. When measured in the laboratory, the specific ^{14}C activity of a marine sample will be less than that of a contemporaneous atmospheric sample. The standard calculation utilizes measured ^{14}C activities and yields an older model age (conventional ^{14}C age) for the marine sample. A reservoir age correction that accounts for the difference in initial ^{14}C activity has to be applied to the model age of the marine sample to make the ages compatible. Unfortunately, the ocean reservoir correction ($^{14}C/^{12}C$ ratio difference between atmosphere and mixed or deep ocean layer) is time dependent (e.g., Stuiver et al. 1993), and changes during the past are not always precisely known. Thus an important uncertainty is introduced when transforming marine calibration data to an atmospheric frame of reference.

The ^{14}C accumulated within a single Earth carbon reservoir (e.g., mixed layer of the ocean, atmosphere, or biosphere) need not be uniformly dispersed. Substantial differences in mixed layer $^{14}C/^{12}C$ ratios exist and are tied, for example, to differences in the ^{14}C supply from deeper (and older) waters. The $^{14}C/^{12}C$ ratios are usually given as $\Delta^{14}C$, which is the relative deviation (in percent or per mil) of the measured ^{14}C activity (after correction for both radioactive decay and isotope fractionation) from the National Bureau of Standards (NBS) oxalic acid ^{14}C activity (Stuiver and Polach 1977). Alternatively, differences in $\Delta^{14}C$ can be given in years, as ^{14}C decays at a rate of nearly 1% per 80 years. Reservoir age corrections of mixed layer subunits may range from 100 years, for samples of a Californian bay to 1200 years for organisms grown in the Antarctic Circumpolar Current (Stuiver et al. 1993). The global average for the mixed layer of the preindustrial Holocene ocean is approximately 400 years.

Relatively fast transport in the troposphere causes atmospheric $^{14}CO_2$ to be fairly uniformly mixed near the Earth's surface. Nevertheless, small regional differences exist, partly due to differences in land–ocean surface areas (e.g., the Northern and Southern Hemisphere differ substantially in oceanic coverage). The meridional atmospheric $\Delta^{14}C$ gradient is mainly forced by latitude-specific atmosphere–ocean CO_2 gas exchange rates that are, in turn, controlled by regional wind strengths, sea-surface temperatures, air–sea pCO_2 differences, and oceanic $^{14}C/^{12}C$ deficiencies (Braziunas et al. 1991). The measured hemispheric difference (~45°N– ~45°S) in atmospheric $\Delta^{14}C$ is approximately 4 per mil (or 32 years) with the Southern Hemisphere being lower (Lerman et al. 1970; Vogel et al. 1986). Within the Northern Hemisphere, the three-dimensional global tracer model of Braziunas et al. (1991) yields up to 2 ± 0.5 per mil regional difference in $\Delta^{14}C$, which is in good agreement with the measured $\Delta^{14}C$ differences of a couple per mil in contemporaneous wood samples of different regions (Stuiver and Quay 1980). For the Southern Hemisphere, there is a stronger latitudinal gradient yielding $\Delta^{14}C$ differences up to 5 ± 2 per mil (excluding Antarctica).

The laboratory measurement procedure adds uncertainties independent of initial reservoir ^{14}C content. It is an unfortunate tradition of the radiocarbon community to underreport the standard error of the age measurement. Age errors are usually based on the Poisson error in the number of counts accumulated during the ^{14}C activity measurement, and this procedure does not account for additional sources of measuring variance. The "error multiplier," K (defined as actual standard error/quoted standard error), differs for individual laboratories and may be as large as a factor of 2 (Scott et al. 1990; Stuiver 1982). Although the sources of variance are additive (resulting in an age dependency of K), K is a convenient expression of the degree to which the quoted error is representative of the overall error in a radiocarbon date. An evaluation of the degree of measuring precision and accuracy of the laboratories involved in time scale calibration can be made relatively easily because wood of the same age is often measured by more than one laboratory.

The assumed constant atmospheric $^{14}C/^{12}C$ ratio is represented in the calculations of conventional radiocarbon ages by the measured $^{14}C/^{12}C$ ratio of the NBS oxalic acid standard. This standard's activity was initially equated to the age-corrected activity of a wood sample of known age. Wood is a preferred material for the calibration of the radiocarbon time scale because the dendroyear counts are considered free of errors. Furthermore, the reservoir corrections to the model ^{14}C ages of terrestrial materials formed in equilibrium with atmospheric $^{14}CO_2$ are, at first order, negligible. Varved sediments formed in marine or freshwater environments are also used for calibration; however, here the estimates of reservoir corrections to be applied to the model ages may introduce errors of century magnitude. Finally, the high precision U/Th dating of corals provides a method where, for the marine environment, model U/Th ages can be compared to model ^{14}C ages.

Three major tree-ring chronologies are available for a test of radiocarbon age conformity. The absolute German Oak (River) chronology covers nearly 10,000 cal years and the Irish Oak (Bog) chronology about 7000 cal years. In the U.S., the Bristlecone Pine chronology extends back to about 8600 cal year B.P. (with 0 yr B.P. = A.D. 1950).

Comparison of ^{14}C Age Results

In this section, I compare the radiocarbon age results (Becker and Kromer 1986; de Jong et al. 1986; Linick et al. 1986; Pearson et al. 1986; Stuiver and Becker 1986; Vogel et al. 1986) from dendrochonologically dated wood of the same age and different origin, as well as those of the same age and same chronology. Of the six laboratories involved, three measured either decadal (Tucson and Seattle) or bi-decadal samples (Belfast). Gröningen, Heidelberg, and La Jolla measured samples grown over shorter time intervals (usually 1 to 3 yr). For comparison purposes, we chose to average the published results over single decades or bi-decades. Usually only part of the ten or twenty years will have been measured, and the "decadal " or "bi-decadal" ^{14}C ages calculated in this manner need not be entirely identical to the ^{14}C ages that would have been obtained by measuring decadal or bi-decadal samples directly. An error multiplier, $K_{Lab\ A\text{–}Lab\ B}$, for interlaboratory comparisons was derived by taking the quotient of the standard deviation σ_{tot} in the observed differences (n = number of comparisons, offset = "a" with positive values when Seattle dates are older) and the average standard deviation σ of the differences calculated from the quoted errors in the ^{14}C determinations.

A test of internal consistency of ^{14}C data of laboratories measuring wood of the same tree chronology (often samples were not supplied during the same year, and wood from identical trees was not necessarily used in different laboratories) provides insight into the sum total of uncertainty tied to the processes of wood allocation, dendro age determination, sample pretreatment, laboratory measurement, and regional $\Delta^{14}C$ distribution between contemporaneous trees (e.g., the area of original growth is not well known for trees collected from alluvial river banks).

Good radiocarbon age agreement of decadal or bi-decadal wood for the German chronology (e.g., Becker and Kromer 1986) is found, for instance, with $K_{Seattle\text{–}Groningen}$ = 1.8 and a = –4 ± 2 yr (3210–3910 B.C., n = 36), $K_{Seattle\text{–}Pretoria}$ = 1.2 and a = 4 ± 2 yr (1930–3350 B.C., n = 72), $K_{Seattle\text{–}La\ Jolla}$ = 1.3 and a = –4 ± 3 yr (2500–5000 B.C., n = 97), and $K_{Seattle\text{–}Belfast}$ = 1.5 and a = –14± 4 yr (5500–6000 B.C., n = 24). Less satisfactory agreement is found in $K_{Seattle\text{–}Belfast}$ =1.8 and a = –39 ± 4 yr (5000–5500 B.C., n = 25) and $K_{Seattle\text{–}Heidelberg}$ = 1.7 and a = –35 ± 4 yr (4075–4925 B.C., n = 43). The reasons for the larger offsets are, as yet, not well understood.

Comparing ^{14}C dates from the German (measured in Seattle and Belfast) and Bristlecone Pine (measured in Tucson and Seattle) chronologies for the 5830–5670 B.C. interval yields $K_{Seattle\text{–}Seattle}$ = 1.3 and a = –6 ± 7 yr (n = 13), $K_{Seattle\text{–}Tucson}$ = 1.8 and a = –3 ± 7 yr (n = 15), and $K_{Tucson\text{–}Belfast}$ = 1.8 and a = –6 ± 7 (n = 7). A comparison of the joint Northwest Pacific and German Chronology measured in Seattle, and the

Irish chronology measured in Belfast, yielded for bi-decadal samples $K_{Seattle-Belfast}$ = 1.6 and $a = 2 \pm 1$ (A.D. 1000–5000 B.C., $n = 302$).

The majority of offsets are in the decade (or less) range, and error multipliers for the age differences are maximally 1.8. Evidently the complex procedure establishing a precise calibration curve back to 10,000 B.P. delivers quality results.

Unfortunately, a precise calibration curve does not remove the uncertainties introduced by the so-called ^{14}C wiggles. During calibration, single ^{14}C dates line up with multiple intercepts, each of which yields a cal date. A continuous spectrum is even possible for the horizontal part of the curve (when atmospheric $^{14}CO_2$ levels decline at the same rate as ^{14}C declines through radioactive decay). In addition, the uncertainty in the calibration curve causes the discrete cal ages to broaden into ranges, even for a ^{14}C age with (hypothetical) zero error. The cal year ranges, obtained from the age calibration of ^{14}C dates between 4000 and 6000 ^{14}C yr B.P. are given in Figure 8.1. The cal yr ranges (time warps) result from the wriggles in the radiocarbon time scale in relation to calendar years and correspond mostly to century-scale variations in atmospheric ^{14}C (e.g., the Maunder and Spörer $\Delta^{14}C$ maxima). To generate the graph, the youngest calibrated age of each range of cal ages obtained for each ^{14}C age was set at zero. For a specific ^{14}C age, the cal year ranges are given by the intercepts of a vertical line drawn through the ^{14}C age and the black areas of Figure 8.1. The ranges can be small (e.g., at 5500 ^{14}C yr B.P.) when a steep linear portion of the calibration

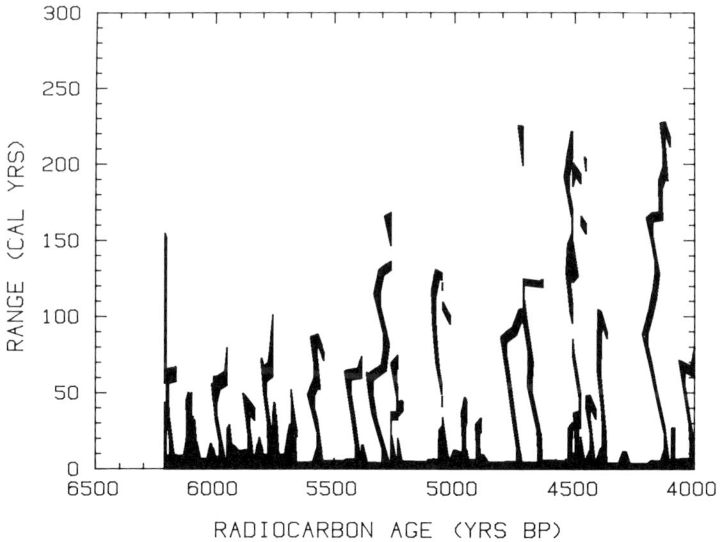

Figure 8.1 The ranges, in cal years, obtained from the calibration of radiocarbon ages (4000–6000 ^{14}C yr B.P. interval). The ranges were produced for an ideal hypothetical case with zero sample standard deviations. The youngest cal age obtained for each radiocarbon age was set at zero. The bi-decadal calibration curve of Stuiver and Pearson (1992) was used. The sample was also assumed to have been formed during a 20-year interval.

curve is encountered, or large when atmospheric ^{14}C content change causes wriggles in the radiocarbon time scale (e.g., at 4500 ^{14}C yr B.P.).

The Figure 8.1 cal age ranges, resulting from the nonunique ^{14}C transformation process, seriously impair the study of century-type climatic information. Clearly it is incorrect for spectral analysis and correlation efforts to remain in the ^{14}C domain. The transformation into cal ages introduces appreciable time scale uncertainty. Improvements can be made by finding suites of contiguous samples for which cal age differences are known (so-called wriggle matching), by selecting intervals with reduced variability (Atwater et al. 1992), or by using samples that have been formed over known cal age intervals that encompass more time. When, for instance, a sample has been formed during one century (assuming a constant growth rate), a 100-year moving average of the calibration curve radiocarbon ages can be used. Averaging the radiocarbon ages reduces the size of the calibration curve wriggles, which results in appreciably smaller time warps (see Figure 8.2). When investigating a detailed time series, the "bulk" samples would have to provide relatively precise marker points whereas the other time determinations would be based on linear interpolation between these points. It should also be noted that AMS dating of single-year macrofossils and seeds in lake sediments erases the reservoir deficiency correction while introducing, at the same time, maximum cal age uncertainty.

The above mitigating procedures (when applicable) reduce the cal age time warps but do not succeed in their complete elimination. To circumvent the time warps introduced by the radiocarbon age calibration entirely, one has to switch to other chronological methods (e.g., tree-ring, varve counting).

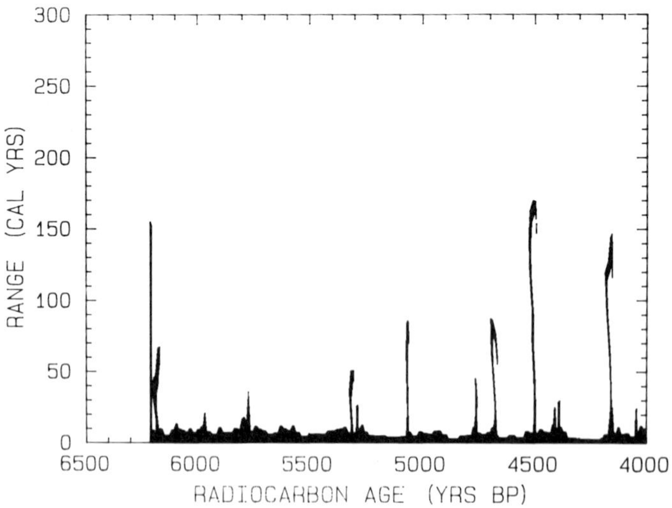

Figure 8.2 Same as Figure 8.1 but for a smoothed (100-yr moving average) calibration curve and a sample formed during 100 years of growth.

PART II: CHRONOLOGY OF THE SEA-LEVEL VARIATIONS DURING THE LAST 300 KYR BY MEANS OF U–TH MASS SPECTROMETRY

Recent development at Caltech, involving the direct determination of ^{230}Th and ^{234}U abundances by mass spectrometry, opened a wide range of new applications for paleoclimate studies, which were previously out of reach of the classical α-counting technique. The typical 2σ precision is now on the order of 50 yr at 10 Ka B.P., 1 Ka at 100 Ka, 5 Ka at 200 Ka, and 10 Ka at 300 Ka. This technique enables a more precise correlation of sea-level highstands, as recorded in coral terraces, with maxima in solar insolation as predicted by Milankovitch's theory (Edwards et al. 1987; Bard et al. 1990a).

Sea-level Reconstruction

The samples analyzed for this study are all composed of the coral species *Acropora palmata*, which is an accurate recorder of the present-day sea level in the Caribbean Sea. The coral specimens were collected on raised terraces from different islands characterized by different rates of tectonic uplift affecting the Caribbean subduction zone. On the island of Curaçao, which is thought to be relatively stable, the Last Interglacial corals are found at about 5–10 meters above the present sea level in accordance with U–Th ages measured on corals collected in stable areas. By contrast, on Haiti (Dodge et al. 1983) and Barbados (Matthews 1973; Fairbanks and Matthews 1978), those corals are found at altitudes ranging between 40 and 60 meters above the present sea level. (On Barbados the uplift rates are even different in the southern, eastern, and western parts of the island [Fairbanks and Matthews 1978]). By assuming that the sea level of the last interglacial was 7 meters above the present sea level, it is possible to calculate sea level for other periods by using linear uplift corrections (Matthews 1973).

Figure 8.3 presents the reconstruction of global sea level based on the TIMS measurements obtained at L–DGO (the data are listed in Fairbanks 1989; Bard et al. 1990a, 1990b, 1991; Hamelin et al. 1991). Superimposed on Figure 8.3 are the sea-level estimates based on a composite benthic $\delta^{18}O$ record considered to be an accurate recorder of the oxygen isotopic composition of the global deep waters (Labeyrie et al. 1987; Vogelsang 1990). The time scale of this record was obtained by tuning the $\delta^{18}O$ variations to an astronomical time scale through a procedure similar to the one described by Shackleton (see below; see also Imbrie et al. 1984). The amplitude of the $\delta^{18}O$ record has been converted into sea-level estimates by normalizing the records for the last glacial maximum.

To a first approximation, there is clear agreement between the U–Th dated sea-level determination and the tuned $\delta^{18}O$ record. This statement is true both for the timing and for the amplitude of the records. However, by looking more closely, it is possible to identify slight discrepancies between the $\delta^{18}O$ curve and the coral results.

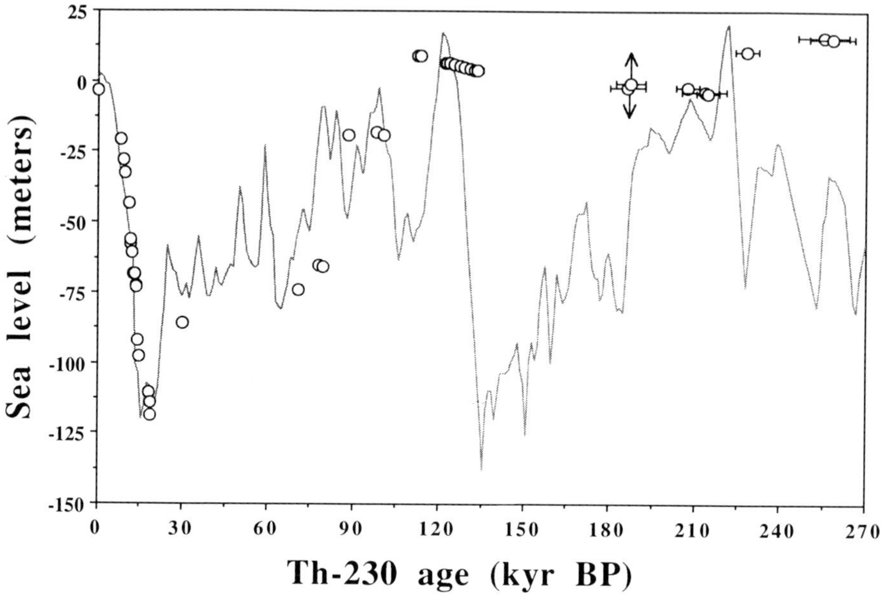

Figure 8.3 *A palmata* paleodepth versus Th–U ages measured by mass spectrometry at L–DGO. Data can be found in Bard et al. (1990a, b, 1991) and Hamelin et al. (1991). To reconstruct this sea-level curve, we have applied linear corrections for uplift by normalizing the last interglacial high stand to 7 meters above the present-day sea level (see text). Solid line: Normalized $\delta^{18}O$ curve determined by Labeyrie et al. (1987) and Vogelsang (1990). The results obtained for a sample corresponding to oxygen isotope stage 7.1 (≈187 Ka) was collected on the southeast coast of Barbados, where the uplift rate is not as well constrained as it is for other parts of the island.

In particular, it appears that coral sea-level estimates obtained for stage 4, 5.1, and 5.5 lead by a few thousand years the $\delta^{18}O$ curve. A first possible explanation to this is that the climatic lag assumed for the astronomical tuning may be overestimated by a few Ka. A second possibility is that the ages obtained by the TIMS technique may not be completely accurate. Although the first explanation is very plausible, some support for the second can be found in the U–Th determinations themselves. This topic will be discussed in the next section.

Another discrepancy has been found between the sea-level estimations obtained for stage 7.5 (correlated with the Kingsland–Aberdare reef complex from Barbados; Radtke et al. 1988). This problem could be caused either by our oversimplifying assumption that tectonic uplift is constant through time or by a non-negligible temperature imprint superimposed on the benthic $\delta^{18}O$ record reconstructed by Vogelsang (1990). Some support for this second hypothesis can be found in the benthic $\delta^{18}O$ reconstruction presented by Martinson et al. (1987), which suggests that during stage 7.5 the sea level was close to the one reached during oxygen isotope stage 7.3.

Isotopic Anomalies of Uranium

The TIMS technique also permits precise determination of the uranium isotopic composition of corals. Initial $^{234}U/^{238}U$ ratios can then be calculated by connecting the measured $^{234}U/^{238}U$ for decay using the ^{230}Th–^{234}U age, a procedure that facilitates comparison with modern seawater.

Chen et al. (1986) showed that $^{234}U/^{238}U$ ratio is constant throughout the ocean with a mean activity ratio of 1.144 (standard deviation of 0.004 on 9 measurements). Recent corals collected in open seawater environments were shown to record accurately the seawater $^{234}U/^{238}U$ ratio, and no anomalous values were reported for such samples from Barbados (Edwards 1988; Bard et al. 1990b), Galapagos (Hamelin et al. 1991), Tobago, or Vanuatu (Edwards 1988; Edwards et al. 1988). By contrast, all corals older than 50 Ka have slightly high initial $^{234}U/^{238}U$ ratios when compared to modern seawater. Hamelin et al. (1991) showed that the distribution of initial $^{234}U/^{238}U$ for 70 to 140 Ka corals is not Gaussian like those for present-day seawater and young corals (30 Ka). Rather, the calculated initial $^{234}U/^{238}U$ of most sample clusters is between 1.150 and 1.165, defining a distribution whose tail is skewed toward much higher values, up to 1.195. Samples older than 150 Ka B.P. are characterized by even greater excesses of ^{234}U (Bard et al. 1991). Assuming that the ^{230}Th-ingrowth ages are accurate, these anomalies translate into high initial $^{234}U/^{238}U$ ratios: about 1.2 at 20 Ka and up to 1.6 at about 450 Ka B.P.

Elsewhere we have proposed that these anomalies result from both diagenetic addition and replacement of U and Th, and possibly from global changes in the $^{234}U/^{238}U$ composition of seawater through time (Bard et al. 1991; Hamelin et al. 1991). However, neither hypothesis is able to explain alone the observed discrepancies. Consequently, as long as these anomalies have not been fully explained, the 2σ precision achieved by TIMS may not be an accurate picture of the actual uncertainty on the age determinations.

In the particular case of the coral samples corresponding to oxygen isotope stage 5.5, the closed-system ages range between 120 and 133 Ka (cf. Figure 8.4). Recent data obtained by the Caltech group suggest even older ages ($\approx$135 Ka) for samples from the Huon Peninsula, New Guinea (Stein et al. 1990). Taken at face value, these ages are in agreement with the iceflow model ages obtained for the Vostok ice core (Lorius et al. 1985; Jouzel et al. 1987), but are in disagreement with the Milankovitch hypothesis since the tuned benthic $\delta^{18}O$ record predicts a glacial maximum at about 135 Ka (cf. Figure 8.3). Nevertheless, the data available also suggest that the oldest samples have anomalously high initial $^{234}U/^{238}U$ ratios, which casts some doubts on the age determination beyond 130 Ka (the New Guinea data are not yet fully published and are thus not represented in Figure 8.4). Although it is clear that much more data are needed to complement the graph presented in Figure 8.4, there is little doubt that a precise and accurate dating of the last interglacial period will require a better understanding of the uranium isotopic anomalies.

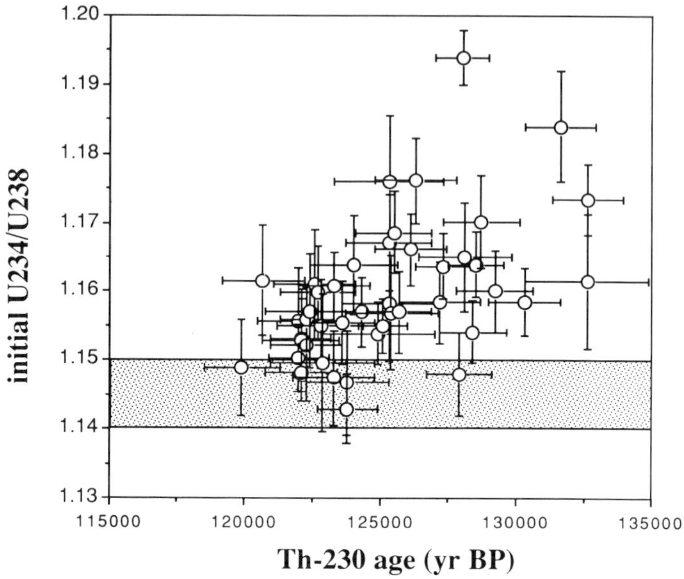

Figure 8.4 Th–U ages versus initial $^{234}U/^{238}U$ obtained by TIMS at Caltech and L–DGO on coral samples from the Last Interglacial Period. The data can be found in Edwards (1988), Bard et al. (1990a, b), Chen et al. (1991), and Hamelin et al. (1991). Note that all samples older than 127 kyr have initial $^{234}U/^{238}U$ ratio significantly higher than the seawater value (shaded area).

PART III: AGES OF THE LAST SIX GEOMAGNETIC REVERSALS BY MEANS OF MILANKOVITCH TUNING

Introduction

The hypothesis that varations in the Earth's orbital geometry are responsible for the timing of climatic change has been under consideration for almost two centuries and was formulated essentially by Milankovitch (1930). However, it was Hays et al. (1976) who provided the first clear documentation that the geological record reveals the independent contribution of the three orbital elements to the determination of past climate. Imbrie et al. (1984) produced a stacked $\delta^{18}O$ record for the past 800 Ka and used astronomical data to fine-tune the geological time scale, and Martinson et al. (1987) developed a computerized system for accomplishing this objective automatically.

Ruddiman et al. (1986) made an attempt to improve the astronomical calibrations proposed by Imbrie et al. (1984) for the interval of time between 620 Ka and 800 Ka. This interval is particularly interesting because the climate system underwent a major change from a regime in which the response was dominated by a 41 Ka periodicity attributed to obliquity forcing (Ruddiman et al. 1986), to a regime characterized by 100 Ka cycles that are loosely associated with the eccentricity cycle (Imbrie et al. 1984).

A reexamination of the astronomical calibrations over this particular interval, in light of the new data available for ODP Site 677, provided the key to a new calibration extending to the six most recent magnetic reversal boundaries.

Results

ODP site 677 was cored using the Advanced Piston Corer (APC) at 1°12′N, 83°44′W in 3461 m water depth (Becker et al. 1988), a location very close to DSDP Site 504. Two holes, 677A and 677B, were recovered with the objective of obtaining a complete Plio-Pleistocene record (Shipboard Scientific Party 1988). The advantage of the sediments at this site is that it has proved possible to generate virtually complete oxygen isotope records for the planktonic species *Globigerinoides ruber* and the benthonic foraminifera (mostly *Uvigerina senticosa*). Analytical data are available in Shackleton and Hall (1989) and additional data are available in Shackleton et al. (1990).

Figure 8.5 shows the oxygen isotope record versus depth and combines the data obtained for Holes 677A and 677B. Oxygen isotope stages as defined by Emiliani (1955), Shackleton and Opdyke (1973), and Ruddiman et al. (1986) are easy to identify and are indicated in Figure 8.5. Of particular importance is the documentation of that part of the record deposited during the Jaramillo subchron (at about 40 m depth in Figure 8.5).

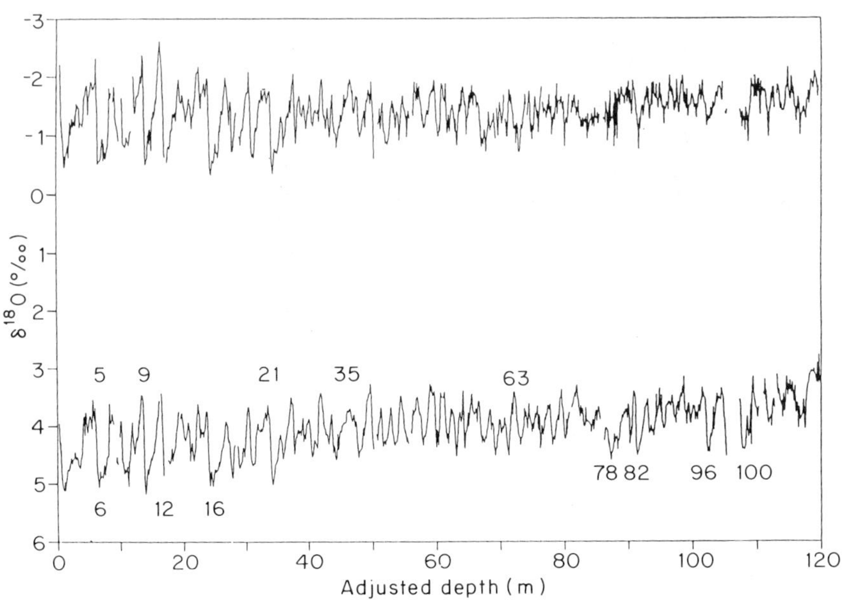

Figure 8.5 Oxygen isotope stratigraphy combining Holes 677A and B: above, planktonic; below, benthonic to 120 m. Selected isotope stages labelled for orinetation.

The detailed procedure for the construction of a tuned time scale for ODP Site 677 is given in Shackleton et al. (1990). It is based on recent recalculations of the variations of solar energy distribution due to changes of Earth's orbital parameters (Berger and Loutre 1991). It should be emphasized, however, that the proposed changes in time scale are not caused by the use of these calculations. Indeed the revision reported here depends mainly on a new interpretation of the $\delta^{18}O$ records over the interval between 600 Ka and 1.2 Ma, where one would obtain almost identical age estimates whether one used the early calculations of Berger (1978) or the newer ones of Berger and Loutre (1991).

The records that were carefully documented by Ruddiman et al. (1986) are very clearly dominated by cycles with a period near 41 Ka. The record from Site 677 is similarly dominated by cycles with a depth wavelength of around 1.7 m over the interval from about 45–65 m. The transitional interval at Site 677 is of particular interest because it displays a secondary cycle with a shorter wavelength of a little under 1 m, which is probably linked with forcing by precession. Because the effect of precession on the seasonal distrbution of solar insolation is modulated by the eccentricity cycle, the precession component provides a better tool for accurate correlation than the more regular obliquity cycle. Spectral analysis of the $\delta^{18}O$ as a function of depth confirms the visual evidence for these cyclicities (Shackleton et al. 1990).

If it is assumed that the 1.7 m cycle is derived from the 41 Ka obliquity cycle, it is apparent that the modulation of the 1 m cycle is consistent with a response to the excentricity cycle that provides the envelope for the climatic precession signal. Consequently, it becomes a straightforward process to generate iteratively an astronomical time scale for the $\delta^{18}O$ record.

In this chapter, it has been assumed that the climate system monitored by the ocean $\delta^{18}O$ record has responded to forcing by insolation variations with the same time-constant throughout the last 2.6 Ma. The model described by Imbrie and Imbrie (1980) has been used to produce a tuning target for correlating the $\delta^{18}O$ record. Figure 8.6 shows the planktonic and benthonic data on a time scale developed so as to align features in the $\delta^{18}O$ record with features in the target, by iteratively assigning age control points and by assuming uniform accumulation between these points.

DISCUSSION

The chronology proposed here may be used to estimate the ages of the last six major reversals of the Earth's magnetic field. Table 8.1 compares our ages with currently accepted geochronological estimates. It is apparent that the "astronomical" ages for the geomagnetic boundaries are systematically older by about 5–7% than the currently accepted values. The same conclusion was reached recently by Hilgen (1991), who used a similar approach by tuning the sapropel records of ODP cores to an astronomical time scale using the record of piston cores to calculate phase relationship.

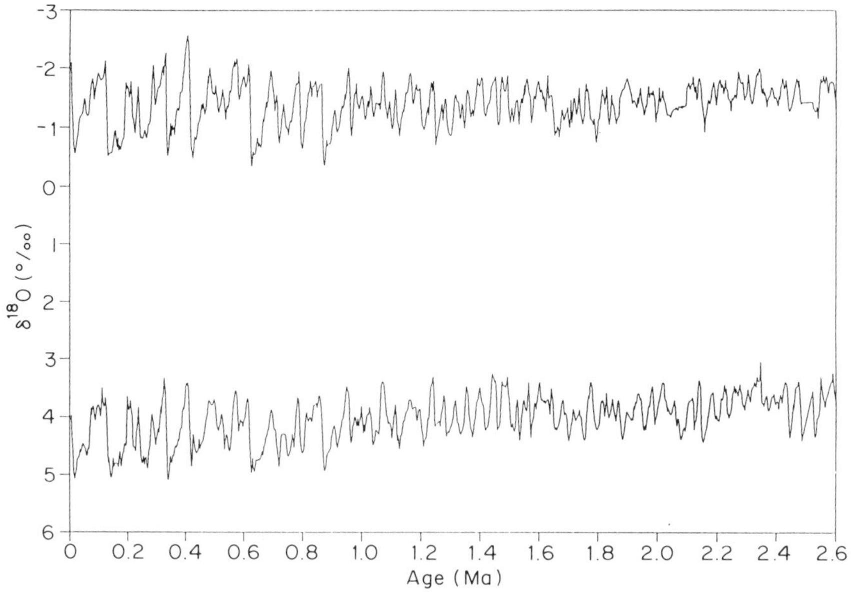

Figure 8.6 Planktonic and benthonic oxygen isotope data for ODP Site 677 for the past 2.6 Ma after tuning the planktonic record to the target described in the text.

Table 8.1 Comparison of ages obtained for selected magnetic boundaries. The ages listed in column 1 are based on K/Ar dating (Berggren et al. 1985; Maniken and Grome 1982). The estimates listed in column 2 are based on the present work.

Isotope Stage	Correlative Reversal	Age 1	Age 2
base 19	base Brunhes	0.73	0.78
mid 27	top Jaramillo	0.92	0.99
mid 31	base Jaramillo	0.98	1.07
base 35	Cobb Mountain	1.10	1.19
base 63	top Olduvai	1.66	1.77
base 71	base Olduvai	1.88	1.95
104	top Gauss	2.47	2.60

This revision of the three most recent reversals could possibly be accommodated by an assumption that the relatively limited number of measurements are systematically in error at the young end of the K/Ar time scale. However, if the older reversals are also older than currently believed, it is also possible for the decay constant generally adopted (Steiger and Jager 1977) to be in error.

Support for the new chronology comes from examination of accumulation rates, especially for the critical interval between Stages 15 and 22. Figure 8.7 compares

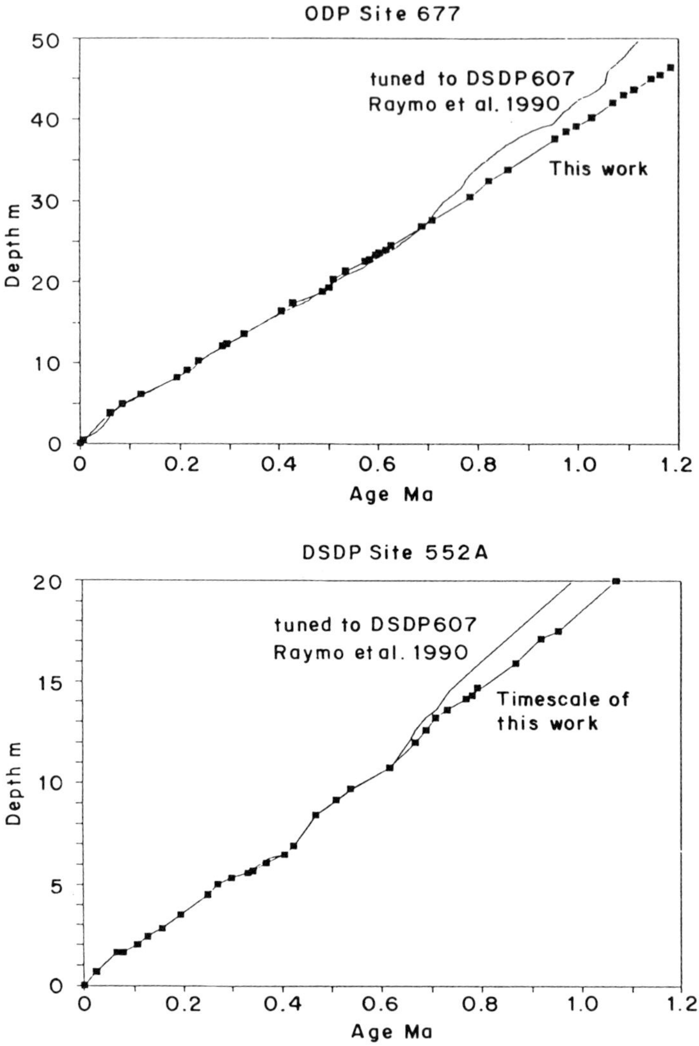

Figure 8.7 Age-depth plots for core ODP 677, V28–238, DSDP522A, and ODP 607 on the time scales of Raymo et al. (1990), SPECMAP (Imbrie et al. 1984), and Ruddiman et al. (1986), compared with age-depth plots using the time scale of this work.

accumulation rates of four cores on a conventional time scale and on the one proposed here. Core V28–238 was used as the reference core in the studies of Imbrie et al. (1984) and of Prell et al. (1986) because it was regarded as the core with the most uniform accumulation of all those for which data were available. DSDP Site 552A is from Rockall at 55°N in the North Atlantic (Shackleton et al. 1984) and DSDP 607 from the mid-Atlantic Ridge at 40°N (Ruddiman et al. 1986). In each case, the previous

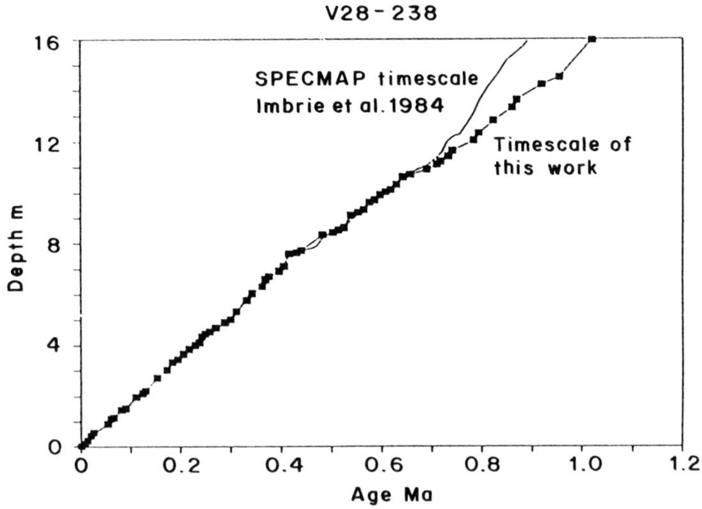

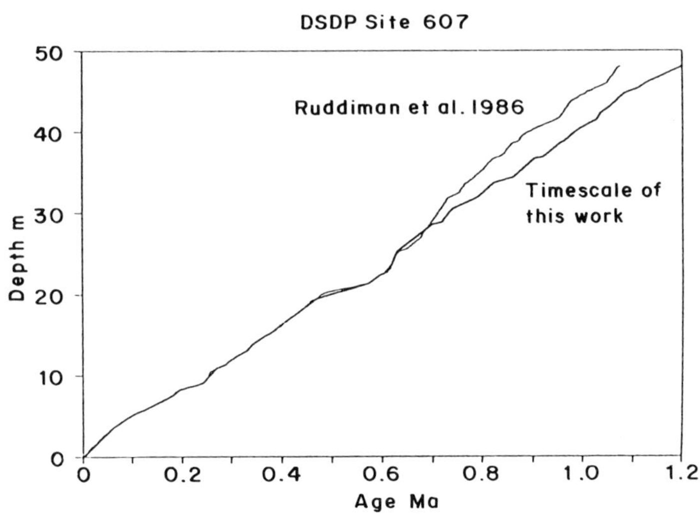

Figure 8.7 (continued)

time scale requires greater changes in accumulation rate than are implied by the time scale proposed in this chapter. Although a systematic change in accumulation rates at the inception of larger glaciation might be expected, a change in accumulation rate in the same direction over a limited time interval in four such different depositional settings in two ocean basins does not seem likely, which in turn gives support for the revised chronology.

Regarding the question of accuracy, there are three issues. First, is the stratigraphic correlation between $\delta^{18}O$ records and the astronomical record correct? This will only be answered definitively by testing the proposed correlation with new data. For the present, one may say that statistically the new correlation is much more convincing than the old one. Second, how accurate is the astronomical record? Here the uncertainties in the interval under discussion are thought to be a few thousand years at most (Berger and Loutre 1991). Third, how constant is the phase relationship between $\delta^{18}O$ and insolation? Here the uncertainty is again limited to a very few thousand years. Thus the overall uncertainty would be less than about 0.50% of age in the Pliocene, assuming that the correlations are correct.

ACKNOWLEDGEMENTS

The ^{14}C work of the Quaternary Isotope Laboratory at Seattle is supported by NSF (Anthropology) grant BNS–9004492. Dr. Becker from the University of Hohenheim, Stuttgart provided samples from the German Oak series, and P.J. Reimer and P.L. Wilkinson gave crucial technical and analytical support.

REFERENCES

Atwater, B.F., M. Stuiver, and D.K. Yamaguchi. 1991. Radiocarbon test of earthquake magnitude at the Cascadia subduction zone. *Nature* **353:**156–158.

Bard, E., R.G. Fairbanks, B. Hamelin, A. Zindler, and C.T. Hoang. 1991. Uranium–234 anomalies in corals older than 150,000 yr. *Geochim. Cosmo. Acta* **55**:2385–2390.

Bard, E., B. Hamelin, and R.G. Fairbanks. 1990a. U/Th ages obtained by mass spectrometry in corals from Barbados: Sea level during the past 130,000 years. *Nature* **346**:456–458.

Bard, E., B. Hamelin, R.G. Fairbanks, and A. Zindler. 1990b. Calibration of ^{14}C time scale over the past 30,000 years using mass spectrometric U–Th ages from Barbados corals. *Nature* **345**:405–410.

Becker, B., and B. Kromer. 1986. Extension of the Holocene dendrochronology by the Preboreal Pine series, 8800 to 10,100 B.P. *Radiocarbon* **28**:961–968.

Becker, K., H. Sakai, et al., eds. 1988. *Proc. ODP Init. Rep.* Pt. A. III, College Station, TX.

Berger, A. 1978. Long-term variations of daily insolation and Quaternary climatic changes. *J. Atmos. Sci.* **35**:2362–2367.

Berger, A., and M.F. Loutre. 1991. Insolation values for the climate of the last 10 million years. *Quat. Sci. Rev.* **10:**297–317.

Berggren, W.A., D.V. Kent, and J.J. Flynn. 1985. Jurassic to Paleogene. II. Paleogene geochronology and chronostratigraphy. In: The Chronology of the Geological Record, ed. N.J. Snelling. *Geol. Soc. Mem.* **10**:141–195.

Braziunas, T.F., I.Y. Fung, and M. Stuiver. 1991. Oceanic and solar forcing of natural geographic variations in atmospheric $\Delta^{14}C$ (abstract). *Radiocarbon* **33**:180.

Chen, J.H., H.A. Curran, B. White, and G.J. Wasserburg. 1991. Precise chronology of the last interglacial period: ^{234}U–^{230}Th data from fossil coral reefs in the Bahamas. *Geol. Soc. Am. Bull.* **103**:82–97.

Chen, J.H., R.L. Edwards, and G.J. Wasserburg. 1986. ^{238}U, ^{234}U and ^{232}Th in seawater. *Earth Planet. Sci. Lett.* **80**:241–251.

de Jong, A.F.M., B. Becker, and W.G. Mook. 1986. High-precision calibration of the radiocarbon time scale 3930–3230 Cal B.C. *Radiocarbon* **28**:939–942.

Dodge, R.E., R.G. Fairbanks, L.K. Benninger, and F. Maurrasse. 1983. Pleistocene sea levels from raised coral reefs of Haiti *Science* **219**:1423–1425.

Edwards, R.L. 1988. High precision thorium–230 ages of corals and the timing of sea level fluctuations in the late Quaternary. Ph.D. thesis, California Institute of Technology.

Edwards, R.L., Chen, J.H., and G.J. Wasserburg. 1987. ^{238}U–^{234}U–^{230}Th–^{232}Th systematics and the precise measurement of time over the past 50,000 years *Earth Planet. Sci. Lett.* **81**:175–192.

Edwards, R.L., F.W. Taylor, and G.J. Wasserburg. 1988. Dating earthquakes with high precision thorium–230 ages of very young corals *Earth Planet. Sci. Lett.* **90**:371–381.

Emiliani, C. 1955. Pleistocene temperatures. *J. Geol.* 63:162.

Fairbanks, R.G. 1989. A 1700 years glacio-eustatic sea level record: Influence of glacial melting rates on the Younger Dryas event and deep-ocean circulation. *Nature* **342**:637–642.

Fairbanks, R.G., and R.K. Matthews. 1978. The marine oxygen isotope record in Pleistocene coral, Barbados, West Indies. *Quat Res.* **10**:181–196.

Hamelin B., E. Bard, A. Zindler, and R.G. Fairbanks. 1991. $^{234}U/^{238}U$ mass spectrometry of corals: How accurate is the U–Th age of the last interglacial period? *Earth Planet. Sci. Lett.* **106**:169–180.

Hays, J.D., J. Imbrie, and N.J. Shackleton. 1976. Variations in the Earth's orbit: Pacemaker of the ice ages. *Science* **194**:121–1131.

Hilgen, F.J. 1991. Astronomical calibration of Gauss to Matuyama sapropels in the Mediterranean and implication for the geomagnetic polarity time scale. *Earth Planet. Sci. Lett.* **1O4**:226–244.

Imbrie, J., J.D. Hays, D.G. Martinson, A. Mclntyre, A. Mix, J.J. Morley, N.G. Pisias, W. Prell, and N.J. Shack1eton. 1984. The orbital theory of Pleistocene climate: Support from a revised chronology of the marine $\delta^{18}O$ record. In: Milankovitch and Climate, ed. A. Berger, J. Imbrie, J.D. Hays, G. Kukla, and B. Saltzman, pp. 269–305. Hingham, MA: Reidel.

Imbrie, J., and J.Z. Imbrie. 1980. Modelling the climatic response to orbital varations. *Science* **207**:943–953.

Jouzel, J., C. Lorius, J.R. Petit, C. Genthon, N.I. Barkov, V.M. Kotlyakov, and V.M. Petrov. 1987. Vostok ice core: A continuous isotope temperature record over the last climatic cycle (160,000 years). *Nature* **329**:403–408.

Labeyrie, L.D., J.C. Duplessy, and P.L. Blanc. 1987. Variations in mode of formation and temperature of oceanic deep waters over the past 125,000 years. *Nature* **327**:477–482.

Lerman, J.C., W.G. Mook, and J.C. Vogel. 1970. C14 in tree rings from different localities. In: Radiocarbon Variations and Absolute Chronology, ed. I.U. Olsson, pp. 275–301. New York: Wiley.

Linick, T.W., A. Long, P.E. Damon, and C.W. Ferguson. 1986. High-precision radiocarbon dating of Bristlecone Pine from 6554 to 5350 B.C. *Radiocarbon* **28**:943–953.

Lorius, C., J. Jouzel, C. Ritz, L. Merlivat, N.I. Barkov, Y.S. Korotkevich, and V.M. Kotlyakov. 1985. A 150,000 year climatic record from Antarctic ice. *Nature* **316**:591–596.

Maniken, E.A., and C.S. Gromme. 1982. Paleomagnetic data from the Coso Range, California and current status of the Cobb Mountain normal geomagnetic polarity event. *Geophys. Res. Lett.* **9**:1272–1282.

Martinson, D.G., N.G. Pisias, J.D. Hays, J. Imbrie, T.C. Moore, and N.J. Shackleton. 1987. Age dating and the orbital theory of the ice ages: Development of a high-resolution 0 to 300,000–year chronostratigraphy. *Quat. Res.* **27**:1–29.

Matthews, R.K. 1973. Relative elevation of late Pleistocene high sea level stands: Barbados uplift rates and their implications. *Quat. Res.* **3**:147–153.

Milankovitch, M. 1930. Mathematische Klimalehre und Astronomische Theorie der Klimaschwankungen. In: Handbuch der Klimatologie I(A), ed. W. Kopppen and R. Geiger, pp. 1–176. Berlin: Gebrüder Borntraeger.

Pearson, G.W., J.R. Pilcher, M.G.L. Baillie, D.M. Corbet, and F. Qua. 1986. High-precision ^{14}C measurement of Irish oaks show the natural ^{14}C variations from AD 1840–5210 B.C. *Radiocarbon* **28**:911–934.

Prell, W., J. Imbrie, D.G. Martinson, J.J. Morley, N.G. Pisias, N.J. Shackleton, and H. Streeter. 1986. Graphic correlation of oxygen isotope stratigraphy, application to the late Quaternary. *Paleoceanog.* **1**:137–162.

Radtke, U., R. Grün, and H.P. Schwarcz. 1988. Electron spin resonance dating of the Pleistocene coral reef tracts of Barbados. *Quat. Res.* **29**:197–215.

Raymo, M.E., W.F. Ruddiman, N.J. Shackleton, and D.W. Oppo. 1990. Evolution of global ice volume and Atlantic–Pacific $d^{13}C$ gradients over the last 2.5 MY. *Earth Planet. Sci. Lett.* **97**:353–368.

Ruddiman, W.F., A. McIntyre, and M.E. Raymo. 1986. Matuyama 41,000-year cycles: North Atlantic Ocean and northern hemisphere ice sheets. *Earth Planet. Sci. Lett.* **80**:117–129.

Scott, E.M., A. Long, and R. Kra, eds. 1990. Proc. International Workshop on Intercomparison of Radiocarbon Laboratories. *Radiocarbon* **32**:253–397.

Shackleton, N.J., A. Berger, and W.R. Peltier. 1990. Astronomical calibration of the lower Pleistocene time scale based on ODP Site 677. *Trans. R. Soc. Edin. Earth Sci.* **81**:251–261.

Shackleton, N.J., and M.A. Hall. 1989. Stable isotope history of the Pleistocene at ODP Site 677. In: *Proc. ODP Init. Rep.* Pt. A, 111, ed. K. Becker, H. Sakai, et al., pp. 253–346. College Station, TX.

Shackleton, N.J., and N.D. Opdyke. 1973. Oxygen isotope and paleomagnetic stratigraphy of equatorial Pacific core V28–238: Oxygen isotope temperatures and ice volumes on a 10^5 and 10^6 year time scale. *Quat. Res.* **3**:39–55.

Shackleton, N.J., et al. 1984. Oxygen isotope calibration of the onset of ice rafting and history of glaciation in the North Atlantic region. *Nature* **307**:620–623.

Shipboard Scientific Party. 1988. Sites 677 and 678. *Proc. ODP*, Sci. Results, ed. K. Becker, H. Sakai, et al., pp. 295–316. College Station, TX.

Steiger, R.H., and E. Jager. 1977. Subcommission on geochronology: Convention on the use of decay constants in geo- and cosmochronology. *Earth Planet. Sci. Lett.* **36**:359–362.

Stein, M., G.J. Wasserburg, P. Aharon P., J.H. Chen, Z.R. Zhu, A.L. Bloom, J. Chappell. 1990. Precise $^{230}Th/^{234}U$ ages of the last interglacial event at Huon Peninsula. Abstract, International Conference on Geochronology. Canberra: ICOG.

Stuiver, M. 1982. A high-precision calibration of the AD radiocarbon time scale. *Radiocarbon* **24**:1–26.

Stuiver, M., and B. Becker. 1986. High-precision decadal calibration of the radiocarbon time scale. *Radiocarbon* **28**:863–910.

Stuiver, M., and G.W. Pearson. 1992. Calibration of the radiocarbon time scale 2500–5000 B.C. In: Four Decades of Radiocarbon Studies: An Interdisciplinary Perspective, ed. R.E. Taylor, A. Long, and R.S. Kra, pp. 19–33. New York: Springer.

Stuiver, M., and T.F. Braziunas. 1993. Modeling atmospheric ^{14}C influences and ^{14}C ages of marine samples to 10,000 B.C. *Radiocarbon* **35:**137–189.

Stuiver, M., and H.S. Polach. 1977. Discussion: Reporting of ^{14}C data. *Radiocarbon* **19**:355–363.

Stuiver, M., and P.D. Quay. 1980. Changes in atmospheric carbon–14 attributed to a variable sun. *Science* **207**:11–19.

Vogel, J.C., A. Fuls, and E. Visser. 1986. Radiocarbon fluctuations during the third millennium B.C. *Radiocarbon* **28**:935–938.

Vogelsang E. 1990. Paläo-Ozeanographie des Europäischen Nordmeeres an Handstabiler Kohlenstoff und Sauerstoffisotope. Ph.D. thesis, University of Kiel. Sonderforschungsbereich 313, Sedimentation im Europäischen Nordmeer 23.

9

Present Means to Define and Estimate Atmospheric "Turbidity" and Its Effects on Radiation Fields at Ground Level: Possible Projections into the Past

F. SIROCKO[1] and E. RASCHKE[2]
[1]Geological Paleontological Institute, University Kiel, Olshausenstr. 40–60, 2300 Kiel, F.R. Germany
[2]Institute of Physics, GKSS Research Center, 2054 Geesthacht, F.R. Germany

ABSTRACT

The turbidity of the atmosphere is a measure for the scattering and absorption of the Sun's radiation in the air and, thus, it is the major control on the mass of energy reaching the Earth's ground. It is caused by gaseous components (mainly water vapor, carbon dioxide, and methane), cloud cover, and aerosols (dissolved or particulate). Concentrations of the gases and dissolved components are uniformly spread over the entire atmosphere; this enables us to reconstruct, from ice cores, their concentrations during former climatic stages. Particulate aerosol components (e.g., volcanic ash and desert dust), however, are not uniformly distributed, but their fossil concentrations in the air can be estimated from their mass-accumulation rates in geological records.

During the last ice age, atmospheric greenhouse gases—water vapor, methane, and carbon dioxide—were reduced, while desert dust aerosols were increased. In the near vicinity (i.e., a few hundred kilometers) of the dust entrainment areas, average seasonal values of aerosol concentration (and thus atmospheric turbidity) should have been twice as high in comparison with present seasonal values, with thermal effects at the ground level to be expected. In the large high-latitude glacial loess regions, the seasonal averages of turbidity must have been similar to seasonal values from modern dust entrainment areas, because dust flux rates in the modern subtropical entrainment area are similar to glacial loess deposition rates. Outside these source regions and their immediate vicinity, dust concentrations during the last glacial were also drastically increased, up to a factor of 20; however, the absolute numbers of particles/volume were still so few that direct effects on the climate can hardly be expected.

Global Changes in the Perspective of the Past
Edited by J.A. Eddy and H. Oeschger

Thus, the glacial atmosphere on a global and annual scale was probably drier and clearer than today, even if a high background aerosol is taken into account.

INTRODUCTION

During past periods of the Earth's history, it has been argued that the atmospheric composition was different than the present one; thus light conditions at the Earth's surface were also different (Petit et al. 1981; Kutzbach and Guetter 1986). These differences may be observable in various geological and historical records. However, at this time, we cannot directly determine the past spectra, amounts, and diffusivities of solar radiation that are so important to life on Earth. The only way to look back into the earlier state of these radiative properties of the atmosphere is through numerical modeling, based on reliable boundary conditions, reconstructed from the geological record, and calibrations versus the present climate state (e.g., Lautenschlager and Herterich 1990). In this chapter we will try to show the state of knowledge about such properties of the Sun's radiation at ground level, based on observable (and computable) quantities. Next we offer a short outline of these relations, describing the dependence of the Sun's radiation field at ground level. Finally, we compile the fossil data base.

PRESENT PROPERTIES OF SOLAR RADIATION AT GROUND LEVEL

At the present time, the Sun's irradiance reaching the Earth's atmosphere amounts to about 1370 Wm^{-2}. Its spectral composition is close to that of radiation from a black body, with a temperature of about 5800 K. Minor changes, e.g., Fraunhofer lines, are due to absorption and emission in the Sun's atmosphere. Major modifications occur when this shortwave radiation, in the spectral interval between 0.1 and 10 μm, penetrates through the atmosphere. The shortwave end (0.1–0.3 μm) is clearly "cut" due to absorption by O_2 and O_3 molecules in the upper atmosphere (above 20 km). Further multiple scattering by molecules and absorption by ozone causes a slight blue shift of the spectrum and reduces the amount reaching the ground. Further spectral changes occur due to strong absorptions by water vapor and carbon dioxide, and causes the solar spectrum to end already at about 3 μm.

Aerosols in the atmosphere, and moreover clouds, change the solar spectrum from bluish to more white and can additionally reduce the downward amount. At the present climate we may receive between 10–80% of the radiation available at the top of the atmosphere, depending on many factors: latitude, season, and area (time of day is not included here). The annual global average may reach values around 50%. Further details are given in many original and review articles (see Goody and Yung 1989 for an excellent summary).

In order to "explain" possible variations in the past, we have to take a closer look at the variations of the atmospheric properties needed to change the solar radiation at ground level.

PARAMETERS AFFECTING THE SUN'S TOTAL AND SPECTRAL IRRADIANCE AT GROUND LEVEL

Astronomical Parameters

Astronomical parameters, i.e., the Earth's orbital characteristics and the position of its axis with respect to the ecliptic plane, show quasiperiodical oscillations capable of causing large changes (about 5–10%) in the present values of the Sun's irradiance reaching the Earth. The magnitude and chronology of these parameters are also well known for the past (Imbrie et al., this volume), their coherency with cold and warm periods is obvious, and they have been successfully used in simulations of the extreme climate states of the Pleistocene and Holocene (e.g., Kutzbach and Guetter 1986).

Aerosols

Aerosols in the atmosphere are particles between the sizes of about 0.01 to 20 μm in diameter. In dust storms, particles of more than 100 μm have been identified. Their amounts—and nature—may change considerably due to the Earth's surface properties and the strength of wind and biological activities (including water supply). At present, the major natural sources are oceans (salt, dimethyl sulfides [DMS]), continental surfaces (soil, dust, seeds, soot, others), volcanic explosions (ash, sulfur), and anthropogenic (mainly over industrial regions; however, like desert dust, it can penetrate even the remote areas of the Arctic). These aerosols tend to act as cloud condensation nuclei, in particular soluble aerosols. Indeed, they diminish the incident radiation and alter its spectrum. Larger particles (> 20 μm) can absorb appreciable amounts of solar radiation, thus altering the heat supply for the atmosphere.

There are various parameters defined to measure the atmospheric turbidity. These values usually relate spectral radiance values at clear but "turbid" skies to some "clean" reference. Since such values were not recorded in the past, only quantitative estimates are possible.

Clouds

Clouds, consisting of water droplets and/or ice crystals, form as a result of various dynamical processes in the atmosphere, of which vertically unstable structures might be the most important source. Thus, if we assume that strong deviations of atmospheric cloudiness existed in the past, we must first look for past atmospheric circulation patterns (regional and global): shift of subtropical high-pressure belts, of the mean location of the Intertropical Convergence Zone, or of the belt of extra-typical disturbances (e.g., the monsoonal regimes or cyclonic depressions in the westwind drifts, etc.). These, in turn, are often related to ocean temperatures and currents (e.g., the present Walker circulations during the El Niño occurrence. Such "looks" are only possible on the basis of numerical model simulations of past circulation systems (e.g., Kutzbach and Guetter 1986; Rind 1987; Joussaume 1990; Lautenschlager and Herterich 1990). However, these have so far not simulated in all detail the cloud cover. Even for the present climate, they are yet unable to simulate the cloud fields and their properties correctly (Cess et al. 1990).

The optical properties of low-level clouds above oceans are assumed to be strongly affected by the supply of DMS aerosols (e.g., Charlson et al. 1987), which, in turn, are result of bioactivities in the near surface water layers and which depend on the water temperature.

Atmospheric Water Vapor and Carbon Dioxide

Atmospheric water vapor and carbon dioxide change the amounts of available solar radiation in the near infrared portions of the spectrum. Drastic changes of CO_2 have been observed in ice cores, which might be due to changes in vegetation on Earth or possibly to the transfer of CO_2 into the deeper ocean currents. Thus, the changes depend on temperature and available moisture. The amount of atmospheric water vapor is clearly nonlinearly related to the mean temperature in the layers near the surface. It is also linked to the presence of clouds, which are even for the present climate still insufficiently treated in any circulation model. Therefore, within the World Climate Research Program, special enhanced efforts have been made to model the present cloud and water vapor distribution.

Ozone

Did strong variations in ozone also occur, which could have affected the shortwave side of the spectrum? Such variations could have been caused by volcanic ash and sulfuric aerosols, reaching the stratosphere. However, lower stratospheric temperatures (dynamics!) could have also caused different ozone amounts as well as of higher or lower amounts of ultraviolet radiation reaching the ground. These variations in the ultraviolet component are assumed to cause selections in various biotic species in oceans and on the continents.

PROJECTIONS INTO THE PAST

How can we estimate past atmospheric turbidity (or better: changes in the total and spectral transmittance of the atmosphere for solar radiation)? Possibly never, quantitatively, but some qualitative speculations appear to be feasible.

Dust from Deserts

Variations in the extent and mass of desert dust have been measured by accumulation rates of eolian particles in the deep oceans, loess deposition at land, and aerosol particles in the ice of the polar regions (for a comprehensive review, see Pye 1987).

The modern dispersal of desert dust over the globe shows clear maxima during summer in the equatorial Atlantic off the western Sahara and in the northern Indian Ocean off the coast of Arabia (McDonald 1938). Another maximum occurs during winter off the southwestern Sahara. In accordance with these observations, maximum accumulation rates of eolomarine dust (up to 100 g/m²/yr) have been reported for these regions (Koopmann 1981; Sirocko et al. 1991). A few thousand kilometers further away, in both polar regions, modern aerosol accumulation is 3–5 orders of magnitude lower (1–100 mg/m²/yr; Windom 1969; Wagenbach et al. 1988) than near the entrainment areas of desert dust (Figure 9.1).

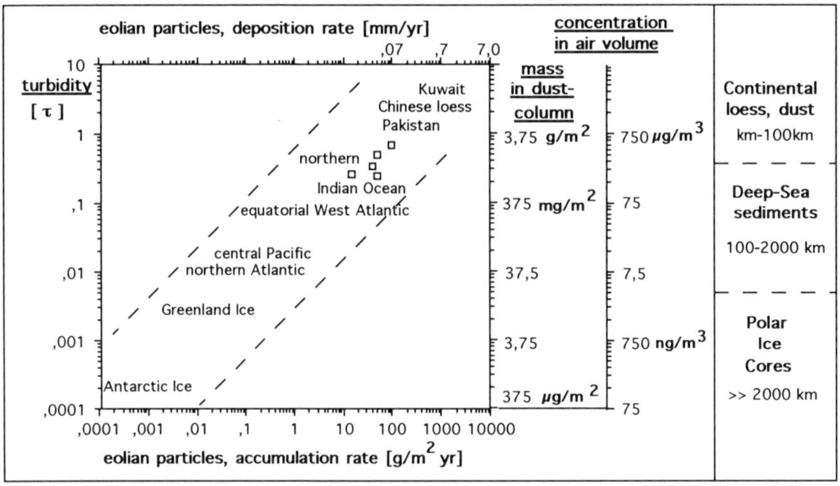

Figure 9.1 Atmospheric turbidity, aerosol content, and eolian sedimentation rates for different parts of the globe. Dots: average annual accumulation rate (g/m²/yr) of wind-transported particles during the Holocene in the deep northern Indian Ocean for five sectors along the coast of Arabia (Sirocko and Sarnthein 1989) versus average atmospheric turbidity (τ) due to dust particles during the dust season (June–August), calculated from GOES satellite images for the same sectors (Sirocko and Sarnthein 1989). Given are also the dust deposition rate (using a dry density of 1.5 g/cm³) and the aerosol mass (g/m²), calculated using the formula of Carlson and Caverly (1977), and the average aerosol concentration in a 5,000 m dust column. The dashed line shows the empirical error bars, which emphasize that this relation is valid only on the scale of two magnitudes. References: Kuwait (Khalaf and Al-Hashas 1983; Foda et a. 1985), Chinese loess (An et al. 1991), warves on the Pakistan shelf (von Stackelberg 1972), Pakistan dust outbreak (Middleton 1986), dust storms on the continents (Pye 1987 and references therein), equatorial west Atlantic (Koopmann 1981; Carlson and Caverly 1977), central Pacific (Janecek and Rea 1985), and Greenland ice, Antarctic ice (Windom 1969; Wagenbach et al. 1988).

For the Saharan and Arabian dust outbreaks, the mass and radiative effect of dust on the turbidity of the atmosphere has been calculated through ground observation and satellite images. On the basis of aircraft measurements, Carlson and Caverly (1977) developed a method to convert τ-values of a Saharan dust layer of 5000 m thickness to mass (M) in the dust column by means of the formula M=3.75t (g/m^2). Similar τ-values have also been obtained from brightness measurements at visible wavelength using the SMS satellite (Norton et al. 1980). In their paper, Norton et al. (1980) described turbidity as the intensity of radiation returned from the Earth–atmosphere system to space, related to the radiation expected from an atmosphere containing a dust layer of specified optical thickness (τ), i.e., the background aerosol is included but clouds, ice, and water vapor are not. Chen (1986) calibrated this algorithm for dust outbreaks from the Arabian desert using visible wavelength data of the GOES satellite, which was located over the equatorial Indian Ocean for one year during the MONEX program in 1979.

Based on the calculation of the dust mass from τ-values 700×10^6t (MT) of dust are derived from Saharan sources (d'Almeida 1987), with 100–200 MT from Arabian sources (Sirocko and Sarnthein 1989). Because these deserts are the major dust sources on the modern globe, the global modern bulk dust discharge might be a value between $1–2 \times 10^9$t (GT) of dust.

To provide a proxy record of the distribution patterns of atmospheric turbidity, we compared the results from the atmosphere over the Arabian Sea with the observed accumulation rates of eolomarine dust particles in the underlying deep-sea sediments of this region (Sirocko and Sarnthein 1989; Figure 9.1). Accordingly, maximum sediment accumulation rates are found in the regions of maximum observed atmospheric turbidity. Figure 9.1 also shows the magnitude of accumulation/deposition rate and average dust concentrations for other parts of the globe. At least in terms of magnitude, the correlation from the Arabian Sea values is substantiated, which shows that the dispersal of aerosol dust mass linearly parallels the accumulation rate of eolian particles. Because the mass is directly related to turbidity, the mass accumulation can be used to quantify the magnitude of past atmospheric turbidity due to eolian particles.

Mass accumulation rates of eolian particles during the last glacial are available for most parts of the glacial (for a review, see Pye 1987). In general, all records agree that glacial dust concentrations clearly exceeded the Holocene values; however, the magnitude of observed changes differs greatly from region to region. In the deep equatorial Atlantic and Northern Indian Ocean, accumulation rates of eolian particles were nearly doubled during the last glaciation (Koopmann 1981; Sirocko et al. 1991). Because the relation between mass and turbidity is linear, glacial turbidity values off of West Africa and Arabia should have doubled in comparison to Holocene t-values (modern average during dust season about 0.4–0.8).

A major difference between the Holocene and the Glacial was the deflation of loess, which occurred in a zone of a few hundred kilometers off the continental ice sheets. The accumulation rate of loess in these regions shows values between 100–1000 $g/m^2/yr$ (An et al. 1991) (i.e., a deposition rate up to 1 mm/yr), which is similar to modern values from the entrainment areas of dust (e.g., dust warves at the coast of Pakistan (von Stackelberg 1972) or Kuwait (Khalaf and Al Hashash 1983), and is thus 1–2 orders of magnitude higher than the accumulation rates of eolomarine sediments. Because the deposition rates of fossil loess (in a dry, cold climate) are so similar to values from modern dust entrainment areas in the dry, hot Subtropics, their mean seasonal τ-values can be expected to have the same magnitude (i.e., τ-values between 1 and 2; Figure 9.1).

In the polar ice-core records of the last glaciation, dust deposition was increased by a factor of 10 to 20 (e.g., Petit et al. 1981). Modern aerosol concentrations in the polar regions reveal values on the order of nanograms/m^3 (i.e., t-values). Thus, even the glacial values were still at least one order of magnitude too low to be of any thermal importance (Figure 9.1). Small thermal effects (<1°C) due to atmospheric dust have been measured in Saharan outbreaks that have t-values between 0.1 and 1 (Carlson and Caverly 1977).

These regional differences in mass of accumulation and amount of change have to be taken into account if we want to estimate the effect of aerosol dust turbidity of the glacial atmosphere. Following the evidence from modern meteorological measurements, dust turbidity has been analyzed by global circulation models when the increased glacial dust loads have caused significant surface temperature anomalies (e.g., Kutzbach and Guetter 1986; Joussaume 1990). Based on the regional scale of large increases in dust mass, we have to question this approach because the areas of subtropic desert dust (where turbidity values must have been about doubled during glacial times) are not large enough to effect climate on a global scale. Thus, the major importance of dust particles on climate might have been in the large loess areas, where τ-values and geographical extent were large enough to cause significant anomalies in the radiation fields at ground level.

In comparison to these changes in the mass of dust, the observed changes in grain size of the aerosols are of second-order importance. Another effect of dust transport near glaciers is the deposition of dust particles on ice, which can change the albedo of the ice. However, this is not an effect related to the turbidity of the atmosphere.

Volcanic Ash

Recent and past volcanic explosions produced large amounts of dust. For an average eruption, the mass of tephra and ash is on the order of a few hundred megatons (e.g., Schmincke 1986), i.e., less than the global annual mass of desert dust (see above). During extreme conditions, however, the mass of ashes may exceed the annual amount transported as desert aerosol (Rampino and Self 1982). Yet only the smallest particles of the ash are able to stay aloft long enough to travel over long distances, provided that they are blown into the middle to upper troposphere or lower stratosphere. In any case, the effect of volcanic ash clouds on the light transmission of the atmosphere will last only a few weeks to a month, but its radiative effect can locally exceed the effect of the desert dusts.

Volcanic Sulfur Emissions

The El Chichon, and recently the Pinatubo, demonstrated drastically that huge amounts of sulfur oxides and other acidic gases can be blown into the atmosphere, where they form aerosol clouds. They are known to spread over large fractions of the globe and may stay there for many months or longer (Hammer et al. 1980). They have been found to lower the temperature at the Earth's surface by more than 1 K (e.g., Tambora in 1815; Rampino and Self 1982). The amount of such gaseous emissions from volcanos has been estimated by three independent approaches even for past volcanic eruptions: (a) Devine et al. (1984) reconstructed the sulfur emission from the chemical composition and mass of tephra; (b) the global concentrations of sulfur in the atmospheric are recorded in ice cores (Raynaud and Siegenthaler, this volume and references therein); (c) Bryson (1989) compiled the available evidence for the frequency of volcanic eruptions during the last 40,000 years and, on this basis, calculated the frequency of sulfur-rich explosions that could have had a globally important magnitude. Accordingly, during glacial times, major events of sulfur emission were less frequent than during the Holocene.

Dimethyl Sulfides (DMS)

DMS particles are due to emissions of sulfide from algae in the sea (Ayers et al. 1991). In ice depositions they can be seen as methane sulfonate (MSA) or non-sea salt sulfates (Legrand et al. 1991). During the glaciations, more than 130% of the amounts were deposited than in the times between. This was most likely not due to higher production rates (which are higher at high temperatures of the surface ocean) but to the fact that DMS can be fixed to dust particles, which were more prominent during glacial times (e.g., Petit et al. 1981); so far, no climate effect has been deduced from such observations. More recent considerations, however, speculate that increased DMS may dampen the greenhouse effect of the atmosphere by decreasing droplet sizes and thus increasing the albedo of clouds (Charlson et al. 1987). Again, we must emphasize that we still do not understand the physics of natural undisturbed clouds well enough to justify the perturbation by DMS particles.

Atmospheric CO_2 and Methane Concentrations

Detailed analyses of past concentrations of these gases have been reconstructed from ice cores (see Raynaud and Siegenthaler, this volume, and references therein). The present concentration of CO_2 (335 ppm) is about 75 ppm higher than values around the preindustrial period, and it may have doubled since the last ice age. Numerical simulations of the climate from the last ice age, with CO_2 concentrations of 180 ppm, estimate surface temperatures to be 1–1.5 K lower than present.

Methane concentrations, presently at 1.7 ppm, have more than tripled its abundance since the last glacial maximum (Chappellaz et al. 1990). This greatly affects longwave infrared (increasing the greenhouse effect) but not the solar spectrum.

These two greenhouse gases are increasing in their abundance due to anthropogenic activities. In the past climate, their abundance was well belanced with the biological activity on Earth. "Simple" predictions and calculations of climatic effects are thus not appropriate, because the response of the biosphere today may be totally different than it was during past climate stages.

Atmospheric Water Vapor

Recently, numerical simulations of future climate states performed with enhanced CO_2 concentration tended to result in higher near-surface temperatures and in more clouds, particularly near the tropopause over all regions of the globe and near the surface over the polar regions. Thus, we can speculate that during the last ice ages, with its cold land surface and ice cover, there was less cloudiness; more clouds may have occurred during the warm interglacials, since the amount of water vapor in the atmosphere is strongly coupled with the mean temperature (Claudius Clapeyson's law). In their simulation of the climate during the ice age, Lautenschlager and Herterich (1990) found slightly less precipitation, in particular over continents. Yet again, we must clearly understand the errors in their model with respect to the hydrological cycle.

CONCLUSIONS

During glacial times, transmittance in the atmosphere was probably better than today, because only desert dust was locally enhanced whereas volcanic ash and sulfur, CO_2, methane, and water vapor were reduced on a global scale. Only sketchy speculations about quantitative changes in atmospheric turbidity during the past thousands of years are possible. There is simply too little evidence available from paleoclimatic analysis to estimate past atmospheric turbidity with an accuracy comparable to that obtained from modern direct measurements. The major problem at this time is to find an approach to reconstruct the paleohydrological budget, especially the distribution patterns of clouds.

REFERENCES

An, Z.S., G. Kukla, S.C. Porter, and X.L. Xiao. 1991. Late Quaternary dust flow on the Chinese Loess Plateau. *Catena* **18**:125–132.

Ayers, G.P., J.P. Ivey, and R.W. Gillet. 1991. Coherence between seasonal cycles of dimethyl sulfide, methanosulphanate and sulphate in marine air. *Nature* **349**:404–406.

Bryson, R.A. 1989. Modeling the NW India monsoon for the last 40000 years. *Clim. Dynam.* **3**:169–177.

Carlson, T.N., and R.S. Caverly. 1977. Radiative characteristics of Saharan dust at solar wavelengths. *J. Geophys. Res.* **82(21)**:3141–3152.

Cess, R.D., et al. 1990. Intercomparison and interpretation of climatic feedback processes in 19 atmospheric general circulation models. *J. Geophys. Res.* **95(D10)**:16601–16615.

Chappellaz, J. 1990. Etude du méthane atmosphérique au cours du dernier cycle climatique. Thesis, University of Grenoble, France.

Charlson, R.J., J.E. Lovelock, M.O. Andreae, G.G. Warren. 1987. Oceanic phytoplancton, atmospheric sulfur, cloud albedo and climate. *J. Geophys. Res.* **89**:9668–9672.

Chen, F.H. 1986. Analyses of a major dust outbreak over the Arabian Sea during Monex. Thesis, University of Wisconsin, Madison.

d'Almeida, G.A. 1987. A model for Saharan dust transport. *J. Clim. App. Meteor.* **24**:903–916.

Devine, J.D., H. Sigurdsson, and A.N. Davies. 1984. Estimates of sulfur and chlorine yield to the atmosphere from vulcanic eruptions and potential climatic effects. *J. Geophys. Res.* **89(B7)**:6309–6325.

Foda, M.A., F.I. Khalaf, and A.S. Kadi-al. 1985. Estimation of dust fallout rates in the northern Arabian Gulf. *Sedimentol.* **32**:595–603.

Goody, R.M., and Y.L. Yung. 1989. Atmospheric Radiation. Oxford: Oxford Univ. Press.

Hammer, C.U., H.B. Clausen, and W. Dansgaard. 1980. Greenland ice sheet evidence of postglacial volcanism and its climatic impact. *Nature* **288**:230–235.

Janecek, T.R., and D.K. Rea. 1985. Quarternary fluctuations in the Northern Hemisphere trade winds and westerlies. *Quat. Res.* **24**:150–163.

Joussaume, S. 1990. Three dimensional simulations of the atmospheric cycle of desert dust particles using a general circulation model. *J. Geophys. Res.* **95(D2)**:1909–1941.

Khalaf, F., and M. Al-Hashash. 1983. Aeolian sedimentation in the northwestern part of the Arabian Gulf. *J. Arid Envir.* **6**:319–322.

Koopmann, B. 1981. Sedimentation von Saharastaub im subtropischen Nordatlantik während der letzten 25,000 Jahre. *Meteor Forsch. Erg.* **C35**:23–59.

Kutzbach, J.E., and P.J. Guetter. 1986. The influence of changing orbital parameters and surface boundary conditions on climate simulations for the past 18,000 years. *J. Atmos. Sci.* **43(16)**:1726–1759.

Lautenschlager, M., and K. Herterich. 1990. Atmospheric response to ice-age conditions: climatology near the Earth's surface. *J. Geophys. Res.* **95(D13)**:22547–22557.

Legrand, M., C. Feniet-Saigne, E.S. Saltzman, C. Germain, N.I. Barkov, and V.N. Petrov. 1991. Ice core record of oceanic emissions of dimethylsulphide during the last climate cycle. *Nature* **350**:144–146.

McDonald, W.F. 1938. Atlas of Climatic Charts of the Ocean. U.S. Dept. Agriculture Weather Bureau, 1, 247, charts 59–62.

Middleton, N.J. 1986. A geography of dust storms in Southwest Asia. *J. Climatol.* **6**:183–196.

Norton, C.C., F.R. Mosher, B. Hinton, D.W. Martin, D. Santek, and W. Kuhlow. 1980. A model for calculating desert aerosol turbidity over the oceans from geostationary satellite data. *J. App. Meteor.* **19(6)**:633–644.

Petit, J.R., M. Briat, and A. Royer. 1981. Ice age aerosol content from east Antarctic ice core samples and past wind strength. *Nature* **293**:391–394.

Pye, K. 1987. Aeolian Dust and Dust Deposits. New York: Academic Press.

Rampino, M.R., and S. Self. 1982. Historic Eruptions ofTambora (1815), Krakatau (1883), and Agung (1963), their stratospheric aerosols, and climatic impact. *Quat. Res.* **18**:127–143.

Rind, D. 1987. Conmponents of the Ice Age circulation. *J. Geophys. Res.* **92(D4)**:4241–4281.

Schmincke, H.U. 1986. Vulkanismus. Darmstadt, Germany: Wissenschaftliche Buchgesellschaft.

Sirocko, F., and M. Sarnthein. 1989. Wind-borne deposits in the Northwestern Indian Ocean: record of Holocene sediments versus modern satellite data. In: Paleoclimatology and Paleometeorology: Modern and Past Patterns of Global Atmospheric Transport, ed. M. Leinen and M. Sarnthein, pp. 401–433. Dordrecht: Kluwer.

Sirocko, F., M. Sarnthein, H. Lange, and H. Erlenkeuser. 1991. The atmospheric summer circulation and coastal upwelling in the Arabian Sea during the Holocene and the last glaciation. *Quat. Res.* **36**:72–93.

von Stackelberg, U. 1972. Faziesverteilung in Sedimenten des indischpakistanischen Kontinentalrandes. *Meteor Forsch. Erg.* **C9**:1–73.

Wagenbach, D., U. Gerlach, K. Moser, and K.O. Münnich. 1988. Coastal antarctic aerosol: the seasonal pattern of its chemical composition and radionuclide content. *Tellus* **40B**:426–436.

Windom, H.L. 1969. Atmospheric dust records in permanent snowfields: Implications to marine sedimentation. *Geol. Soc. Am. Bull.* **80**:761–782.

10

Use of General Circulation Models to Follow Climatic Tracers on a Global Scale

J. JOUZEL[1,2], S. JOUSSAUME[1,3], and R. KOSTER[4]
[1]Laboratoire de Modélisation du Climat et de l'Environnement, DSM, Bat 709, CEN Saclay, 91191 Gif sur Yvette, France
[2]Laboratoire de Glaciologie et Géophysique de l'Environnement, B.P. 96, 38402 St. Martin d'Hères Cedex, France
[3]Laboratoire d'Océanographie Dynamique et Climatologie, Université Pierre et Marie Curie, 75252 Paris Cedex 5, France
[4]Hydrological Sciences Branch, Code 974, NASA, Goddard Space Flight Center, Greenbelt, MD 20771, U.S.A.

ABSTRACT

The atmospheric cycles of water isotopes and desert dust can be investigated with general circulation models (GCMs). Indeed, through simulations of present-day and glacial climates, GCM tracer studies can help interpret the nature of important climate–tracer relationships, such as those between temperature and isotope content of precipitation and between atmospheric circulation and dust deposition. GCM tracer diagnostics can also be applied to studies of the hydrological cycle by establishing the origins of precipitating water.

INTRODUCTION

General circulation models (GCMs) contribute to our understanding of current, past, and future climates by simulating the time evolution of various global, discretized atmospheric fields (e.g., wind speed, temperature, surface pressure, specific humidity) through integration of the relevant basic equations. In classical applications, only air and normal water ($H_2{}^{16}O$) are advected through the GCM's atmosphere. GCMs, however, are now being used to study the global behavior of atmospheric tracers such

Global Changes in the Perspective of the Past
Edited by J.A. Eddy and H. Oeschger

as stable water isotopes (HDO and $H_2^{18}O$)[1] and desert dust particles. This is accomplished by modeling the relevant tracer source and sink processes in each grid box.

The modeling of water isotopes and desert dust in GCMs is motivated by their important changes with climate, especially between interglacial and glacial periods, as documented from ice cores (e.g., Jouzel, Lorius et al. 1987; Petit et al. 1981). The long-term objectives of the climatic tracer studies are (1) to allow a direct comparison between model results and observations for model verification, and (2) to facilitate the translation of paleodata (such as δD, $\delta^{18}O$ or dust content) into information on paleoclimate, e.g., information on ancient temperatures and atmospheric circulations. A third objective concerns determining the origin of the water vapor contributing to a given region's precipitation. This has important hydrological implications as well as implications for paleoclimate studies: the temperature and relative humidity in a moisture source region can influence the deuterium excess of precipitation (Jouzel et al. 1982).

This chapter provides an overview of this relatively new direction of research, which is spearheaded by groups working with the LMD (Laboratoire de Météorologie Dynamique, Paris) GCM, and the NASA/GISS (Goddard Institute for Space Studies, New York) GCM. Each section below focuses on the scientific objectives and features specific to each tracer type. They contain a synthesis of the key published results, which focus on studies of modern climate and the climate of the Last Glacial Maximum (LGM, 18 Ka B.P.).

DEUTERIUM AND OXYGEN-18 CYCLES

Numerous examples in the literature over the last 40 years have demonstrated that deuterium and oxygen-18 distributions in water can be successfully applied to studies of cloud physics, climatology, hydrology, and paleoclimatology. Prior to the pioneering GCM work of Joussaume et al. (1984), water isotope modeling was essentially limited to simple Rayleigh-type models, which form precipitation from isolated air masses (see Jouzel 1986 for a review). These models cannot adequately account for the complexity of dynamical and microphysical processes leading to the formation of individual precipitation events. A GCM, on the other hand, can.

The two key questions addressed in the modeling studies are (1) can the GCMs reproduce the main features of the δD and $\delta^{18}O$ distributions in global precipitation and water vapor for present-day climate, and (2) how do such distributions change under different (e.g., ice age) climate conditions, and how do they support our interpretations of paleodata? A third important objective, which is internal to the modeling approach, is to understand how the GCM results relate to those obtained with the simpler isotopic models.

[1] HDO and $H_2^{18}O$ concentrations are denoted hereafter in δ units, or differences per mille, i.e., $\delta = 1000$ (R_{sample} $-R_{SMOW}$ /R_{SMOW}, where R_{sample} and R_{SMOW} are the isotopic concentrations of the sample and of SMOW (Vienna Standard Mean Ocean Water).

Water isotopes are transported in the GCM via the same processes that transport nonisotopic model water, i.e., surface evaporation and dewfall, water vapor transport and diffusion, atmospheric condensation, subsequent evaporation of falling precipitation, and ground hydrology. Isotopic fractionation is modeled at every change of phase in the water cycle due to the slightly different saturation vapor pressures and molecular diffusivities of $H_2{}^{16}O$ and the isotopic species. The LMD and NASA/GISS tracer models employ similar modeling philosophies; the main differences in the two models, however, stem from differences in their dynamical and physical parameterizations (Joussaume et al. 1984; Jouzel, Russell et al. 1987; Koster et al. 1988).

Simulations of the water isotope cycles have been performed with both the LMD and GISS GCMs for both modern (Joussaume et al. 1984; Jouzel, Russell et al. 1987) and LGM climatic conditions (Joussaume and Jouzel 1987; Joussaume 1989b; Joussaume and Jouzel 1993; Jouzel et al., future manuscript). CLIMAP (1981) data sets for sea surface temperature, sea ice, and ice sheets were used for the LGM boundary conditions.

Figure 10.1 and Table 10.1 briefly summarize the simulation results, focusing on the annual mean relationship between the isotope content of precipitation and the surface air temperature. The linear regressions for temperatures above and below 15°C are provided. The higher range (T > 15°C) encompasses tropical and equatorial sites, for which $\delta^{18}O$ in precipitation is controlled mostly by precipitation amount and not by temperature (Dansgaard 1964); the observed behavior is reproduced well in both GCMs. Below 15°C, the models reproduce the observed linear relation between $\delta^{18}O$ and temperature, with excellent agreement between the observed and predicted gradients. Furthermore, the linear relationship between δD and $\delta^{18}O$ seen in IAEA (1981) observations and suggested by Craig (1961) is reproduced by both GCMs. These results convincingly demonstrate the ability of the GCMs to reproduce the main features of δD and $\delta^{18}O$ in global precipitation.

For the ice-age climate, isotope concentrations in precipitation exhibit, in middle to high latitudes, a dependency with temperature quite similar to that simulated for present-day conditions (Figure 10.1). In high latitudes, the simulated difference (LGM present day) in $\delta^{18}O$ is in good agreement with ice-core data. The variability in the predicted temperature–isotope relationship, however, calls for some caution in applying regional GCM results to the interpretation of paleodata.

Table 10.1 Comparison between the observed and predicted present-day and ice age $\delta^{18}O$ surface air temperature, T_a, relationships ($T_a < 15°C$) for the LMD and GISS GCMs (annual means). The correlation coefficents, r, are included.

	Present-day	Ice Age
Observations	slope 0.64; r = 0.96	
LMD GCM	slope 0.57; r = 0.88	slope 0.65; r = 0.92
GISS GCM	slope 0.59; r = 0.96	slope 0.63; r = 0.97

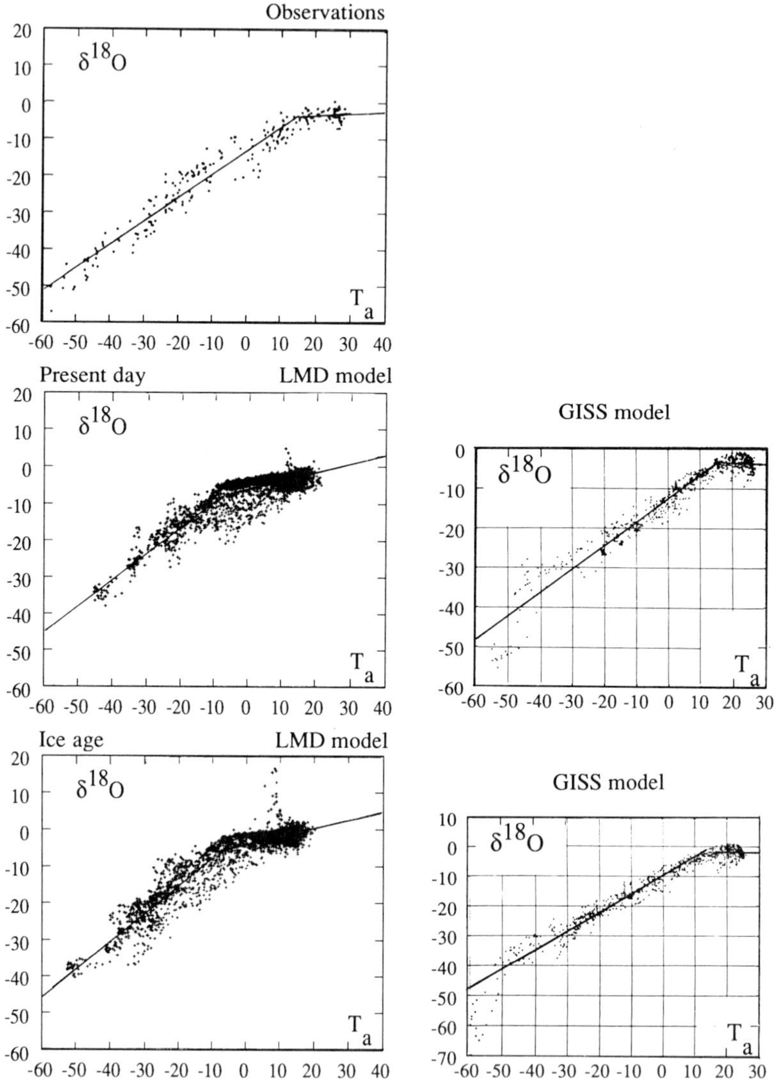

Figure 10.1 Annual mean $\delta^{18}O$ in precipitation as a function of mean surface air temperature from observations (adapted from Jouzel et al. 1987) and from the present-day and ice age simulations performed with the LMD and GISS GCMs. Regression lines are also shown (see Table 10.1).

DESERT DUST

GCMs are appropriate tools for examining the global-scale pick-up and fallout of desert dust, given the long distances traveled by dust particles. Much of the initial motivation for incorporating dust tracers into the LMD GCM came from analyses of

Greenland and Antarctic ice-core data, which show a strong increase in the deposition of continental dust during Last Glacial Maximum.

Joussaume (1990) fully describes the modeling of the desert dust cycle with the LMD GCM. Dust is picked up by simulated surface winds over all desert areas, with the extent of the dust source modeled as a function of soil moisture. The dust is diffused and transported by the model winds, and it is finally removed from the atmosphere by gravitational settling, interception at the surface via turbulence, or rainout. This first experiment considered only one size range of particles (around 1 μm) and did not include interactions with radiation and cloud physics.

Simulations with dust tracers have been performed for both present-day (Joussaume 1990) and LGM (Joussaume 1993) climates under both February and August conditions. One of the main findings is a strong seasonal behavior in the dust plumes, with twice as much airborne dust in August as in February. This is essentially due to seasonality in the vertical distribution of dust; in the summer season, the uplift of dust is stronger due to higher vertical instability, leading to a reduced efficiency of dry deposition. Some of the features in the present-day simulated dust fields, such as the long-range transport of Saharan dust over the northern tropical Atlantic Ocean in August, are in qualitative agreement with the few available observations. Other simulated features, however, disagree with the observations, perhaps due to deficiencies in the simulated climate itself, particularly in the wind fields.

For the LGM simulation, the model predicts only a small increase in globally averaged dust source strength (+ 15%) and atmospheric dust content (+ 8%) compared to present-day conditions. Larger changes in atmospheric dust content (up to a twofold increase) are simulated over such regions as the tropical Atlantic Ocean and Europe due to changes in the atmospheric circulation. The simulated change in dust deposition over East Antarctica (Figure 10.2) is much too weak: the dust removal flux in the model is reduced by half which, when scaled by the reduced snow accumulation, leads to no change in the dust concentration of snow, whereas observations indicate a ten- to twentyfold increase (Petit et al. 1981; De Angelis et al. 1987). Similarly, in Greenland, the model predicts a fourfold increase in dust concentration, which is much less than the hundredfold increase observed at Camp Century (Thompson 1977).

Joussaume (1993) suggests that these discrepancies might result from errors in the simulation of source areas. In the GCM simulations, dust falling over East Antarctica essentially comes from Australia for both present-day and LGM climates; the weak change during the LGM reflects weak changes in this source. Mineralogical studies of dust samples extracted from ice cores (Gaudichet et al. 1992), however, suggest a change in the source location during the LGM. A more recent study of measured strontium and neodime isotopes identifies an Argentine origin for LGM dust (Grousset et al. 1992). Thus, the GCM's problems may be related to its inability to simulate a strong South American source (Figure 10.2) during the LGM.

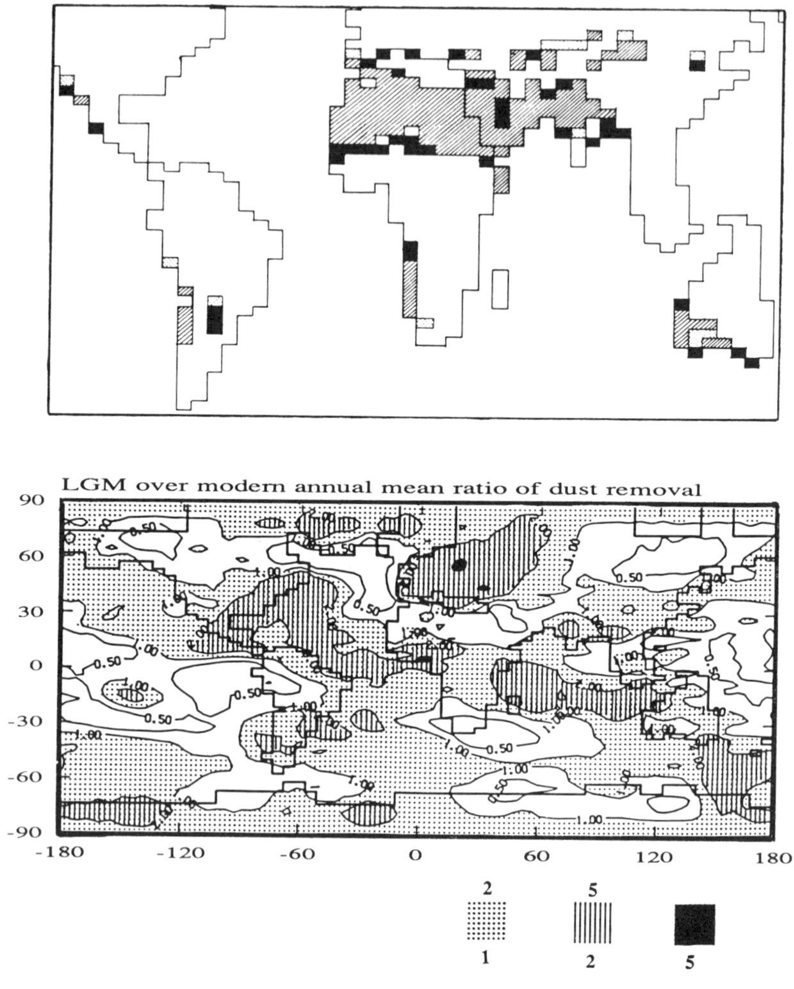

Figure 10.2 Simulations of the desert dust cycle with the LMD GCM for present-day and LGM conditions. Above: simulated extent of the desert source regions—dotted areas for present-day only; black areas for LGM only; and hatched areas for both climates. Below: total annual mean (August and February) dust removal—ratio of simulated ice age values to simulated present-day values. Isolines are at 1/2, 1, 2, and 5 (reproduced from Joussaume 1989b).

THE ORIGIN OF PRECIPITATION

Precipitation is one of the most important manifestations of climate, yet it is difficult to reconstruct the history and origin of air masses producing precipitation in a given area from measurements. The GCM tracer approach, however, is especially well-suited to the "origins of precipitation" problem.

In GCM studies, a set of grid squares on the Earth's surface is defined as a tracer source region, and the tracer is evaporated from the region at the calculated water evaporation rate. The tracer moves through the atmosphere via the same processes that transport the model water. The distribution of tracer precipitation indicates, after a suitable start-up time, the final destinations of the evaporated tracer water and thereby quantifies the lateral scales of vapor transport in the GCM. Naturally, the atmospheric part of the model is relatively simple, since such processes as isotopic fractionation and gravitational deposition need not be considered.

Using this approach, Joussaume et al. (1986) and Koster et al. (1986) studied the contribution of various source regions to local precipitation and the degree to which oceanic moisture penetrates continental regions. Koster et al. (1986), who focused on precipitation in the Northern Hemisphere, found that the percent contribution of an important continental source to mid-latitude or high-latitude precipitation is usually largest in the summer and smallest in the winter, in phase with the seasonal cycle of continental evaporation. This trend is lost in the Subtropics and even reversed in the Sahel. The global study of Joussaume et al. (1986) demonstrated the important contribution of continental evaporation to July precipitation over North America and Eurasia. Both studies (Joussaume et al. 1986; Koster et al. 1988) demonstrated the strong contribution of Atlantic water vapor to precipitation in central America, the southeastern United States, West central Africa, and the Atlantic coast of Europe (Figure 10.3). The comparison of the two sets of results exhibits, however, a more extensive lateral influence of oceanic vapor in the LMD model which may come from the use of a more diffusive advection scheme for the tracers in this model and/or differences in the wind fields. The strong difference over South America only reflects a difference in the source region which is restricted to the Northern Hemisphere in the GISS simulation but not in the LMD one.

A study with the GISS tracer model (Druyan and Koster 1989) examined the sources of Sahelian precipitation in wet and dry years. It was found that the dry conditions were caused by Northern African moisture convergence patterns and by low recycling rates, not by a lack of moisture advection from ocean to continent. The GISS model was also used to establish an average source temperature of 12°C for the GCM's Antarctic precipitation (Koster et al. 1992). This study suggests that differences in source temperature do not explain much of the scatter observed in the δD–temperature relationship.

CONCLUDING REMARKS

Rather than establishing a list of work to be done with paleoclimatic tracers in GCMs (the list would be quite long), we would like to draw attention to two key aspects of tracer modeling. First, a proper tracer transport scheme is critical to tracer modeling

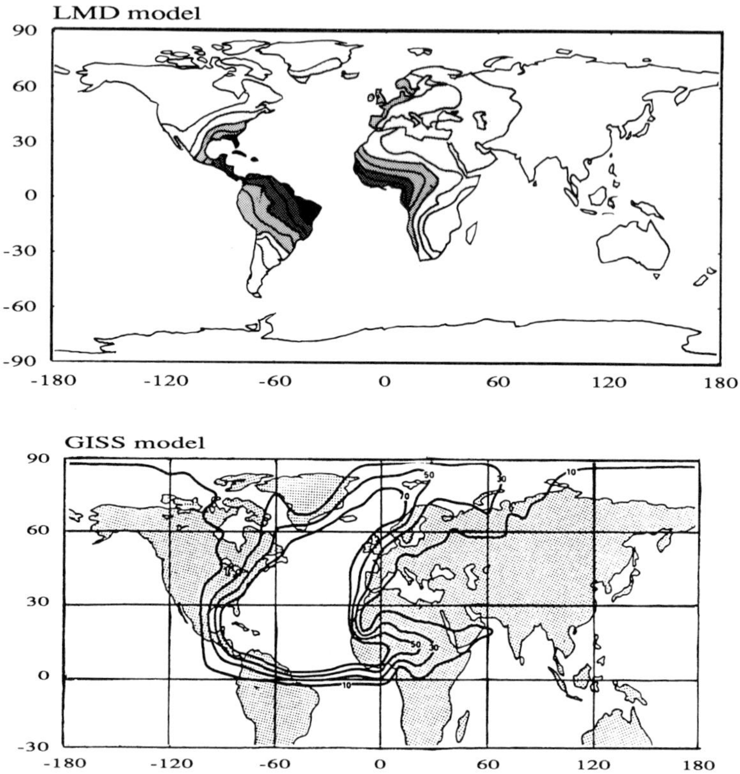

Figure 10.3 Percentage of local July precipitation derived from the Atlantic source region in the LMD (above) and GISS (below) GCM experiments. The Atlantic source region is restricted to the Northern Hemisphere in the GISS simulation but expands down to 30°S in the LMD simulation. In the latter, isolines are at every 10%: light shading 10–30%; medium shading 30–50%; heavy shading 50–70%; black more than 70% (adapted from Joussaume et al. 1986). GISS experiments isolines 10, 30, 50, and 70% (adapted from Koster et al. 1988).

(Joussaume et al. 1984). A recent study by Jouzel et al. (1991) clearly demonstrates that the $\delta^{18}O$ fields produced over Antarctica under different transport schemes are significantly different. Second, it is difficult to define proper source regions for dust and to parameterize dust mobilization. Joussaume (1990) discusses this in full and also draws attention to the need for separating the dust into various size classes. This is now being done with the NASA/GISS model (C. Genthon, pers. comm.).

In conclusion, climatic tracers in GCMs can be used effectively in studies of various aspects of past global changes, as illustrated by the above examples. We note that this approach has also been applied to studies of tritium (HTO) geochemistry (Koster et al. 1989) and may be extended to other tracers such as volcanic aerosols and cosmogenic isotopes (^{10}Be, ^{36}Cl).

ACKNOWLEDGEMENTS

We would like to thank all of the scientists at LMD, GISS, and other institutions who have participated in the studies and reviewed in this article. The project with the LMD model was supported by the Programme National d'Étude de la Dynamique du Climat and the computing work performed at the Centre de Calcul Vectoriel pour la Recherche.

REFERENCES

CLIMAP. 1981. Seasonal reconstructions of the Earth's surface at the Last Glacial Maximum. Map and Chart Series MC–36. Geological Society of America.

Craig, H. 1961. Isotopic variations in meteoric waters. *Science* **133**:1702–1703.

Dansgaard, W. 1964. Stable isotopes in precipitation. *Tellus* **16**:436–468.

De Angelis, M., N.I. Barkov, and V.N. Petrov. 1987. Aerosol concentrations over the last climatic cycle (160 kyr) from an Antarctic ice core. *Nature* **325**:318–321.

Druyan, L.M., and R. Koster. 1989. Sources of Sahel precipitation for simulated drought and rainy season. *J. Climate* **2**:1438–1446.

Gaudichet, A., M. de Angelis, S. Joussaume, J.R. Petit, Y.S. Korotkevitch, and V.N. Petrov. 1992. Comments on the origin of dust in East Antarctica for present and ice-age conditions. *J. Atm. Chem.* **14:**129–142.

Grousset, F.E., P.E. Biscaye, M. Revel, J.R. Petit, K. Pye, S. Joussaume, and J. Jouzel. 1992. Antarctic (Dome C) ice-core dust at 18 k.y. B.P.: Isotopic constraints on origins. *Earth Planet. Sci. Lett.* **111:**175–182.

International Atomic Energy Agency (IAEA). 1981. Statistical treatment of environmental isotope data in precipitation. Technical report series 206. Vienna: IAEA.

Joussaume, S. 1989a. Desert dust and climate: An investigation using an atmospheric general circulation model. In: Paleoclimatology and Paleometeorology: Modern and Past Patterns of Global Atmospheric Transport, ed. M. Leinen and M. Sarnthein, pp. 253–263. Dordrecht: Kluwer.

Joussaume, S. 1989b. Simulations du climat du dernier maximum glaciaire à l'aide d'un modèle de circulation générale de l'atmosphère incluant une modélisation du cycle des isotopes de l'eau et des poussières d'origine détertique. Thèse d'Etat, Université Paris VI.

Joussaume, S. 1990. Three-dimensional simulations of the atmospheric cycle of desert dust particles using a general circulation model. *J. Geophys. Res.* **95**:1909–1941.

Joussaume, S. 1993. Paleoclimatic tracers: An investigation using an atmospheric general circulation model under ice-age conditions. 1. Desert dust. *J. Geophys. Res*. **98:**2767–2805.

Joussaume, S., and J. Jouzel. 1987. Simulations of paleoclimatic tracers using atmospheric general circulation models. In: Abrupt Climatic Changes: Evidence and Implications, ed. W.H. Berger and L.D. Labeyrie, pp. 369–381. NATO ASI series. Dordrecht: Reidel.

Joussaume, S., and J. Jouzel. 1993. Paleoclimatic tracers: An investigation using an atmospheric general circulation model under ice-age conditions. 2. Water isotopes. *J. Geophys. Res.* **98:**2807–2830.

Joussaume, S., J. Jouzel, and R. Sadourny. 1984. A general circulation model of water isotope cycles in the atmosphere. *Nature* **311**:24–29.

Joussaume, S., R. Sadourny, and C. Vignal. 1986. Origin of precipitating water in a numerical simulation of the July climate. *Ocean-Air Int.* **1**:43–56.

Jouzel, J. 1986. Isotopes in cloud physics multiphase and multistage condensation processes. In: Handbook of Environmental Isotope Geochemistry, the Terrestrial Environment, ed. P. Fritz and J.C. Fontes, vol. 2, pp. 61–112. New York: Elsevier.

Jouzel, J., C. Lorius, J.R. Petit, C. Genthon, N.I. Barkov, V.M. Kotlyakov, and V.M. Petrov. 1987. Vostok ice core: A continuous isotope temperature record over the last climatic cycle (160,000 years). *Nature* **329**:403–408.

Jouzel, J., L. Merlivat, and C. Lorius. 1982. Deuterium excess in an East Antarctic ice core suggests higher relative humidity at the oceanic surface during the last glacial maximum. *Nature* **299**:688–691.

Jouzel, J., G.L. Russell, R.J. Suozzo, R.D. Koster, J.W.C. White, and W.S. Broecker. 1987. Simulations of the HDO and $H_2^{18}O$ atmospheric cycles using the NASA/GISS general circulation model: The seasonal cycle for present-day conditions. *J. Geophys. Res.* **92**:14739–14760.

Jouzel, J., R.J. Suozzo, R.D. Koster, J.W.C. White, and W.S. Broecker. 1991. Simulations of the HDO and $H_2^{18}O$ atmospheric cycles using the NASA/GISS general circulation model: Sensitivity experiments for present-day conditions. *J. Geophys. Res.* **96**:7495–7507.

Koster, R., W. Broecker, J. Jouzel, R. Suozzo, G. Russell, D. Rind, and J. White. 1989. The global geochemistry of bomb-produced tritium: General circulation model compared to available observations and traditional interpretations. *J. Geophys. Res.* **94**:18305–18326.

Koster, R.D., P.S. Eagleson, and W.S. Brocker. 1988. Technical Report 317, Ralph M. Parsons Laboratory, Dept. of Civil Engineering. Cambridge, MA: MIT.

Koster, R., J. Jouzel, R. Suozzo, and G. Russell. 1992. Effect of evaporative source temperature on HDO concentrations in Antarctic precipitation. *Clim. Dynam.* **7**:195–203.

Koster, R., J. Jouzel, R. Suozzo, G. Russell, W. Broecker, D. Rind, and P. Eagleson. 1986. Global sources of local precipitation as determined by the NASA/GISS GCM. *Geophys. Res. Lett.* **43**:121–124.

Petit, J.R., M. Briat, and A. Royer. 1981. An ice-age aerosol content from East Antarctic ice-core samples in past wind strength. *Nature* **293**:391–394.

Thompson, L.G. 1977. Microparticles, ice sheets and climate. Report 64, OSU Institute of Polar Studies.

Standing, left to right:
J. White, G. Farquhar, N.J. Shackleton, M. Lautenschlager, B. Stauffer, R. Bradley

Seated, left to right:
E. Bard, B. Molfino, S. Joussaume, F. Sirocko

11

Group Report: Evaluating Strategies for Reconstructing Past Global Changes—What and Where Are the Gaps?

R. BRADLEY, Rapporteur

E. BARD, G. FARQUHAR, S. JOUSSAUME,
M. LAUTENSCHLAGER, B. MOLFINO,
E. RASCHKE, N.J. SHACKLETON,
F. SIROCKO, B. STAUFFER, J. WHITE

WHAT PALEODATA ARE MOST URGENTLY NEEDED?

Although great strides have been made in our understanding of past global changes, it is clear that many uncertainties remain. In assessing where the gaps are in our knowledge of past climates, it is instructive to first begin with a review of what *is* generally known. To this end, Appendix 11.1 provides an overview of the primary proxy data sources, together with some comments on reliability of the paleodata and the limitations of each approach. The list is not exhaustive and includes only those records which can provide a continuous stratigraphy. Thus, records of episodic events (such as moraines, shorelines, etc.) are not considered. Although some types of information studied may not provide a very definitive paleoclimatic signal, when considered in combination with other parameters studied in the same record, confidence in the interpretation is often considerably enhanced. Multiparameter reconstructions are therefore desirable whenever possible.

Appendix 11.1 reveals the breadth of effort currently being applied to the reconstruction of past global change. However, such a table cannot provide a sense of how the spatial coverage of data varies through time. Regardless of the period of interest, there are always large areas of the globe which are underrepresented or simply missing in any data set. Often these gaps reflect the difficulty of obtaining appropriate proxies

Global Changes in the Perspective of the Past
Edited by J.A. Eddy and H. Oeschger © 1993 John Wiley & Sons Ltd.

for paleoclimatic reconstruction in those regions rather than a lack of adequate sampling. For example, in spite of considerable efforts, there have been few cases where an unequivocal climatic signal has been extracted from trees in the lowland intertropical zone, yet this is where the bulk of the world's forests are found (Jacoby 1989; Jacoby and D'Arrigo 1990). Consequently, there are very few high-frequency records from this important region. The need for extending geographic coverage to as many parts of the world as possible requires that all possible efforts be made to investigate potentially useful new methods that might be applied in such areas.

Elsewhere, increasing the density of known proxies must continue to be a high priority. This is particularly important for the reconstruction of past global changes at higher frequencies, where natural variability may be subsumed within the noise level of each proxy record. There is a scale-dependency such that, in order to examine the higher-frequency climate signals of global significance, one requires data from a large number of records with a wide geographical distribution to separate out signals that are only of regional significance.

One variable that it might be possible to add to the inventory of paleoclimatic reconstructions is humidity, based on stable hydrogen (δD) and oxygen ($\delta^{18}O$) isotopes in plant cellulose. δD values are more strongly affected by the phase-change fractionation relative to the diffusion fractionation during the evaporation of leaf water. This situation is reversed for $\delta^{18}O$ values, which respond more strongly to the isotopic fractionation induced by diffusion. As the diffusional gradient is partly imposed by the humidity gradient from the leaf water to the air, humidity is a stronger signal in the $\delta^{18}O$ record than in the δD record of plant cellulose. By combining the two isotopic records, and using appropriate scaling between the two isotopes, it should be possible to focus attention primarily on the humidity signal while decreasing the impact of other signals, such as the isotopic composition of the source water for the plant. One important caveat in this approach is the assumption that the isotope record in cellulose responds to changes in the isotopic composition of leaf water. There is some debate concerning the compartmentalization of water in a leaf (and hence which compartment to model for isotopes) and also whether the isotopic signal is reset via equilibration with stem water (equivalent to source water) outside of the leaf. We recommend further research on this matter to try and resolve the uncertainties.

WHAT ARE THE MAJOR UNCERTAINTIES IN UNDERSTANDING PROCESSES OF GLOBAL CHANGE?

How Has the Hydrological Cycle Changed over Time?

The hydrological cycle is the crucial link between the various components of the climate system. It affects the radiation budget through direct water vapor greenhouse effects and the formation of clouds and, in addition, ice-sheet buildup and ocean dynamics through surface hydrological budgets. It is also the major source of uncertainty for modeling future climate, especially the question of cloud feedbacks in atmospheric general circulation models (GCMs) and the net freshwater flux in the ocean GCMs (Cubasch and Cess 1990; Duplessy and Maier-Reimer, this volume).

Unfortunately, we have very few tools to measure directly any of these variables in the past. Paleovegetation can be used to get an estimate of precipitation changes on the continents (Frenzel 1991; Bartlein et al. 1984). In addition, the calculation of paleosalinity by Duplessy et al. (1991) will provide a data base to localize changes in river runoff, and the patterns of precipitation minus evaporation. However, much more research on these problems is needed.

One of the most significant unknowns in our understanding of past climate is the distribution of cloudiness and its changes through time. Clouds play such an important role in the global energy balance that information on changes in this parameter would be extremely useful. At present, however, there is no way of obtaining such information. There may be some biochemical indicators in plant material (such as violoxanthin and zeaxanthin, which protect plant leaves from photoinhibition by excess light), but whether these can be related to cloudiness and/or radiation is not really known; furthermore, it is not known whether such an indicator could be derived from paleorecords. Another possible approach might be the $\delta^{13}C$ composition in corals and forams that live with endosymbiotic algae (Fairbanks and Dodge 1979; Patzold 1984) and obviously respond to the availability of light. However, nutrients will exert first-order control, so this approach may not prove useful. Any progress in reconstructing cloudiness or radiation conditions would be welcome, though it may be an insoluble problem. The best solution at present is to address more broadly the question of the hydrological cycle by seeking information on paleo-precipitation, evaporation, and run-off regimes. In particular, we need much more coring and, especially, precise dating of sediment cores from the shelves and upper continental slopes of the major world rivers to recontruct paleowater discharge and its effect on the salinity of the ocean surface.

Is There a Significant Direct Effect of Changed Atmospheric CO_2 (Past or Future) on Natural Terrestrial Ecosystems?

Plant physiological theory suggests that CO_2 "fertilization" effects should be strongest in water-limited environments, but should also recur where light is limiting, and at high temperatures. The theory also suggests that effects should be least at cool temperatures and also where nutrients are most limiting.

The extrapolation from theory to experiment has been made in the short term (say, months) for enhanced CO_2 in glasshouse studies and, in a few cases, in outdoor studies. As expected, doubling of CO_2 had little long-term effect on growth or carbon accumulation in tundra vegetation, but citrus trees in Arizona and salt-marsh plants in North Carolina showed large and continuing effects. There have only been a limited number of chamber studies at lower than present levels of CO_2 (e.g., Masle et al. 1990) and none in natural environments. Since there were much lower CO_2 levels during the last glacial maximum, it would be useful to conduct more of these studies at low CO_2 concentrations.

Despite the thousands of reports on growth enhancement of indivdual plants at high CO_2, especially in herbaceous plants, there are no unambiguous reports of the effects on ecosystems of the ~80 ppm increase in atmospheric concentration that has occurred since industrialization. In part this may be because of the other extensive changes which humans have wrought (land use, pollution, etc.). Alternatively, this may be because of methodological problems, especially the difficulties of maintaining ecological change. Possibly the only measure that has been tested is tree-ring width. Although the most recent studies of trees growing in "clean" environments—Huon pines in western Tasmania—reveal the largest growth increments of the last 1500 years occurred in the last few decades, it is difficult to separate these effects from those of increasing temperature, since Tasmanian summers have become warmer this century.

The seasonal variation in atmospheric CO_2 has also been increasing. Again, it is unclear whether this is caused by increasing terrestrial activity or whether this is related to CO_2 concentration alone. If significant direct CO_2 effects on ecosystem functions do occur, then a modern climate–plant relationship cannot be used to infer (a) future or past vegetation or (b) future climate. Greater emphasis needs to be given to physiological and paleobotanical observations of low CO_2 environments and to the modeling of ecosystems functioning under such conditions. More specifically, were vegetation patterns at the last glacial maximum determined in part by low CO_2 concentration, and, if so, how large were these effects compared to those of climate?

What Is the Climatic Significance of Isotopic Changes in Plant Material?

Considerable uncertainty surrounds the physiological effects on the isotopic composition of plants, particularly during large-scale pCO_2 changes, such as during the last glacial–interglacial transition and during the last 100 years. How plants respond, for example, to increased pCO_2, whether by increasing growth (fertilization) or maintaining an assimilation rate dictated by other limiting factors, will influence the $\delta^{13}C$ of the plants. This could be particularly important to the $\delta^{13}C$ interpretation of the last glacial in foraminifera, etc., as the average $\delta^{13}C$ of plants could change by up to a few permille. To resolve this problem, a better characterization of the $\delta^{13}C$ of terrestrial plants during the last glacial is needed; we do not view this as a particularly difficult task.

How Well-preserved Is the Climatic Signal in Marine Sediments?

Although the process of carbonate dissolution is well known qualitatively, it must be quantified. The circulation of ocean deep water determines gradients in carbonate dissolution that vary in time and space. The effect of the gradients is to dissolve the sediment fauna partially, thereby distorting the signal of what actually lived at the sea surface. Ideally, one should quantify the dissolution in order to estimate the original faunal composition. Dissolution paleoindicators should be independent from the measured parameters of choice (Appendix 11.1). An example of this can be found in the recent work of Matsuoka et al. (1991) who use a fragmentation index of *Calcidiscus leptoporus*, a coccolithophorida, to infer when sediments are well and poorly preserved. The indicator is applicable in all oceans, from approximately 45°S to 60°N. It has been tested temporally against the *G. menardii* (a foraminifera) fragmentation index and variations in calcium carbonate, and spatially in a suite of surface sediment samples from the Equatorial Atlantic, which vary over depth. The advantage of the *C. leptoporus* index is that it was based on a phytoplankton, while the majority of isotopic and faunal paleotemperature estimates from the marine sediments are based on zooplankton (Appendix 11.1). Thus this is an independent measure which, once calibrated, can be used to unravel the effect of dissolution on paleoclimatic signals.

How Can Data and Models Be Combined to Understand Past Changes in Climate?

To understand the processes involved during the glacial to interglacial transition better, coupled atmosphere–ocean–ice sheet–biosphere global models are required. Such models are under development, with some simplified versions already available (e.g., two-dimensional models). In the coming years, global atmospheric models will, however, still be used for snapshot studies to understand past changes in atmospheric circulation better and to test their ability to reproduce climatic changes. In this regard, good spatial coverage of paleodata for snapshot periods in the past are still needed, either for boundary conditions (e.g., ice sheets, SSTs) or for validation of model outputs (e.g., precipitation, land–surface–air temperature). In the future, when the use of coupled atmosphere–ocean models becomes more appropriate, some of the fields originally used for boundary conditions will become validation tools, requiring a less dense global coverage.

The study of climate over the last 500 years using global models is more difficult since the forcings are not clearly known. This problem can be addressed either through sensitivity experiments (e.g., solar constant or SST changes) or by global-coupled atmosphere–ocean models, over several centuries, to study the internal variability of the coupled system and to compare simulations with data in that time window.

HOW CAN DIFFERENT TYPES OF RECORDS BE CORRELATED?

It is well known that one of the impediments to using geological data to make global reconstruction of past conditions is the difficulty of correlating records from different environments and regions. It has become possible to correlate sedimentary records from almost all areas of the ocean basins reliably, primarily through the marine $\delta^{18}O$ record. Thus the challenge is usually seen in terms of correlating terrestrial and marine sequences. Several categories of approach to this problem may be distinguished:

1. using events external to the climate system;

2. seeking terrestrial environmental indicators in sediments of the adjacent ocean;
3. assuming similar climatic controls in the records on the continent and nearby sea surface;
4. using comparable dating methods to obtain chronologies for all sequences.

Some methods for Pts. (1) and (2) are given in Tables 11.1 and 11.2 below. As regards Pt. (3), a few unconventional examples are given in Table 11.3; more straightforward geological approaches such as relating low sea level, river incision, and glacial till deposits to glaciation are not. Dating methods are outlined in Table 11.4. We recommend that more attempts be made to link terrestrial and marine sequences using approaches along the lines of those suggested in Tables 11.2 and 11.3.

Table 11.1 Correlation tools external to the climate system.

	Precision	Limitations	Recording Environment
Volcanic ash layers	~1 yr	distribution pattern different from each event; not global	general (ice, marine, and terrestrial sediment)
^{10}Be production peaks		at present only clearly identified in ice cores	ice, some sedimentary environments
Magnetic reversals	~10^3 yr	rare; least well established >780,000 B.P.	sediments containing magnetic particles
Magnetic excursion	~10^3 yr	record poorly validated at present	sediments containing magnetic particles
Secular variations	~10 yr	only useful if determined in continuous sequence with <10 yr resolution	sediments containing magnetic particles

WHAT SAMPLING STRATEGIES SHOULD BE EMPLOYED IN ANALYZING PALEODATA?

In a diverse but limited group of workers, it is dangerous to make general recommendations regarding sampling strategies for improving our understanding of paleoclimate. However, two aspects deserve mention. First, a large proportion of the available records are undersampled, with three consequences: there is a general loss of information; the bandwidth accessible for any time series analysis is reduced, and the true noise level cannot be assessed. Geologists often feel that it is only important to reveal the big signal, but any analytical approach requires optimal resolution.

Table 11.2 Continental indicators for correlating to marine sedimentary sequences.

	Continental Event Recorded	Limitations
Terrestrial dust	Aridity and/or windiness	In ocean sediments only detectable near sources and in appropriate wind trajectory, but easily measured in ice
Pollen from terrestrial continental vegetation	Broad-scale vegetation	Only measurable near margins; differential transport/preservation
Iceberg meltout	Glaciation	Distribution largely controlled by marine environment
Lake diatoms	Aridity in previously wetter environments	Unclear whether peaks represent driest times or times at which lakes are retreating fastest
Plant fragments, especially phytoliths	Wind, vegetation type	Only detectable near sources and in appropriate wind trajectories
River-controlled salinity	Pluvial events, deglaciation	Sediments near major rivers unstable; locus of salinity; minimum changes with sea-level changes
Ocean $\delta^{18}O$	Integrated global glaciation	Very generalized
Ocean $\delta^{13}C$	Integrated global vegetation and soil	Very generalized and poorly understood

Table 11.3 Climate-dependent tools for correlation.

	Example of Applications	Caveats
Air temperature	Ice D/H to marine SST; pollen-estimated temperature to marine SST	Misuse could obscure mechanisms of climate change
Atmospheric pCO_2	Ice-core pCO_2 to marine $\delta^{13}C$ to plant stomatal concentration	Plant fossil measures of pCO_2 need investigation
$\delta^{13}C$ in atmospheric CO_2	Plant $\delta^{13}C$ to ice-core $\delta^{13}C$ to marine plankton	Ice-core $\delta^{13}C$ record not available; controls on plant $\delta^{13}C$ and marine plankton $\delta^{13}C$ complex and need further work
Atmospheric methane	Ice-core CH_4 to low-latitude wetness	Natural CH_4 variations not understood, low-latitude wetness may not be the control
$\delta^{18}O$	$\delta^{18}O$ of atmospheric O_2 in ice sheets to ocean water $\delta^{18}O$	Mediated through marine photosynthesis; time constant $\sim 10^3$ years

Table 11.4 Approaches to dating paleorecords.

Sources	Measured Signals	Reliability	Range	References
Trees	annual rings	excellent	0–11,500 yr	Becker et al. 1991
Corals	annual bands	±1 yr	1000 yr	Patzold 1984
Lakes	varves	± a few %	0–13,000 yr	Stuiver 1971
Ice	annual layers		0–40,000 yr	Hammer 1989
Organic matter carbonates	^{14}C	±0.5% problems of calibration	0–40,000 yr	Bard et al., this volume
Carbonates	^{230}Th	problems of open system ±1%	0–300,000 yr	Bard et al. 1990
Carbonates	^{234}U	±10%	300,000–10^6	Ludwig et al. 1991
Volcanic deposits	K/Ar	±1%	0–∞	Van den Bogard et al. 1989
Burned minerals, loess	TL	±10%	0–10^6	Valladas et al. 1988
Teeth, loess carbonates	ESR *in situ*	±10%		Wintle 1990
Silicate of rocks	*in situ* cosmogenic nuclides	±20%	depends on the isotope; under promising development	Radtke et al. 1988; Grun 1990; Phillips et al. 1990

The second aspect is that we recommend more effort be devoted to those sections in which several different parameters can be estimated. It is often impossible to achieve this; however, we can learn from the example of ice cores, where of necessity a great number of records have been derived from each core because so few cores are available.

With limited capacities for analytical measurements, the question arises whether a recovered core should be analyzed continuously in equally spaced increments, or if special emphasis should be given to certain especially interesting sections.

There was general agreement that a survey with at least part of the available analytical techniques should be performed along the entire core, especially to look for special events (signals from volcanic erruptions) and possible time markers. However, detailed investigations are additionally needed in special sections of all available records. This matter has been discussed in some detail in previous reports (see Working Group Recommendations in Bradley 1991 and Eddy 1992), but we believe two "time windows" deserve particular attention: the last 500–1000 years and the transition from the last glaciation to the Holocene (about 16,000–10,000 yrs B.P.). Very detailed knowledge about the climate and its regional changes during the last 500 years will enable us to decide whether deviations of climatic events from mean values are within natural variablity or are a significant signal of anthropogenic influence.

The climatic change that will be faced within about the next 100 years is the same order of magnitude as during the transition from the last glaciation to the Holocene. The processes and the time sequence of anthropogenically caused change in the future will be different from the glacial to interglacial transition, but some basic processes will be the same. Thus, this interval is of particular relevance today and should be given a high research priority.

Appendix 11.1 Records of past global change. (Part 1: Ocean proxy data. Part 2: Nonmarine proxy data.) Criterion for inclusion of a measured parameter is that it provides a continuous signal suitable for time-series analysis. Limitations include spatial, temporal, or notable measure constraints. References are recent review articles or articles where reliability is derived. Compiled by B. Molfino with help from N.J. Shackleton, M. Sarnthein, L. Labeyrie, E. Bard, F. Sirocko, R. Bradley, J. Cole, A. Zwick, and F. Schweingruber.

Part 1: *Ocean proxy data*. Reliability refers to the statistical or analytical error of the proxy signal, if the measured parameter has been calibrated; otherwise, reliability refers to the analytical or statistical error of the measured parameter. Under reliability, "counting error" specifies the approximate 95% confidence interval for binomial proportions between 30–70% based upon the given number of samples, n; this statistical error in conjunction with the overall range of variation of the measured parameter offers some constraint on the reliability.

I. Ocean Proxy Data

	Source	Measured Parameter	Proxy Signal	Reliability	Limitations	References
A. Biological						
1. Biotal census (%)	Forams	Assemblages	SST	Standard error of estimate for the ocean-wide equations are: ±1.2°C for the Atlantic, ±1.3°C for the Indian, ±2.0°C for the Pacific, which explains 85–95% of the variance	Equations are ocean-calibration specific; in the Atlantic, paleoestimates suspect if monospecific polar assemblage dominates sample, since summer temperatures cannot be accurately predicted below 6°C	CLIMAP 1981
			Surface salinity	Standard error of estimate ±0.8‰ for Bay of Bengal, ±0.4‰ for the North Atlantic, which explains ~72% of the variance.	Spatially restricted in application	Cullen 1981; Kipp 1976
			Productivity	Standard error of estimate ±12 (for total Atlantic) to ±13.7 (for truncated Atlantic) $gCm^{-2}yr^{-1}$, which explains ~60% of the variance	Underestimates higher productivities (~ >120g C $m^{-2}yr^{-1}$)	Mix 1989; Sarnthein et al. 1992

Source	Measured Parameter	Proxy Signal	Reliability	Limitations	References
		Photic zone hydrography	Uncalibrated	Study restricted to tropical Atlantic (20°N–20°S) sediment samples; relationship "tested" using LGM tropical Atlantic data base and GCM model results of double wind stress run	Ravelo et al. 1990
	Assemblage dominance shift	Meridonal position of subtropical convergence	Dependent on spacing of cores	Documented for the STC of the Indian and South Atlantic; subpolar assemblage species more susceptible to dissolution than subtropical assemblage species	Prell et al. 1979
	G.pachyderma left-coiling	Polar front movement	Dependent on spacing of cores; counting error (n = 350) is ±5% and percentages range from 0–100%	Actual polar front position is based upon faunal (G. pachyderma accounting for 45–65% of fauna), flora (no coccoliths), and sediment (high ice-rafted detritus) evidence	Ruddiman and McIntyre 1977
	G. bulloides	Monsoonal upwelling	Uncalibrated; counting error (n = 300) is ±5% and percentages range from 0–50%	Resticted to Arabian Sea? Documented for last 30 Ka	Prell 1984
radiolaria	assemblages	SST	Standard error of estimate ±1.4°C for South Atlantic equation, ±1.9°C for Southern Indian Ocean, ±1.5°C for Antarctic Ocean	Each regional equation uses a different set of species; each set has eliminated deep-dwellers and combined species counts of easily misidentified genus	Morley 1979, 1989; CLIMAP 1981
	C. davisiana	Presence of winter sea-ice formation	Uncalibrated; counting error (n = 300) is ±5% and percentages range from 0–60%	High-latitude, deep-living species; relationship to Sea of Okhotsk modern conditions (low salinity in summer & fall; sea ice in winter and early spring) inferred, since C. davisiana is ∼5% in modern sediment samples, except for Sea of Okhotsk where it is >20%	Morley & Hays 1979

continues

I. Ocean Proxy Data (*continued*)

	Source	Measured Parameter	Proxy Signal	Reliability	Limitations	References
	Coccoliths	Assemblages	SST	Standard error of estimate ±1.8°C for the Atlantic Ocean	Applicable for 0–70 Ka only due to evolutionary changes in genus Emiliania and Gephyrocapsa	Molfino et al. 1982
		F. produnda	Nutricline dynamics	Uncalibrated; counting error (n = 500) is ±4% and percentages range from 10–70%	Restricted to tropical–subtropical waters; documented in equatorial Atlantic for the last 200 Ka, though species does extend to the Miocene	Molfino & McIntyre 1990
	Diatoms	Assemblages	SST	Standard error of estimate of ±1°C for Southern Ocean summer (0–10°C)	Equation bias if estimated silica loss exceeds ~70%, equating to poor temperature estimates of values >14°C	Pichon et al. 1993
			Opal dissolution	Standard error of estimate ±20%	Applicable in Southern Ocean with northern limit the subtropical convergence (approximately 12°C summer isotherm)	Bareille et al. 1992
		Melosira (per gm)	Aridity (winds)	Uncalibrated; counting precision 8% for abundances $> 5\ 10^6\ g^{-1}$, 20% for abundances $< 5\ 10^6\ g^{-1}$	Advective signal of freshwater diatoms from exposed lake beds; documented for low-latitude Atlantic for the last million years	Pokras & Mix 1987; deMenocal 1991
		Phytoliths (per gm)	Presence of tall-grass savanna (winds)	Uncalibrated	Advective signal of silicified grass cells; documented for low-latitude Atlantic for the last million years	Pokras & Mix 1985; deMenocal 1991
	Dinoflagellates	Assemblages	Salinity	•	Under development	•
2. Biogenous sediment (weight percent)	Bulk sediments	Organic carbon	Productivity	Analytical precision ±0.3% of measured value; range of values 0.1–6.0%	All production has potential for destruction overprint; limited range of values in marine sediment, but has no spatial restriction	Verardo et al. 1990

	Source	Measured Parameter	Proxy Signal	Reliability	Limitations	References
		Opal	Productivity	Analytical precision ±4.0% of measured value; range of values 3.0–100.0%	All production has potential for destruction overprint; spatially limited to high latitudes and equatorial divergence; measurement lower limit of 3% opal (values within 0–3% range due to silica leaching from clay)	Mortlock & Froelich 1989
		Nitrogen	Productivity	Analytical precision ±1.6% of measured value; range of values 0.1–6.0%	All production has potential for destruction overprint; limited range of values in marine sediment, but has no spatial restriction	Verardo et al. 1990
		Calcium carbonate	General preservation and positioning of CCD	Precision ±1.0% gasometric; precision ±0.1% coulometry	Records combination of production, dilution, and dissolution	Berger 1973; Verardo 1991
(fluxes)	Bulk sediments	Organic carbon	Productivity measure	Standard error estimate 9.7 gC m^{-2} yr^{-1}, which explains ~80% of the variance	Quality of statistical fit dependent on ability to determine flux	Sarnthein et al. 1988, 1992
3. Trace Elements	Benthic forams	Oxygen isotopes	Global ice volume	~ ±10 m sea level equivalent	Difficult to separate temperature effect	Shackleton & Opdyke 1973; Shackleton 1987
			Deep-water temperature	~ ±0.5°C	Difficult to separate ice volume effect	Labeyrie et al. 1987, 1992; Zahn & Mix 1991
			Deep-water salinity	~ ±0.5‰	Needs thermohaline hypothesis to separate temperature and salinity	Labeyrie et al. 1987, 1992; Zahn & Mix 1991

I. Ocean Proxy Data (*continued*)

Source	Measured Parameter	Proxy Signal	Reliability	Limitations	References
	Carbon isotopes	AOU (global average deep-water circulation)	$\sim \pm 10\%$ of saturation value	Affected by local carbon flux	Curry et al. 1988; Duplessy et al. 1988
	Carbon isotopes, ocean gradients	% NADW production	Uncalibrated	Concept of NADW endmember not always clear	Raymo et al. 1990
	Cadmium–calcium ratio	Phosphate in deep water	$\sim \pm 10\%$ of index value	•	Boyle 1990, 1992
	Barium–calcium ratio	Productivity	•	•	Lea & Boyle 1990
Planktic forams	Oxygen isotopes	Global ice volume	$\sim \pm 10$ m sea-level equivalent	Sensitive to temperature and salinity effects and, in particular, to meltwater spikes	Shackleton & Opdyke 1973, 1976; Shackleton 1987
		Surface temperature	$\sim \pm 1°C$	Difficult to separate salinity effect	Broecker 1986
		Surface salinity	$\sim \pm 0.5‰$	Difficult to separate temperature effect; salinity derived using biotic transfer functions estimates for SST	Duplessy et al. 1991
	Carbon isotopes	Surface ocean $\delta^{13}C$	$\sim \pm 0.1‰$	Local contaminating effects may confuse global signal	Curry & Crowley 1987; Sarnthein & Winn 1990
Benthic & planktic forams	Carbon isotope difference	Atmospheric CO_2	$\sim \pm 20$ ppm	True uncertainty much greater than analytical	Shackleton & Pisias 1985

	Source	Measured Parameter	Proxy Signal	Reliability	Limitations	References
		^{14}C difference	Ocean mixing ratio	$\sim \pm 200$ yr	Bioturbation is a very serious limitation	Shackleton et al. 1988; Andree et al. 1986
	Organic matter	Carbon isotopes	pCO_2 in surface waters	Calibrated relationship has r = 0.65	Useful in fully pelagic areas only	Rau et al. 1991; Westerhausen 1991
		^{37}UK (long-chain alkenones)	SST	$\sim \pm 1°C$	Applicable from 0–500 Ka	Brassel et al. 1986; Prahl & Wakeham 1987
	Corals	Oxygen isotopes	Temperature	$\pm 1°C$	Complications with salinity and rainfall	McConnaughey 1989
		^{14}C anomalies	Upwelling	•	•	Druffel & Suess 1983
		Cadmium	Nutrient level	•	•	Shen et al. 1987; Shen & Boyle 1988
		Barium	Nutrient level	•	•	Lea et al. 1989
		Strontium	Temperature	•	•	Smith et al. 1979
		$\delta^{18}O$	Rainfall, Southern Oscillation	•	Complications related to temperature effects	Cole & Fairbanks 1990
		Fluorescence	Continental run-off		•	Isdale 1984
B. Nonbiological	Bulk sediment	Terrigenous dust	Continental aridity	•	Pelagic or hemipelagic sediment only	Tiedemann 1991

continues

I. Ocean Proxy Data (*continued*)

Source	Measured Parameter	Proxy Signal	Reliability	Limitations	References
	Trace element isotopes	Position of specific wind systmes	Depends on element and isotope	Need ocean-wide maps during distinct time snapshots to observe clear gradients that point towards the source area of the element which allows us to distinguish fluvial from eolian proportions	Grousset et al. 1988
Bulk sediment	Grain-size parameter	Wind speed and direction	•	•	Sarnthein et al. 1981
	Magnetic susceptibility	Dust	Analytical error <<1.0% of measure value	Diagenesis (the chemical changes that occur within sediments as they become compacted after burial), i.e., magnetic susceptibility measures atmospheric dust and oceanic basalt-derived clays	Bloemendal & deMenocal 1989
	Quartz shards	Winds	•	Inferred to be eolian in origin if found in pelagic sediments undisturbed by turbidity currents and far from land	Leinen et al. 1986
	Pollen	Seasonal wind direction	•	•	Hoochiemstra et al. 1987
Bulk sediment	Ice-rafted detritus	Iceberg flow pattern	Errors estimate to be ±3% on mapped values (averaged values in isotopic stages)	Sand-sized volcanic ash in North Atlantic sediments during isotopic stages 2, 3, 4, and 5 (130 Ka) inferred to be delivered to the area by ice-rafting	Ruddiman 1977; Ruddiman & McIntyre 1977
Bulk sediment	% clay > modal grain size	windborne sediments	Qualitative (on–off)	If % clay < grain size, fluvial discharge dominant	Koopmann 1981

II. Nonmarine Proxy Data

	Source	Measured Parameter	Proxy Signal	Reliability	Limitations	References
II. Nonmarine Proxy Data *A. Lake Sediments* 1. Biological Content						
1.1 Biotal Census	Diatoms	assemblages	Trophic status; acidification	Very good but qualitative; certain species are excellent indicators	Limited understanding of ecological constraints	Agbeti & Dick-man 1989; Smol et al. 1986
			ice cover	Unknown	New approach requiring verification; restricted to polar lakes	Smol 1988
			temperature, salinity	Good, very good	Closed lakes	Fritz et al. 1991
			salinity/alkalinity	●	●	Gasse & Tekaia 1982; Battarbee 1986
	pollen	assemblages	temperature, precipitation	Very good	●	Webb et al. 1987; Hoochiemstra 1987
	ostracodes	assemblages	productivity, water chemistry	Good but qualitative	Limited understanding of ecosystem relationships	Carbonal et al. 1988
	rotifer (eggs)	assemblages	water temperature	Tendency of changes in environment only	Transfer functions very poorly understood	
	chrysophycean cysts	assemblages	●	●	Uncertain transfer functions	Smol et al. 1986

continues

II. Nonmarine Proxy Data (*continued*)

	Source	Measured Parameter	Proxy Signal	Reliability	Limitations	References
	chironomids cladocera	assemblages	water quality, pH, trophic state, & temperature	Excellent	•	Boucherle & Zully 1983; Hofmann 1986; Frey 1986
1.2 Biogenous Sediment	bulk sediment	plant pigments	productivity	Very good	Chlorophyll not stable; signal uncertain	Bengtsson & Enell 1986
1.3 Stable Isotopes	diatoms	$\delta^{18}O$	water balance temperature	Qualitative only		Siegenthaler & Eicher 1986
	lake marl, ostracodes	$\delta^{18}O$	precipitation $\delta^{18}O$, temperature, water balance	Semiqualitative; indicates tendency toward change	Transfer function subject to uncertainty; vital affect important	Siegenthaler & Eicher 1986
	lake marl, ostracodes	$\delta^{13}C$	productivity	Largely qualitative; indicate tendency toward change	Processes not well understood; many possible complications	
2. Nonbiological Content	bulk sediment	clay minerals	weathering conditions in catchments; temperature	Qualitative only	Bedrock control may swamp environ-mental signal	Weingarten et al. 1990
		carbonate	productivity	Good	Uncertainty of carbonate-forming processes during loss on ignition measurements	
		phosphorus	productivity	Good	Mobility of phosphorus in the sediment under different conditions	Williams et al. 1976
		nitrogen	indicator of nutrient content	Good	•	
		magnesium, aluminum, iron	indicators of erosion	Good	Dependence on redox condition	

	Source	Measured Parameter	Proxy Signal	Reliability	Limitations	References
		manganese, iron	oxygen condition	Tendency	•	
B. Eolian Sediments on Land	loess	magnetic susceptibility	processes of pedogenesis	Of measurements, very high; interpretation uncertain and qualitative	Debatable as to whether signal reflects accumulation rates or pedogenetic processes, or partly both	Kukla et al. 1988
		$\delta^{18}O$ in carbonate; $\delta^{13}C$ in carbonate and molluscs	precipitation	•	•	Goodfriend 1988
C. Cave Deposits	speleothems	$\delta^{18}O$	precipitation ^{18}O; temperature	•	•	Thompson et al. 1976; Harmon et al. 1978
		$\delta^{13}C$	vegetation cover, precipitation D/H	•		Schwarcz et al. 1976
D. Terrestrial Biota	tree rings	ring widths	temperature, precipitation/water balance	Signal strength varies with location	Restricted to areas where trees are climatically limited, but extreme events may also be identified in other cases	Cook et al. 1991; Duvick & Blasing 1981
			extremes of cold, drought			La Marche & Hirshboeck 1984
		late wood density	summer temperature	Varies with location; generally $\pm 0.5°C$	Restricted to high latitudes and high altitudes	Schweingruber et al. 1991; Briffa et al. 1990

continues

II. Nonmarine Proxy Data (*continued*)

	Source	Measured Parameter	Proxy Signal	Reliability	Limitations	References
		$^{13}C/^{12}C$	drought, humidity, precipitation, temperature and/or atmospheric $^{13}C/^{12}C$	Varies with location and complicating factors	Ratio measures physiological proesses that are related to proxies listed in a complex manner; sites must be selected where sources of variation are largely limited to the proxy of interest	Farquhar et al. 1989; Stuiver & Braziunas 1987
		$^{18}O/^{16}O$, $^{2}H/^{1}H$	temperature and humidity	Currently qualitative	Uncertainties due to physiological effects	
	packrat middens	pollen, macrofossils	precipitation	Very good where climatic limits on plant distribution are well known	Restricted to arid and semiarid environments	Betancourt et al. 1990
		$^{13}C/^{12}C$	atmospheric $^{13}C/^{12}C$	$\pm 0.1‰$	Relies on availability of C_4 cellulose material with no climate overprint	
	peat	pollen, macrofossils	temperature, precipitation	Excellent		
		stable isotope	temperature	Qualitative		
E. Continental Ice	Ice	stable isotopes $\delta^{18}O$, $\delta^{2}H$	surface temperature, accumula-tion rate	$\pm 1°C$; for decadal mean accumulation rates $\pm 1\%$	Dry snow zone and upper percolation zone; competing SST signal	Jouzel & Merlivat 1984
		deuterium excess	SST, humidity, wind speed at moisture source area	Unknown	Incomplete understanding of relative importance of SST, humidity, and wind	Johnsen et al. 1989
		electrical conductivity	Acidity of ice (volcanic signals) accumulation rate	Information is primarily qualitative; accumulation rates $\pm 1\%$	Dry snow zone; uncertain signal source	Hammer 1989

Source	Measured Parameter	Proxy Signal	Reliability	Limitations	References
	particles (amount and size distribution)	Storminess, aridity of surrounding land masses accumulation rate	Information is primarily qualitative; accumulation rates ±1%	Dry snow zone; percolation zone	Clausen & Langway 1989
	major ions (SO_4^{2-}, NH_3+, Cl^-,...)	volcanic signals, sea ice extension accumulation rate	Information is primarily qualitative; accumulation rates ±1%	Dry snow zone; upper percolation zone	Delmas & Legrand 1989
	reactive gases (H_2O_2, HCHO, ...)	oxidation capacity of the atmosphere	Allows at present only rough estimates	Dry snow zone	Delmas & Legrand 1989
	organic acids	transport of air masses, biomass burning	Only estimates at present	Dry snow zone	
	physical properties melt-layers, ...)	summer temperature accumulation rate	Information is primarily qualitative; accumulation rates ± 1%	Upper percolation zone	Shoji & Langway 1989
	radioisotopes (^{10}Be, ^{36}Cl, ...)	solar activity accumulation rate	Primarily estimates	Accumulation rate variability, residence time in atmosphere?	Lorius et al. 1989
air bubbles	gas composition (CO_2, CH_4, ...)	atmospheric composition	CO_2: ±10 ppm; CH_4: ±30 ppb	Dry snow zone	Siegenthaler, this volume
	gas content	altitude at ice surface	Currently unknown	Dry snow zone	
	isotopic composition of gases	carbon and water cycle	?	Dry snow zone	

REFERENCES

Agbeti, M., and M. Dickman. 1989. Use of lake fossil diatom assemblages to determine historical changes in trophic status. *Can. J. Fish. Aq. Sci.* **46**:1013–1020.

Andree, M., H. Oeschger, W.S. Broecker, N. Beavan, M. Klas, A.C. Mix, G. Bonani, H.J. Hofmann, M. Suter, W. Wölfli, and T.H. Peng. 1986. Limits on the ventilation rate for the deep ocean over the last 12,000 years. *Clim. Dynam.* **1**:53-62.

Bard, E., B. Hamelin, and R.G. Fairbanks. 1990. U/Th ages obtained by mass spectrometry in corals from Barbados sea level during the past 130,000 years. *Nature* **346**:456–458.

Bareille, G., M. Labracherie, L.D. Labeyrie, J.J. Pichon, and J.L. Turon. 1992. Biogenic silica accumulation rate during the Holocene in the Southeastern Indian Ocean. *Marine Chem.*, in press.

Bartlein, P.J., T. Webb III, and E.C. Fleri. 1984. Holocene climatic change in the northern Midwest: Pollen-derived estimates. *Quat. Res.* **22**:361–374.

Battarbee, R.W. 1986. Diatom analysis. In: *Handbook of Holocene Palaeoecology and Palaeohydrology*, ed. B.Berglund, pp. 527-570. Chichester: Wiley.

Becker, B., B. Kroner, and P. Trimborn. 1991. A stable isotope tree-ring timescale of the Late Glacial/Holocene boundary. *Nature* **353**:647–649.

Bengtsson, L., and M. Enell. 1986. Chemical Analysis. In: Handbook of Holocene Palaeoecology and Palaeohydrology, ed. B. Berglund, pp. 423-451. Chichester: Wiley.

Berger, W.H. 1973. Deep-sea carbonates: Pleistocene dissolution cycles. *J. Foram. Res.* **3**:187-195.

Betancourt, J.L., T.L. Van Devender, and P.S. Martin. 1990. Packrat Middens: The Last 40,000 Years of Biotic Change. Tucson: Univ. of Arizona Press.

Bloemendal, J., and P.B. deMenocal. 1989. Evidence for a change in the periodicity of tropical climate cycles at 2.4 Myr from whole-core magnetic susceptibility measurements. *Nature* **342**:897-899.

Boucherle, M.M., and H. Zully. 1983. Cladoceran remains as evidence of change in trophic state in three Swiss lakes. *Hydrobiol.* **103**:141-146.

Boyle, E.A. 1990. Quaternary deep-water paleoceanography. *Science* **249**:863-870.

Boyle, E.A. 1992. Cadmium and $\delta^{13}C$ paleochemical ocean distributions during the stage 2 glacial maximum. *Ann. Rev. Geophys.*, in press.

Bradley, R.S., ed. 1991. Global Changes of the Past. Boulder, CO: Univ. Corp. for Atmospheric Research.

Brassel, S.C., G. Eglinton, I.T. Marlowe, U. Pflaumann, and M. Sarnthein. 1986. Molecular stratigraphy: A new tool for climatic assessment. *Nature* **320**:129-133.

Briffa, K.R., T.S. Bartholin, D. Ekstein, P.D. Jones, W. Karlen, F.H. Schweingruber, and P. Zetterberg. 1990. A 1400-year tree-ring record of summer temperatures in Fennoscandia. *Nature* **346**:434-439.

Broecker, W.S. 1986. Oxygen isotope constraints on surface ocean temperatures. *Quat. Res.* **26**:121-134.

Carbonel, P., J.-P. Colin, D. Danielopol, H. Löffler, and J. Nuestreva. 1988. Paleoecology of limnic ostracodes: A review of some major topics. *Palaeogeog. Palaeoclim. Palaeoecol.* **62**:413-461.

Clausen, H.B., and C.C. Langway, Jr. 1989. The ionic deposits in polar ice cores. In: The Environmental Record in Glaciers and Ice Sheets, ed. H. Oeschger and C.C. Langway, Jr., pp. 225-247. Dahlem Konferenzen. Chichester: Wiley.

CLIMAP. 1981. Seasonal reconstructions of the Earth's surface at the last glacial maximum. *Geol. Soc. Am. Map Chart Ser.* **36**.

Cole, J.E., and R.G. Fairbanks. 1990. The Southern Oscillation recorded in the ^{18}O of corals from Tarawa atoll. *Paleoceanog.* **5**:669-683.

Cook, E., T. Bird, M. Peterson, M. Barbetti, B. Buckley, R. d'Arrigo, R. France, and P. Tans. 1991. Climatic change in Tasmania inferred from a 1089-year tree-ring chronology of Huon Pine. *Science* **253**:1266-1268.

Cubasch, U., and R.D. Cess. 1990. Processes and modeling. In: Climate Change: The IPCC Assessment, ed. J.T. Houghton, G.J. Jenkins, and J.J. Ephraums, pp. 69–91. Cambridge: Cambridge Univ. Press.

Cullen, J.L. 1981. Microfossil evidence for changing salinity patterns in the Bay of Bengal over the last 20,000 years. *Palaeogeog. Palaeoclim. Palaeoecol.* **35**:315-356.

Curry, W.B., and T. Crowley. 1987. The $\delta^{13}C$ of equatorial Atlantic surface waters: implications for ice age pCO_2 levels. *Paleoceanog.* **2**:489-518.

Curry, W.B., J.C. Duplessy, L.D. Labeyrie, and N.J. Shackleton. 1988. Quaternary deep-water circulation changes in the distribution of $\delta^{13}C$ of deep water ΣCO_2 between the last glaciation and the Holocene. *Paleoceanog.* **3**:317-342.

Delmas, R.J., and M. Legrand. 1989. Long-term changes in the concentration of major chemical compounds (soluble and insoluble) along deep ice cores. In: The Environmental Record in Glaciers and Ice Sheets, ed. H. Oeschger and C.C. Langway, Jr., pp. 319-341. Dahlem Konferenzen. Chichester: Wiley.

deMenocal, P.B. 1991. Pliocene-Pleistocene evolution of tropical aridity. PhD Thesis, Columbia Univ.

Druffel, E.M., and H.E. Suess. 1983. On the radiocarbon record in banded corals: exchange parameters and net transport of $^{14}CO_2$ between atmosphere and surface ocean. *J. Geophys. Res.* **88**:1271-1280.

Duplessy, J.C., L. Labeyrie, A. Juillet-LeClerc, F. Maitre, J. Duprat, and M. Sarnthein. 1991. Surface salinity reconstruction of the North Atlantic Ocean during the last glacial maximum. *Oceanol. Acta* **14**:311–324.

Duplessy, J.C., N.J. Shackleton, R.G. Fairbanks, L.D. Labeyrie, D. Oppo, and N. Kallel. 1988. Deepwater source variations during the last climatic cycle and their impact on the global deepwater circulation. *Paleoceanog.* **3**:343-360.

Duvick, D.N., and T.J. Blasing. 1981. A dendroclimatic reconstruction of annual precipitation amounts in Iowa since 1680. *Water Res. Res.* **17**:1183-1189.

Eddy, J.E. 1992. The Pages Project: Proposed Implementation Plans for Research Activities. IGBP Global Change Report, vol. 19. Stockholm: International Geosphere–Biosphere Programme.

Fairbanks, R.G., and R.E. Dodge. 1979. Annual periodicity of the $^{18}O/^{16}O$ and $^{13}C/^{12}C$ ratios in the coral *Montastrea annularis*. *Geochim. Cosmo. Acta* **43**:1009–1020.

Farquhar, G.D., J.D. Ehleringer, and K.T. Hubick. 1989. Carbon isotope discrimination and photosynthesis. *Ann. Rev. Plant Phys. Mol. Biol.* **40**:503-537.

Frenzel, B. 1991. The history of flora and vegetation during the Quaternary. *Prog. Botany* **52**:359–380.

Frey, D.G. 1986. Cladocera analysis. In: Handbook of Holocene Palaeoecology and Palaeohydrology, ed. B. Berglund, pp. 667-692. Chichester: Wiley.

Fritz, S.C., S. Juggins, R.W. Batterbee, and D.R. Engstrom. 1991. Reconstruction of past changes in salinity and climate using a diatom-based transfer function. *Nature* **352**:706-708.

Gasse, F., and F. Tekaia. 1983. Transfer functions for estimating palaeoecological conditions (pH) from East African diatoms. *Hydrobiol.* **103**:85-90.

Goodfriend, G.A. 1988. Mid-Holocene rainfall in the Negev desert from ^{13}C of land snail shell organic matter. *Nature* **333**:757-760.

Grousset, F.E., P.E. Biscaye, A. Zindler, J.M. Prospero, and R. Chester. 1988. Neodymium isotopes as tracers in marine sediments and aerosols: North Atlantic. *Earth Planet. Sci. Lett.* **97**:367-378.

Grun, R. 1990. Dose response of the paramagnetic centre at g:2.007 in corals. *Ancient TL* **8**:20–22.

Hammer, C.U. 1989. Dating by physical and chemical seasonal variation and reference horizons. In: The Environmental Record in Glaciers and Ice Sheets, ed. H. Oeschger and C.C. Langway, Jr., pp. 99–121. Dahlem Konferenzen. Chichester: Wiley.

Harmon, R.S., P. Thompson, H.P. Schwarcz, and D.C. Ford. 1978. Late Pleistocene paleoclimates of North America as inferred from stable isotope studies of speleothems. *Quat. Res.* **9**:54-70.

Hofmann, W. 1986. Chironomid analysis. In: Handbook of Holocene Palaeoecology and Palaeohydrology, ed. B. Berglund, pp. 715-727. Chichester: Wiley.

Hoochiemstra, H., A. Bechler, and H.-J. Beug. 1987. Isopollen maps for 18,000 B.P. of the Atlantic offshore of Northwest Africa: evidence for paleowind circulation. *Paleoceanog.* **2**:561-582.

Isdale, P. 1984. Fluorescent bands in massive corals record centuries of coastal rainfall. *Nature* **310**:578-579.

Jacoby, G.C. 1989. Overview of tree-ring analysis in tropical regions. *IAWA Bull.* **10**:99–108.

Jacoby, G.C., and R.D. D'Arrigo. 1990. Teak (*Tectona Grandis* L.F.), a tropical species of large-scale dendroclimatic potential. *Dendrochron.* **8**:83–98.

Johnsen, S.J., W. Dansgaard, and J.C. White. 1989. The origin of arctic precipitation under present and glacial conditions. *Tellus* **41B**:452-468.

Jouzel, J., and L. Merlivat. 1984. Deuterium and oxygen-18 in precipitation: Modeling of the isotopic effect during snow formation. *J. Geophys. Res.* **89**:11749-11757.

Kipp, N.G. 1976. New transfer function for estimating past sea-surface conditions from sea-bed distribution of planktonic foraminiferal assemblages in the North Atlantic. In: Investigation of Late Quaternary Paleoceanography and Paleoclimatology, ed. R.M. Cline and J.D. Hays. *Geol. Soc. Am. Mem.* **145**. Boulder, CO: Geological Society of America.

Koopmann, B. 1981. Saharan dust deposition in the northern subtropical Atlantic during the last 25,000 years. *Meteor Forsch. Erg.* **C35**:23-59.

Kukla, G.J., F. Heller, L.X. Ming, X.T. Chun, T.S. Liu, and Z.S. An. 1988. Pleistocene climates in China dated by magnetic susceptibility. *Geology* **16**:811-814.

Labeyrie, L.D., J.C. Duplessy, and P.L. Blanc. 1987. Variations in mode of formation and temperature of oceanic deep waters over the past 125,000 years. *Nature* **327**:477-482.

Labeyrie, L.D., J.C. Duplessy, J. Duprat, A. Juillet-Leclerc, J. Moyer, N. Kallel, and N.J. Shackleton. 1992. Changes in the vertical structure of the North Atlantic Ocean between glacial and modern times. *Quat. Sci. Rev.*, in press.

LaMarche, V.C. and K.K. Hirschboeck. 1984. Frost rings in trees as records of major volcanic eruptions. *Nature* **307**:121-126.

Lea, D.W., and E.A. Boyle. 1990. A 210,000-year record of barium variability in the deep northwest Atlantic Ocean. *Nature* **347**:269-272.

Lea, D.W., E.A. Boyle, and G.T. Shen. 1989. Coralline barium records temporal variability in equatorial Pacific upwelling. *Nature* **340**:373-376.

Leinen, M., D. Cwienk, G.R. Heath, P.E. Biscaye, V. Kolla, J. Thiede, J.P. Dauphin. 1986. Distribution of biogenic silica and quartz in recent deep-sea sediments. *Geology* **14**:199-203.

Lorius, C., G. Raisbeck, J. Jouzel, and D. Raynaud. 1989. Long-term environmental records from Antarctic ice cores. In: The Environmental Record in Glaciers and Ice Sheets, ed. H. Oeschger and C.C. Langway, Jr., pp. 343-361. Dahlem Konferenzen. Chichester: Wiley.

Ludwig, K.R., B.J. Szabo, J.G. Moore, and K.R. Simmons. 1991. Crustal subsidence rates off Hawaii determined from $^{234}U/^{238}U$ ages of drowned coral reefs. *Geology* **19**:171–174.

Masle, G., D. Farquhar, and Gifford. 1990. *Aust. J. Plant Physiol.* **17**:465–487.

Matsuoka, H., A. McIntyre, B. Molfino, and D. Verardo. 1991. A sensitive dissolution indicator *Calcidiscus leptoporus*: Concordance with climate-forced dissolution at orbital time scales. *EOS* **72**:271.

McConnaughey, T.A. 1989. C-13 and O-18 isotopic disequilibria in biological carbonates. I. Patterns. *Geochim. Cosmo. Acta* **53**:151-162.

Mix, A.C. 1989. Pleistocene paleoproductivity: evidence from organic carbon and foraminiferal species. In: Productivity of the Ocean: Present and Past, ed. W. H. Berger, V.S. Smetacek, and G. Wefer, pp. 313-340. Dahlem Konferenzen. Chichester: Wiley.

Molfino, B., N. Kipp, and J. Morley. 1982. Comparison of foraminiferal, coccolithophorid, and radiolarian paleotemperature equations: Assemblage coherency and estimate concordancy. *Quat. Res.* **17**:279-313.

Molfino, B., and A. McIntyre. 1990. Precessional forcing of nutricline dynamics in the equatorial Atlantic. *Science* **249**:766-769.

Morley, J.J. 1979. A transfer function for estimating paleoceanographic conditions based on deep-sea surface sediment distribution of radiolarian assemblages in the South Atlantic. *Quat. Res.* **12**:381-395.

Morley, J.J. 1989. Radiolarian-based transfer functions for estimating paleoceanographic conditions in the south Indian Ocean. *Marine Micropal.* **13**:293-307.

Morley, J.J., and J.D. Hays. 1979. *Cycladophora davisiana*: A stratigraphic tool for Pleistocene North Atlantic and interhemispheric correlation. *Earth Planet. Sci. Lett.* **44**:383-389.

Mortlock, R.A., and P.N. Froelich. 1989. A simple method for the rapid determination of biogenic opal in pelagic marine sediments. *Deep Sea Res.* **36**:1415-1426.

Patzold, J. 1984. Growth rhythms recorded in stable isotopes and density bands in the reef coral *Porites Lobata* (Cebu, Philippines). *Coral Reefs* **3**:87–90.

Phillips, F.M., M.G. Zreda, S.S. Smith, D. Elmore, P.W. Kubik, and P. Sharma. 1990. Cosmogenic chlorine-36 chronology for glacial deposits at Bloody Canyon, eastern Sierra Nevada. *Science* **248**:1529–1532.

Pichon, J.J., L.D. Labeyrie, G. Bareille, M. Labracherie, J. Duprat, and J. Jouzel. 1992. Surface water temperature changes in the high latitudes of the Southern Hemishphere over the last glacial-interglacial cycle. *Paleoceanog.*, in press.

Pokras, E.M., and A.C. Mix. 1985. Eolian evidence for spatial variability of late Quaternary climates in tropical Africa. *Quat. Res.* **24**:137-149.

Pokras, E.M., and A.C. Mix. 1987. Earth's precession cycle and Quaternary climatic change in tropical Africa. *Nature* **326**:486-487.

Prahl, F.G., and S G. Wakeham. 1987. Calibration of unsaturation patterns in long-chain ketone compositions for paleotemperature assessment. *Nature* **330**:367-369.

Prell, W.L. 1984. Variation of monsoonal upwelling: A response to changing solar radiation. In: Climate Processes and Climate Sensitivity, ed. J.E. Hansen and T. Takahashi. *Geophys. Monog.* **29**:48-57. Washington, D.C.: American Geophysical Union.

Prell, W.L., W.H. Hutson, and D.F. Williams. 1979. The subtropical convergence and late Quaternary circulation in the southern Indian Ocean. *Marine Micropal.* **4**:225-234.

Radtke, U., R. Grun, and H.P. Schwarcz. 1988. Lectron spin resonance dating of the Pleistocene coral reef tracts of Barbados. *Quat. Res.* **29:**197–215.

Rau, G.H., P.N. Froelich, T. Takahashi, and D.J. DesMarais. 1991. Does sedimentary organic $\delta^{13}C$ record variations in Quaternary ocean [CO_2 (aq)]? *Paleoceanog.* **6**:335-347.

Ravelo, A.C., R.G. Fairbanks , and S.G.H. Philander. 1990. Reconstructing tropical Atlantic hydrography using planktonic foraminifera and an ocean model. *Paleoceanog.* **5**:409-431.

Raymo, M.E., W.F. Ruddiman, N.J. Shackleton, and D.W. Oppo. 1990. Evolution of Atlantic-Pacific $\delta^{13}C$ gradients over the last 2.5 m.y. *Earth Planet. Sci. Lett.* **97**:353-368.

Ruddiman, W.F. 1977. North Atlantic ice-rafting: a major change at 75,000 years before the present. *Science* **196**:1208-1211.

Ruddiman, W.F., and A. McIntyre. 1977. Late Quaternary surface ocean kinematics and climatic change in the high-latitude North Atlantic. *J. Geophys. Res.* **82**:3877-3887.

Sarnthein, M., U. Pflaumann, R. Ross, R. Tiedemann, and K. Winn. 1992. Transfer functions to reconstruct ocean paleoproductivity, a comparison. In: Evolution of Upwelling systems since the Miocene, ed. C.P. Summerhays, W.L. Prell, and K.C. Emeis. London: Blackwell.

Sarnthein, M., G. Tetzlaff, B. Koopmann, K. Wolter, and U. Pflaumann. 1981. Glacial and interglacial wind regimes over the eastern subtropical Atlantic and North-West Africa. *Nature* **293**:193-196.

Sarnthein, M., and K. Winn. 1990. Reconstruction of low and mid latitude export productivity, 30,000 yr B.P. to present: Implications for global carbon reservoirs. In: Climate–Ocean-Interactions, ed. M. Schlesinger, pp. 319-342. Dordrecht: Kluwer.

Sarnthein, M., K. Winn, J.C. Duplessy, and M.R. Fontugue. 1988. Global variations of surface ocean productivity in low and mid latitudes: influence on CO_2 reservoirs of the deep ocean and atmosphere during the last 21,000 years. *Paleoceanog.* **3**:361-399.

Schwarcz, H.P., R.S. Harmon, P. Thompson, and D.C. Ford. 1976. Stable isotope studies of fluid inclusions in speleothems and their paleoclimatic significance. *Geochim. Cosmo. Acta* **40**:657-665.

Schweingruber, F.H., K.R. Briffa, and P.D. Jones. 1991. Yearly maps of summer temperatures in western Europe from A.D. 1750 to 1975 and western North America from 1600 to 1982: results of a radiodensitometrical study on tree rings. *Vegetatio* **92**:5-71.

Shackleton, N.J. 1987. Oxygen isotopes, ice and sea level. *Quat. Sci. Rev.* **6**:183-190.

Shackleton, N.J., J.C. Duplessy, M. Arnold, P. Maurice, M.A. Hall, and J. Cartlidge. 1988. Radiocarbon age of last glacial Pacific deep water. *Nature* **335**:708-711.

Shackleton, N.J., and N.D. Opdyke. 1973. Oxygen isotope and palaeomagnetic stratigraphy of equatorial pacific core V28-238: Oxygen isotope temperatures and ice volumes on a 10^5 and 10^6 year scale. *Quat. Res.* **3**:39-55.

Shackleton, N.J., and N.D. Opdyke. 1976. Oxygen-isotope and paleomagnetic stratigraphy of pacific core V28-239 late Pliocene to latest Pleistocene. In: Investigation of Late Quaternary Paleoceanography and Paleoclimatology, ed. R.M. Cline and J.D. Hays. *Geol. Soc. Am. Mem.* **145**. Boulder, CO: Geological Society of America.

Shackleton, N.J., and N.G. Pisias. 1985. Atmospheric carbon dioxide, orbital forcing, and climate. In: The Carbon Cycle and Atmospheric CO_2: Natural Variations Archean to Present, ed. E.T. Sundquist, and W.S. Broecker. *Geophys. Monog.* **32**:303-317. Washington, D.C.: American Geophysical Union.

Shen, G.T., and E.A. Boyle. 1988. Determination of lead, cadmium and other trace metals in annually banded corals. *Chem. Geol.* **67**:47-62.

Shen, G.T., E.A. Boyle, and D.W. Lea. 1987. Cadmium in corals as a tracer of historical upwelling and industrial fallout. *Nature* **328**:794-796.

Shoji, H., and C.C. Langway, Jr. 1989. Physical property reference horizons. In: The Environmental Record in Glaciers and Ice Sheets, ed. H. Oeschger and C.C. Langway, Jr., pp. 161-175. Dahlem Konferenzen. Chichester: Wiley.

Siegenthaler, U., and U. Eicher. 1986. Stable oxygen and carbon isotope analysis. In: Handbook of Holocene Palaeoecology and Palaeohydrology, ed. B. Berglund, pp. 407-422. Chichester: Wiley.

Smith, S.V., R.W. Buddemeier, R.C. Redalje, and J.E. Houck. 1979. Strontium-calcium thermometry in coral skeletons. *Science* **204**:404-407.

Smol, J.P. 1988. Paleoclimate proxy data from freshwater arctic diatoms. *Verh. Int. Verein. Limnol.* **23**:837-844.

Smol, J.P., R.W. Battarbee, R.B. Davis, and J. Merilainen, eds. 1986. Diatoms and Lake Acidity. Dordrecht: Junk.

Stuiver, M. 1971. Long-term ^{14}C variations. In: Radiocarbon Variations and Absolute Chronology, ed. I.U. Olsson, pp. 197–213. New York: Wiley.

Stuiver, M., and T.F. Braziunas. 1987. Tree cellulose $^{13}C/^{12}C$ isotope ratios and climatic change. *Nature* **328**:58-60.

Thompson, P., H.P. Schwarcz, and D.C. Ford. 1976. Stable isotope geochemistry, geothermometry and geochronology of speleothems from West Virginia. *Geol. Soc. Am. Bull.* **87**:1730-1738.

Tiedemann , R. 1991. Climatic history of northwest Africa during the last 8 million years and paleoceanography of the adjacent Atlantic: High resolution time series from ODP-Sites 658-661. PhD Thesis, Geol. Palaeont. Inst., University of Kiel.

Valladas, H., J.L. Reyss, J.L. Joron, G. Valladas, O. Bar-Josef, and B. Vandermeersch. 1988. Thermoluminescence dating of Mousterian "Proto-Cro-Magnon" remains from Israel and the origin of modern man. *Nature* **331**:614–616.

Van den Bogard, P., C.M. Hall, H.U. Schminke, and D. York. 1989. Precise single grain $^{40}Ar/^{39}Ar$ dating of a cold to warm climate transition in Central Europe. *Nature* **342**:523–525.

Verardo, D.V. 1991. Late Pleistocene biogenic flux in the tropical Atlantic: A response to climate change. PhD Thesis, City University of New York.

Verardo, D.V., P.N. Froelich, and A. McIntyre. 1990. Determination of organic carbon and nitrogen in marine sediments using the Carlo Erba NA-1500 Analyzer. *Deep Sea Res.* **37**:157-165.

Webb, T. III, P.J. Bartlein, and J.E. Kutzbach. 1987. Climatic change in eastern North America during the past 18,000 years: comparison of pollen data with model results. In: The Geology of North America, Vol. K-3, North America and Adjacent Oceans during the Last Deglaciation, ed. W.F. Ruddiman and H.E. Wright, pp. 447-462. Boulder, CO: Geological Society of America.

Weingarten, B., R. Yuretich, R. Bradley, and M.L. Salgado-Labouriau. 1990. Environmentally controlled variations in clay mineral assemblages from lake sediments in the Venezuelan Andes. In: Proc. 9th International Clay Conference, ed. V.C. Farmer and Y.Tardy. *Sci. Geol. Mem.* **88**:53-61.

Westerhausen, L. 1991. Organische sedimente im aquatorialen Ostatlantik: Einflusse von Herkunft, Transportmustern, Diagenese und Klimaschwankungen. PhD thesis, University of Kiel.

Williams, J.D.H., J.-M. Jacquet, and R.L. Thomas. 1976. Forms of phosphorus in the surficial sediments of Lake Erie. *J. Fish. Res. B. Can.* **33**:413-429.

Wintle, A. 1990. A review of current research on TL dating of loess. *Quat. Sci. Rev.* **9**:385–397.

Zahn, R., and A.C. Mix. 1991. Benthic foraminiferal $\delta^{18}O$ in the ocean tempeature salinity density field: constraints on ice-age thermohaline circulation. *Paleoceanog.* **6**:1-20.

12

Role of Trace Gases: The Problem of Lead and Lag

D. RAYNAUD[1] and U. SIEGENTHALER[2]
[1]Laboratoire de Glaciologie et Géophysique de l'Environnement Rue Molière, Domaine Universitaire, B.P. 96, 38402 St.-Martin-d'Hères Cédex, France
[2]Physikalisches Institut, Universität Bern, Sidlerstrasse 5, 3012 Bern, Switzerland

ABSTRACT

Polar ice cores and deep-sea sediments offer a unique opportunity to reconstruct the records of important climatic parameters, such as temperature, continental ice volume, and atmospheric greenhouse gases over the last climatic cycle. In general, the records indicate a remarkable correlation between CO_2 and CH_4 concentrations and the orbitally forced climate. This chapter discusses different evidence for lead and lag between changes in radiatively active trace gases and climatic variations. During deglaciations, both CO_2 and CH_4 increased approximately in phase with temperature in high southern latitudes, but the gases started to change ahead of any significant melting of continental ice. The radiative forcing due to greenhouse gases therefore preceded the climate forcing due to the vanishing ice sheets, and the initial greenhouse-gas increase was not a consequence of sea level change. During the initiation of the glaciation, CH_4 decreased also approximately in phase with Southern Hemisphere temperature, but CO_2 lagged by several thousand years (Ka) behind. Finally, abrupt climatic changes (on a time scale of centuries) not correlated with orbital forcing are observed during the ice age-Holocene transition and during the ice age (around 40 Ka B.P.). CO_2 or CH_4 variations associated with these abrupt changes are also observed in the ice record. Up to now, only the CH_4 oscillation found during the glacial–interglacial transition can be interpreted with confidence as a real atmospheric signal, most likely linked to the Younger Dryas. More detailed information is needed about the role of greenhouse gases in abrupt climatic events.

INTRODUCTION

The occurrence of large fluctuations of atmospheric greenhouse gases during the last climatic cycle is by now well established. Two fundamental questions have arisen: (a) what are the mechanisms at the origin of these fluctuations, and (b) what is the contribution of such atmospheric changes on climate variability? The modeling of

Global Changes in the Perspective of the Past
Edited by J.A. Eddy and H. Oeschger

biogeochemical cycles, atmospheric chemistry, and climate are key tools for providing global answers to these questions, but the model outputs are constrained by the available palaeo data and have to account for the leads and lags observed in the different records. The aim of this chapter is to review the lead and lag evidence from the proxy data covering the last climatic cycles and to compare these observations with some model results. We focus the discussion on the leads and lags between greenhouse gases (specifically CO_2 and CH_4) and important climatic indicators such as temperature and continental ice volume. Given our fields of activity, the emphasis is on proxy data from ice cores, although there is clearly a wealth of information on the lead-lag problem contained in deep-sea sediment records.

RELATION BETWEEN ATMOSPHERIC COMPOSITION AND CLIMATE

Spectral analysis of palaeoclimatic records from the past 700,000 years has shown that in addition to the dominant 100-Ka cycle, the orbital periods of ca. 20 and 40 Ka are strongly present. This indicates that orbital variations provide the external forcing that drives glacial–interglacial cycles. However, climatic variations are not linearly related to astronomical forcing: the relatively fast terminations cannot be explained by a linear relationship between insolation and climate, and the 100-Ka cycle is not seen with significant power in the insolation curves of the past. Relatively strong nonlinear feedback mechanisms obviously act within the Earth's climate system to amplify and modify the original external forcing. For the lead-lag problem, we have to distinguish between rapid and slow feedbacks. Typical rapid feedbacks are changes in atmospheric water vapor concentration and in sea-ice cover after a change in temperature, which contribute to the greenhouse effect and affect the Earth's albedo, respectively. For our purpose, they can be regarded as immediate. The slow feedbacks most relevant to the lead-lag problem are changes in albedo and in the parameters of large ice sheets caused by their waxing and waning, with a typical response time of 10,000 years, and those in vegetation, with response times of one to a few centuries. Another important feedback that can influence both global heat and atmospheric composition are climate-induced changes in ocean circulation, with a response time of less than a century (Bryan and Spelman 1985; Maier-Reimer and Mikolajewicz 1989).

The atmospheric composition can change in direct or indirect response to climatic processes. On the other hand, the concentration of greenhouse gases affects the Earth's radiation balance and, therefore, its climate. We will consider these two links next.

Greenhouse Gases as Climatic Factors

The atmospheric CO_2 concentration during the last glacial maximum was about 70% of its Holocene value, which must have created a reduced greenhouse effect, compared with the present situation. Simulations of the ice age climate have been made using

General Circulation Models. According to Boccoli and Manabe (1987), the CO_2 change was responsible for about 40% of the global temperature change, as well as the rapid feedback effects induced by the original greenhouse change, in particular the reaction of atmospheric water vapor, clouds, and snow and sea-ice albedo. The reduced methane concentration during the ice age directly contributed about 15% as much as CO_2 to the temperature change (Chappellaz et al. 1990). In addition, its indirect effects due to feedbacks on atmospheric chemistry (stratospheric water vapor, tropospheric ozone) may have been significant (Raynaud et al. 1988; Chappellaz et al. 1990). Therefore, these two greenhouse gases together could have been responsible for half the temperature change. In harmony with these model results, a multivariate analysis of the Vostok ice-core data has estimated the greenhouse-gas contribution to have been 50 ± 10% to the temperature change (Lorius et al. 1990). Clearly, atmospheric composition has been a significant factor in large climatic cycles. If there were a sudden change in atmospheric CO_2, the global temperature would not immediately establish a new equilibrium because the warming-up of the ocean takes time. The time scale for reaching the new equilibrium temperature is on the order of 100 years (e.g., Hansen et al. 1984). Compared to this, the natural greenhouse-gas changes have been relatively slow, so that a lag between atmospheric composition changes and the corresponding climate response would hardly be noticeable. A possible exception are the rapid CO_2 oscillations during the ice age observed in Greenland ice cores (see below).

CLIMATIC CHANGE AS THE CAUSE OF ATMOSPHERIC COMPOSITION CHANGES

There is general agreement that the cause of the the glacial–interglacial CO_2 variations should be sought in the ocean, which is the Earth's largest carbon-exchanging reservoir. The terrestrial biomass most likely was reduced, which by itself would have produced higher ice age CO_2 levels. Different hypotheses for the specific mechanisms involved have been put forward, a short review of which is provided in Broecker and Peng (1989). Changes in atmospheric CO_2 concentration are most directly caused by modifications in the chemical composition of surface water; that is, in its total CO_2 concentration (ΣCO_2) or alkalinity. This can be brought about by either (a) a redistribution of geochemical properties within the ocean or (b) a modification in the whole-ocean composition due to external input or sedimentation. These two categories of mechanisms are differ with respect to the response times of the atmospheric composition. The first category includes changes in ocean circulation or in biological activity, which together influence the ΣCO_2 in surface water and, therefore, the partial pressure of CO_2, pCO_2. There is clear evidence that the circulation of the glacial ocean differed from the present one; a reduced North Atlantic Deep Water (NADW) formation caused the deep Atlantic more closely to resemble today's deep Pacific (higher concentrations of ΣCO_2 and nutrients, lower oxygen concentration and lower

$\delta^{13}C$ values). Ocean circulation is therefore a good candidate for causing CO_2 variations. For example, a weakening of the vertical circulation in high latitudes would result in ΣCO_2 and fewer nutrients being mixed up from depth to the surface per unit time. Under conditions of constant biological activity, this would lead to a better utilization, and therefore to a depletion, of ΣCO_2 and nutrients in high-latitude surface ocean waters, and consequently to a decrease in atmospheric CO_2 (e.g., Wenk and Siegenthaler 1985). The circulation of the global ocean can change rapidly, within a century or less, in response to external forcing. After an abrupt change in ocean circulation, the carbon in the ocean and in the atmosphere would reach a new equilibrium distribution on a time scale of about 200 years (Wenk and Siegenthaler 1985). Other mechanisms involving a redistribution of oceanic properties would be a decrease in biological activity in high latitudes (slower biological pump), a change in the Redfield ratio C/P utilized by marine organisms, and a rise in alkalinity of surface waters as a consequence of water mass rearrangements connected to a reduced NADW formation (Broecker and Peng 1989). The overall delay of change in atmospheric CO_2 concentration in response to ocean-internal forcing is of the order of a few centuries.

A possible explanation for the lower ice age CO_2 levels would be that the average oceanic concentration of nutrients (P, N) was greater, leading to a more active biological carbon pump and therefore to a stronger surface depletion of ΣCO_2. This mechanism is part of the shelf sediment hypothesis of Broecker (1982), according to which the sea level changes are accompanied by deposition or erosion of organic-rich shelf sediments. If governed by such a mechanism, atmospheric CO_2 should closely follow the buildup or the melting of the continental ice sheets. Another mechanism is a change in the cycle river input-sedimentation of $CaCO_3$ that could lead to a whole-ocean change in the alkalinity (and the concentration of CO_3^{2-} ions) and therefore in surface-water pCO_2. Boyle (1988) postulated such an effect as a consequence of a vertical rearrangement of ΣCO_2 and nutrients in the glacial ocean. The $CaCO_3$ compensation involved in this mechanism has a time constant of several 1000 yr (Broecker and Peng 1989). Such a time lag should be discernible in adequate records.

The major natural source of atmospheric methane is anaerobic terrestrial environments, i.e., wetlands; and its main sink is reaction with OH radicals in the atmosphere (e.g., Cicerone and Oremland 1988). The atmospheric lifetime of methane is about 10 years, so its concentration would, on the time scales of the proxy data we consider here, react immediately to changes in the sink mechanism, while modifications in the extent of wetlands and in their strength as a CH_4 source after some climatic event could occur with a response time of centuries or less.

THE RECORDS

Deep-sea sediments and polar ice contain proxy data for both atmospheric greenhouse gases and climatic parameters (temperature, ice volume, atmospheric transport, oceanic circulation, etc.), as indicated in Table 12.1. They are complementary.

Table 12.1 Global atmospheric/climatic aprameters.

Global Atmospheric/Climatic Parameters	Proxy Data from Ice Cores and Deep-sea Sediements
Atmospheric Trace Gases (CO_2, CH_4, ...)	• Ice cores: Air enclosed in ice • Deep-sea sediments: Difference ($\Delta\delta^{13}C$) between planktonic and benthic foraminiferal $\delta^{13}C$ as a proxy for CO_2
Temperature and Precipitation	
Antarctic or Greenland air temperature	• Isotopic composition ($\delta^{18}O$ or Δ) of polar ice
Snow accumulation rate	• ^{10}Be in ice
Ocean temperature	• $\delta^{18}O$ and composition of deep-sea sediments
Ice Volume	• Deep-sea sediments: $\delta^{18}O$ of forams • Ice cores: $\delta^{18}O$ of air enclosed in ice cores
Atmospheric and Oceanic Circulation	• Ice cores: concentration of dust and aerosols for atmospheric circulation • Deep-sea sediments: $\delta^{13}C$ and Cd/Ca in forams for oceanic circulation; dust for atmospheric circulation

The deep-sea cores provide long-term records which are interpreted for the period of the late Pleistocene in terms of continental ice volume, temperature, circulation, and chemistry of the ocean and atmospheric transport. Of special interest here are the $\delta^{13}C$ differences between planktonic and benthic foraminifera ($\Delta\delta^{13}C$), which have been used by Shackleton et al. (1983) to deduce an atmospheric CO_2 record for the last climatic cycle and by Shackleton and Pisias (1985) to extend this record over the past 340 Ka. Further $\Delta\delta^{13}C$ data have been discussed along this line by Curry and Crowley (1987). These authors point out that $\Delta\delta^{13}C$ not only reflects pCO_2 so that the use of $\Delta\delta^{13}C$ as a proxy for atmospheric CO_2 should be reevaluated. In addition, a reorganization of the deep ocean circulation affected $\delta^{13}C$ strongly in some ocean basins (particularly in the Atlantic Ocean), so that any individual $\Delta\delta^{13}C$ record may deviate significantly from the global mean.

The ice cores provide a shorter record, which today is limited to the past 160,000 years (one full climatic cycle). The time resolution is in general better than in the deep-sea record. The ice-core data are interpreted in terms of Antarctic or Greenland temperature and accumulation rates, of aerosols of various origins reaching the polar atmospheres, of continental ice volume, and of atmospheric composition. An important particularity of ice cores is that they actually contain two types of record:

temperature, accumulation rate, and aerosols, which are deduced from measurements performed on the ice phase itself, and greenhouse gases and proxy for ice volume, which are obtained by analyzing the gas phase which is incorporated as air bubbles in the ice well up after the snow deposition at the surface of the ice sheet. This means that the air will be younger than the ice itself. Focusing on the greenhouse-gas record, the ice provides a quasi-direct atmospheric record (in contrast to the oceanic $\Delta\delta^{13}C$ proxy). This has been documented for CO_2 and CH_4, but which could also be extended to other trace gases such as N_2O.

Information on leads and lags is obtained by comparing the different records contained in deep-sea sediments and ice cores. The main sources of uncertainty are: (a) the time resolution of the records, (b) dating problems that arise when data from different records are compared, and (c) the signal-to-noise ratio in a given record when determining the time at which the signal starts.

Because of current uncertainties in dating, it is generally not possible to assess leads and lags by directly comparing a record preserved in ice with one measured in deep-sea sediments. There is some hope that such uncertainties can be reduced by searching for comparable horizons such as dust in ice and marine records or $\delta^{18}O$ of atmospheric oxygen trapped in ice and $\delta^{18}O$ of benthic forams as proxies for continental ice volume.

For practical purposes, we first present the data concerning the general trends in the climatic cycle (deglaciations, entering a glaciation, phase relationship in the Milankovitch cycles) and we then discuss abrupt changes including those associated with the Younger Dryas event.

The Last Climatic Cycle

Phase Relationship between CO_2 and Temperature

The best documented CO_2 records from ice cores arise from Byrd (West Antarctica) for the last deglaciation (Neftel et al. 1988; Staffelbach et al. 1991; see Figure 12.1) and from Vostok (East Antarctica) for the penultimate deglaciation and the following inception of the glaciation (Barnola et al. 1987; Barnola et al. 1991; see also Figure 12.2). They indicate that during each deglaciation the warming experienced in Antarctica (as inferred from isotopic measurements on ice performed on the same cores) was roughly in phase with the CO_2 increase. It is of special interest to infer from such records which started first, the CO_2 increase or the warming. The answer is unclear due to uncertainties linked with the determination of the age difference between the enclosed air and the ice itself, the time resolution of the records (especially at Vostok), and the difficulty of locating precisely where the CO_2 and temperature signals start to emerge from the background noise of the full glacial conditions. The last point is clearly illustrated in the detailed results for Byrd (Figure 12.1), where there is an uncertainty of about 1.5 Ka for CO_2 and 2 Ka for temperature in determining the precise timing of the start of the systematic change. On the other hand, the best guesses

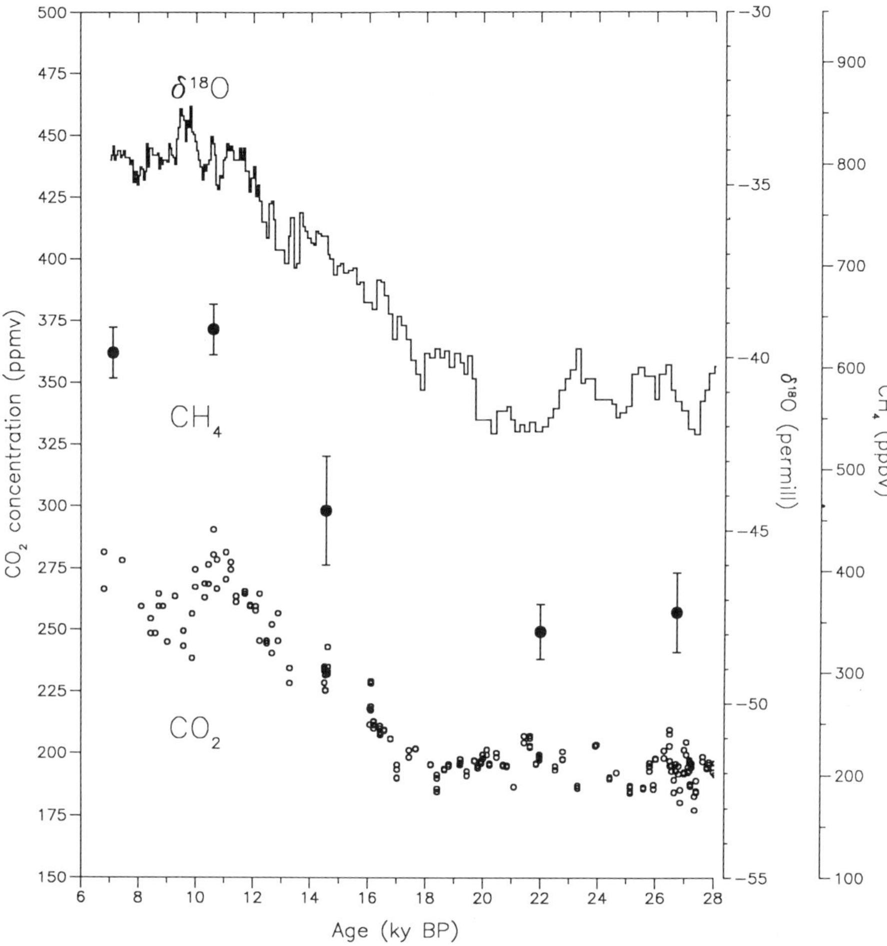

Figure 12.1 Bryd ice core: changes of CO_2, CH_4, and $\delta^{18}O$ of ice (used as a proxy for temperature) during the last deglaciation. Age scale: C. Hammer, pers. comm. After Staffelbach et al. (1991) and Stauffer et al. (1988).

concerning the penultimate deglaciation, as recorded by the Vostok core (Barnola et al. 1991), are that either both temperature and CO_2 started to increase in phase or that the CO_2 increase lagged slightly (by less than 1 Ka) behind the warming phase. No reliable data from ice cores are available for investigating the phase relationship between CO_2 and Greenland temperature over a deglaciation period (see, for instance, Staffelbach et al. 1991 for a discussion of the measurements performed on the Dye 3 core).

In contrast to deglaciation, the CO_2 change (decrease), when entering the glacial period at the end of the interglacial, lags behind Anarctic temperatures by several thousand years (see Figure 12.2).

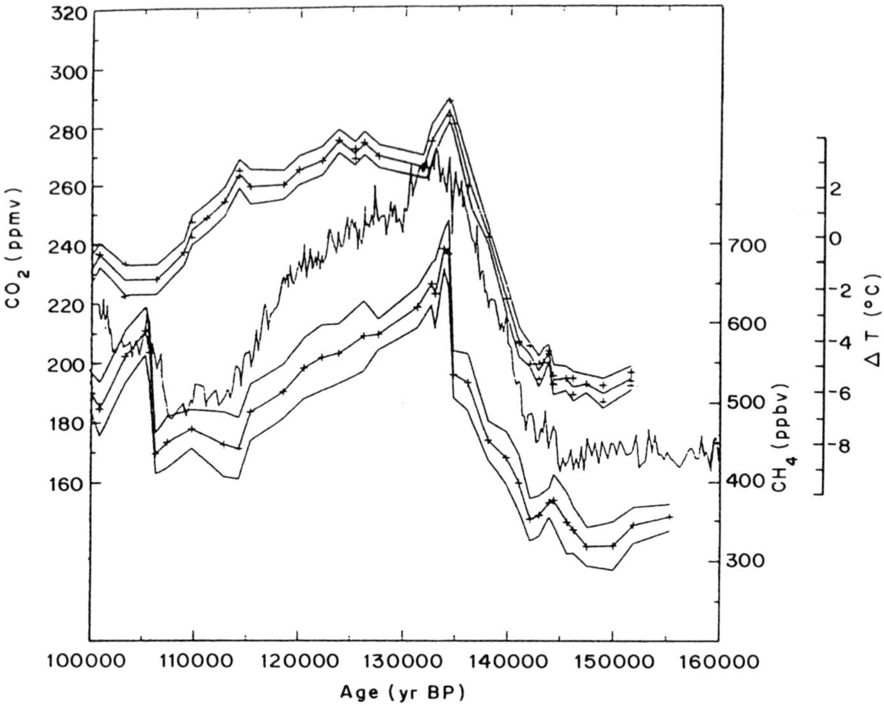

Figure 12.2 Vostok ice core: changes of CO_2 (top), CH_4 (bottom), and Antarctic temperature (middle curve, deduced from ice isotopic composition) during the 100–160 Ka B.P. period, including the penultimate deglaciation, the interglacial, and the inception of the following glaciation. From Chappellaz (1990) and Barnola et al. (1991).

The detailed record from Byrd shows a CO_2 decrease after an initial maximum, with values near 250 ppmv between 8 and 10 Ka B.P. (see Figure 12.1). No clear correlation with the isotopic temperature record is observed. The cause of this reduction remains to be identified. Oeschger (1991) suggested that it might be connected to the early Holocene buildup of the land biomass.

Phase Relationship between CO_2, Ice Volume, and Temperature

The first evidence of the relationship between CO_2 and ice volume arose from the deep-sea record with the comparison between $\Delta\delta^{13}C$ and $\delta^{18}O_{foram}$ used as proxies for atmospheric CO_2 and ice volume changes respectively (Shackleton and Pisias 1985; Curry and Crowley 1987; Imbrie et al. 1989). In the Milankovitch frequency bands, $\Delta\delta^{13}C$ leads $\delta^{18}O_{foram}$ over the past 300 Ka. As we will see below, information on the CO_2-ice volume phase relation for the penultimate deglaciation is available from the ice record. For that period, the oceanic record suggests that CO_2 began to increase before continental ice volume started to decrease.

Concerning sea-surface temperature and ice volume, analyses of the marine record have shown that SST at high southern latitudes leads ice volume changes in general (see for instance Imbrie et al. 1989). The second piece of evidence arises from the $\delta^{18}O$ record of atmospheric oxygen, as obtained from ice-core measurements. Atmospheric oxygen is cycled through the global photosynthesis-respiration cycle with a turnover time of 2000–3000 yr, and via marine photsynthesis, the isotopic composition of O_2 is determined by $\delta^{18}O_{ocean}$. Therefore, $\delta^{18}O_{atm}$ represents a proxy for changes in $\delta^{18}O_{ocean}$ and hence in continental ice volume. It is a more complex proxy for ice volume than $\delta^{18}O_{foram}$ but has the advantage that it can be measured on the same air sample extracted form ice as CO_2. A recent and detailed investigation of the penultimate deglaciation period along the Vostok core (Sowers et al. 1991) indicates unambiguously that the $\delta^{18}O_{atm}$ signal began to shift after the CO_2 started to increase (Figure 12.3). After deconvolving the $\delta^{18}O_{atm}$ signal into a simulated $\delta^{18}O_{ocean}$ curve, Sowers et al. estimate that atmospheric CO_2 began to increase 4 to 7 Ka before any significant input of glacial meltwater to the oceans, and that it had risen already by about 35 ppmv above glacial values before the initial reduction of continental ice.

Phase Relationship between CO_2 and CH_4

The record of atmospheric CH_4 is interesting not only because this is another greenhouse gas than CO_2, but also because the geochemical processes governing the two gases are different. CH_4 concentrations during the past should mainly have been controlled by continental sources (wetlands) and by changes in the oxidizing capacity (OH concentration) of the atmosphere, whereas atmospheric CO_2 changes were driven mainly by the ocean.

The two last deglaciation periods show well-marked increases in CH_4, from about 350 to 650 ppbv (Stauffer et al. 1988; Raynaud et al. 1988; Chappellaz et al. 1990). In terms of leads and lags, the Vostok record indicates that CH_4, like CO_2, is roughly in phase with the glacial to interglacial warmings (Chappellaz et al. 1990). The CO_2–CH_4 comparison is facilitated by the fact that both are measured on the air phase. For the penultimate deglaciation (Figure 12.2), it is difficult to decide whether CH_4 started to increase simultaneously with or 1000–2000 years prior to CO_2 and Antarctic temperature. The CH_4 change could reflect a response of the natural CH_4 sources at low latitudes. Alternatively, a decrease of the atmospheric sink could have been involved.

In contrast to CO_2, the methane signal is in phase with Antarctic temperature at the time of the glaciation inception (see Figure 12.2). Also, the spectral analysis of the CH_4 record indicates periodicities in rather good agreement with orbital frequencies, and cross spectral analysis between CH_4 and temperature signals shows no significant lag in the phase relationship (Chappellaz et al. 1990).

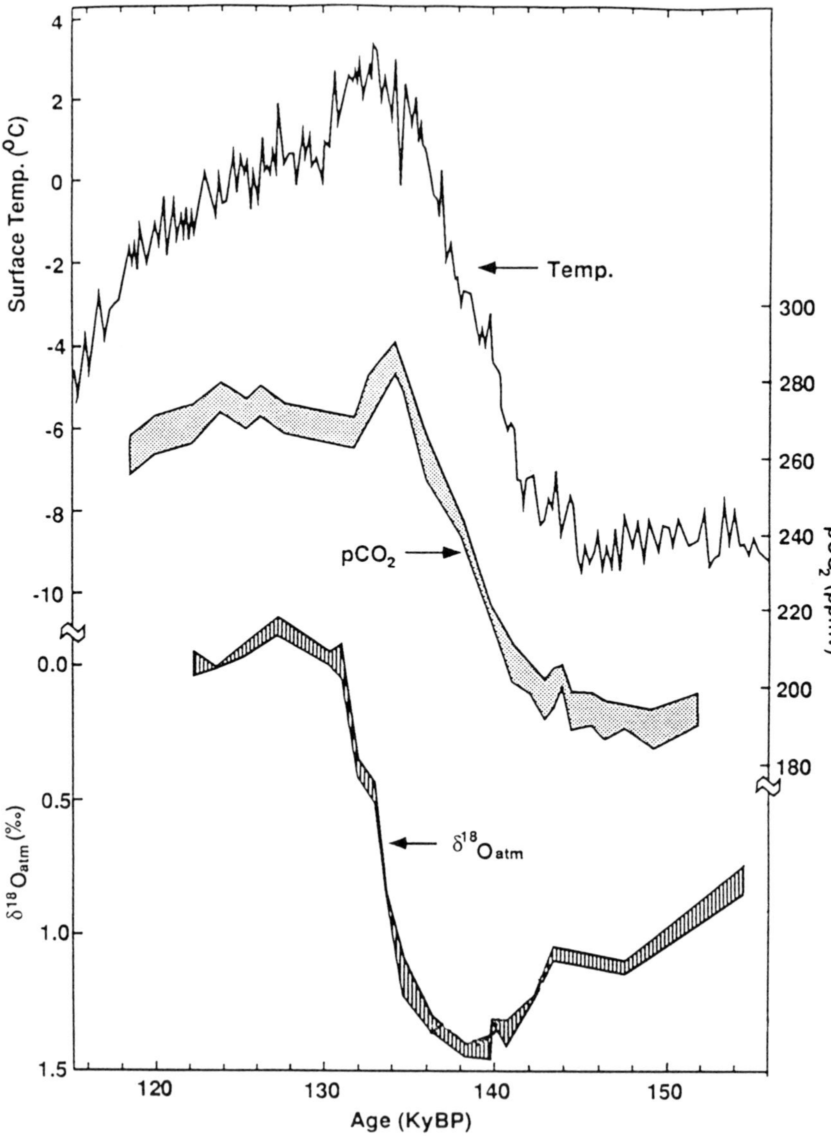

Figure 12.3 Vostok ice core: Changes in Antarctic temperature (as deduced from ice isotopic composition), CO_2, and $\delta^{18}O$ of atmospheric oxygen during the penultimate deglaciation. From Sowers et al. (1991).

Greenhouse Gas Concentrations during Rapid Climate Changes

In the deep ice cores from Greenland, rapid oscillations of $\delta^{18}O$ and other parameters have been observed during the ice age, indicating the existence of a bistable climatic system, at least in the North Atlantic region. Stauffer et al. (1984) and Staffelbach et al. (1988) found that in the Dye 3 and Camp Century cores, the CO_2 concentration changes strictly parallel to $\delta^{18}O$ during such oscillations. The CO_2 changes amount to ca. 50 ppm, and they are extremely rapid; if an undisturbed stratigraphy is assumed, they took place within one to a few decades. Since the atmosphere is mixed within a few years, the interhemispheric CO_2 difference cannot amount to more than a few ppm, and the rapid changes observed in the Greenland ice should also be seen in Antarctic ice cores. However, studies on the Byrd ice core (Antarctica) did not show similar CO_2 oscillations during the ice age (Neftel et al. 1988), so the question arises whether those observed in Greenland do indeed reflect atmospheric changes. The abruptness of the changes observed in Greenland poses some problems. There are no known mechanisms that could produce atmospheric CO_2 changes of 50 ppmv within less than a century. Also, if an atmospheric CO_2 change occurred simultaneously with a climatic change, the $\delta^{18}O$ shift would be expected at less deep levels in the ice than the CO_2 shift because of the age difference between the trapped air and the surrounding ice. However, the transitions are observed at exactly the same depth for $\delta^{18}O$ and for CO_2. Staffelbach et al. (1988) concluded that the stratigraphy at those depths is disturbed in such a way that the transitions are artificially shortened in the ice. It seems probable that the ice age $\delta^{18}O$ records around 40 Ka B.P. from Greenland do reflect rapid climatic oscillations of the Younger Dryas type, typically lasting a few 1000 years with transitions occurring within ca. 100 years. Whether the observed CO_2 variations represent actual atmospheric changes or rather artifacts due, for example, to summer melting during the warm periods, cannot be definitely decided at present. The results from the ongoing GRIP and GISP 2 deep drillings in Greenland should help to resolve this problem.

A review of the available CO_2 and CH_4 ice-core data associated with the Younger Dryas recently has been made by Jouzel et al. (1992; see Figure 12.4). One difficulty is that records of the last deglaciation general exist at depths in the ice sheet where the ice is brittle and consequently less appropriate for obtaining reliable gas measurements, especially with respect to relatively short events like the Younger Dryas. In the Dye 3 core (Greenland), the ice corresponding to the last part of the deglaciation and the Holocene may have been affected by melt layers, which, especially for CO_2, will modify the initial atmospheric signal. For Greenland, a good high-resolution CO_2 record thus is still lacking. Such a record should be provided by the new GISP II and GRIP ice cores. For Antarctica, the compilation of the available data from Vostok (Barnola et al. 1987; Barnola et al. 1991), Dome C (Raynaud and Barnola 1985), and Byrd (Neftel et al. 1988; Staffelbach et al. 1991) suggest that the CO_2 increase linked with the last deglaciation occurred in two steps, with a stage when the rise of CO_2 was interrupted (see Figures 12.1 and 12.4). It remains to be determined if this stage is

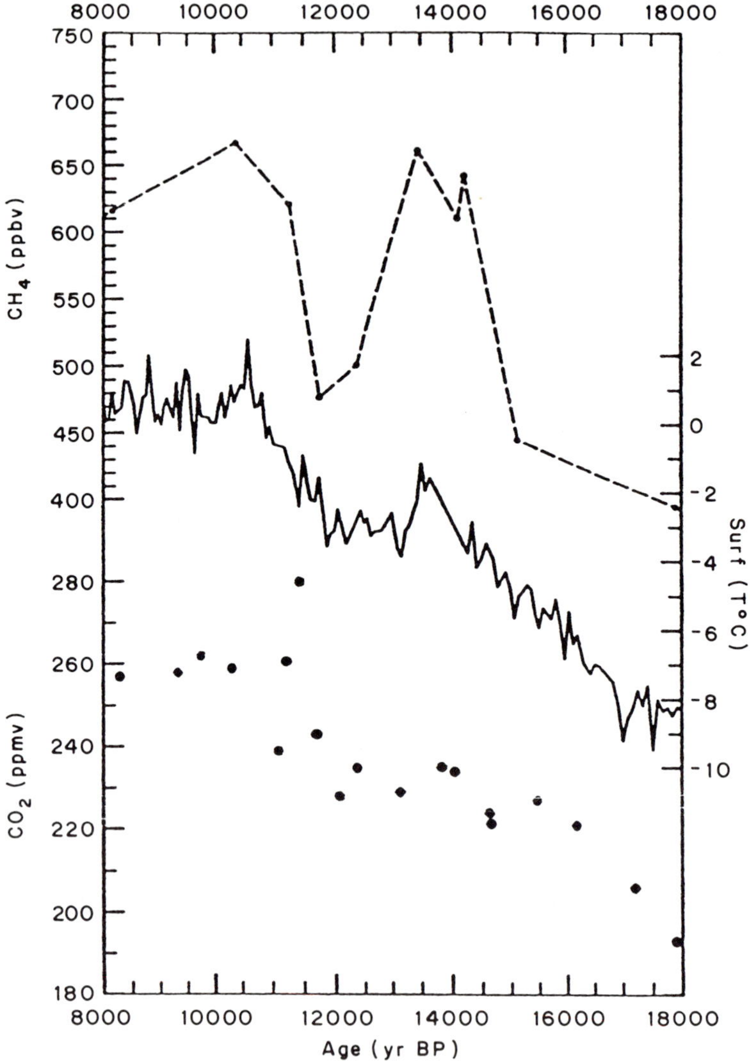

Figure 12.4 Variations of CO_2 and CH_4 concentrations during the last deglaciation as deduced from gas measurements: CH_4 from the Vostok core (dashed line) and CO_2 from the Vostok and Dome C cores) dots). The smoothed deuterium Vostok profile translated in terms of temperature is shown for comparision (solid line). From Jouzel et al. (1992).

connected with the Younger Dryas climatic event, as has been shown to be the case in the Northern Hemisphere.

The CH_4 signal is much clearer. The Antarctic Vostok record reveals a large CH_4 oscillation from about 650 to 480 ppbv, roughly at the time when the glacial–interglacial warming at Vostok was interrupted by a slight cold reversal (Chappellaz et al.

1990; see also Figure 12.4). However, we can not assess the phase relationship between CH_4 and temperature because of the uncertainty linked with the relative dating of air and ice and the limited time resolution of the methane measurements. Stauffer et al. (1988) also noted a low value in the measurements performed on the Greenland ice (Dye 3 core), which falls into the Younger Dryas event.

CONCLUSIONS

The data reviewed here lead to the following conclusions concerning the role of greenhouse gases as climatic factors and the causes of changes in greenhouse-gas concentration:

1. Atmospheric methane changed in phase with southern temperature during the last climatic cycle. This is true for the initiation and the termination of glaciation, as well as for the whole glacial period. This is consistent with the idea that the atmospheric concentration of methane was controlled by the extent or activity of wetlands (as a source of this trace gas), in parallel and with little delay (centuries) to climatic change.
2. During the last two deglaciations, CO_2 changed approximately simultaneously with temperature in high southern latitudes. The data are consistent with synchronous response, but also with a CO_2 lag of up to 1 Ka. During the initiation of the last glaciation, CO_2 lagged by several Ka behind southern temperature (and CH_4). The phase relationship between CO_2 and southern temperature was therefore not the same for initiation and termination of a glaciation. Spectral analysis of the deep-sea records, which does not distinguish between ice buildup and deglaciation, indicates that $\Delta\delta^{13}C$ (as a proxy for CO_2) leads SST.
3. Based on ice-core results, the phase relations between CO_2, CH_4, and ice volume during the penultimate deglaciation indicate that greenhouse-gas concentrations began to increase several Ka before a detectable change in the volume of continental ice. The deep-sea records consistently show that ice volume systematically lagged behind southern hemisphere SST and CO_2 (Imbrie et al. 1989) and also support a CO_2 increase starting before the continental ice volume change during the penultimate deglaciation (Shackleton and Pisias 1985; Curry and Crowley 1987). The Earth's radiation balance (i.e., the greenhouse effect) therefore started to change ahead of the ice volume changes.
4. Based on the above, we conclude that the initial greenhouse-gas increase during deglaciation was not caused by sea-level change. The marine sedimentary record indicates that Cd/Ca in the South Atlantic, which can be considered as an indicator of water mass distribution led the carbon cycle changes ($\Delta\delta^{13}C$) (Imbrie et al. 1989). This, together with the synchroneity of CO_2 and southern temperature, favors a mechanism of ocean internal forcing of the CO_2 changes.

In contrast, the CO_2 lag of several Ka behind southern temperature at the start of the glaciation is more difficult to use for constraining the mechanisms of the CO_2 change for that period, since there are uncertainties about the phase relationship between the various climatic factors.

5. North Atlantic SST lagged behind Southern Hemisphere SST, $\Delta\delta^{13}C$ and Cd/Ga and, for the 23-Ka cycle, behind ice volume (Imbrie et al. 1989). There seems to be a problem with the relative timing of these events: since North Atlantic SST and NADW formation seem to have varied synchronously during the last deglaciation, the deep-water chemistry of the South Atlantic (Cd/Ca) would be expected to change in phase with North Atlantic SST.
6. Around the Younger Dryas cold phase, the atmospheric level of methane decreased significantly after having increased with the first deglacial warming. The deglacial CO_2 increase seems to have been interrupted during or before the Younger Dryas. The cooling of North Atlantic surface waters is suspected to have triggered the Younger Dryas. The methane oscillation could be linked to this cooling which also affected Northern Hemisphere temperature and the hydrological cycle and therefore the continental methane sources.

During the glaciation, around 40 Ka B.P., rapid oscillations are recorded in Greenland ice, which are not directly related to orbital forcing. In parallel to these oscillations, the CO_2 content of trapped air changed abruptly. These changes have not been confirmed in Antarctic ice cores. If they were real, they would pose serious problems for our understanding of the global carbon cycle and its relation to climate.

The Younger Dryas-type oscillations with their abrupt climatic shifts provide the opportunity to check the leads and lags in detail. It is therefore very important to study the greenhouse-gas variations during these periods carefully, in particular to assess whether the rapid glacial CO_2 oscillations were real and to determine the phases of CO_2 and CH_4 with respect to the climatic parameters.

ACKNOWLEDGEMENTS

Thanks are due to J. Chappellaz, M. McElroy, and B. Stauffer for valuable comments on the manuscript. Our work was supported by the commission of the European Communities, the Programme National pour l'Etude de la Dynamique du Climat and the Commissariat à l'Energie Atomique, on the French Side, as well as by the Swiss National Science Foundation.

REFERENCES

Barnola, J.M., D. Raynaud, Y.S. Korotkevich, and C. Lorius. 1987. Vostok ice core: A 160,000-year record of atmospheric CO_2. *Nature* **329**:408–414.

Barnola, J.M., P. Pimienta, D. Raynaud, and Y.S. Korotkevich. 1991. CO_2-climate relationship as deduced from the Vostok ice core: A reexamination based on new measurements and on a reevaluation of the air dating. *Tellus* **43B**:83–90.

Boyle, E. 1988. *J. Geophys. Res.* **93**:15701–15714.

Broccoli, A.J., and S. Manabe. 1987. The influence of continental ice, atmospheric CO_2, and land albedo on the climate of the last glacial maximum. *Clim. Dynam.* **1**:87–99.

Broecker, W.S. 1982. Ocean chemistry during glacial time. *Geochim. Cosmo. Acta* **46**:1689–1705.

Broecker, W.S., and T.-H. Peng. 1989. The cause of the glacial to interglacial atmospheric CO_2 change: A polar alkalinity hypothesis. *Glob. Biogeochem. Cyc.* **3**:215–239.

Bryan, K., and M.J. Spelman. 1985. The ocean's response to a CO_2-induced warming. *J. Geophys. Res.* **90**:11679–11688.

Chappellaz, J. 1990. Etude du méthane atmosphérique au cours du dernier cycle climatique. Thesis, University of Grenoble.

Chappellaz, J., M.M. Barnola, D. Raynaud, Y.S. Korotkevich, and C. Lorius. 1990. Ice core record of atmospheric methane over the past 160,000 years. *Nature* **345**:127–131.

Cicerone, R.J., and R.S. Oremland. 1988. Biogeochemical aspects of atmospheric methane. *Glob. Biochem. Cyc.* **2**:299–327.

Curry, W.B., and T.J. Crowley. 1987. The $\delta^{13}C$ of equatorial Atlantic surface waters: Implications for ice age pCO_2 levels. *Paleoceanog.* **2**:489–517.

Hansen, J., A. Lacis, D. Rind, G. Russell, P. Stone, I. Fung, R. Ruedy, and J. Lerner. 1984. Climate sensitivity: Analysis of feedback mechanisms. In: Climate Processes and Climate Sensitivity, ed. J.E. Hansen and T. Takahashi. *Geophys. Monog.* **29**:130–163. Washington, D.C.: American Geophysical Union.

Imbrie, J., A. McIntyre, and A. Mix. 1989. Oceanic response to orbital forcing in the late Quaternary: Observations and experimental strategies. In: Climate and Geo-Sciences, ed. A. Berger et al., pp. 121–164. Dordrecht: Kluwer.

Jouzel, J., J.R. Petit, N.I. Barkov, J.M. Barnola, J. Chappellaz, P. Ciais, V.M. Kotlyakov, C. Lorius, V.N. Petrov, D. Raynaud, and C. Ritz. 1992. The last deglaciation in Antarctica: Further evidence of a "Younger Dryas" type climatic event. In: The Last Deglaciation: Absolute and Relative Dating, ed. E. Bard and W.S. Broecker.

Lorius, C., J. Jouzel, D. Raynaud, J. Hansen, and H. le Treut. 1990. The ice-core record: Climate sensitivity and future greenhouse warming. *Nature* **347**:139–145.

Maier-Reimer, E., and U. Mikolajewicz. 1989. Experiments with an OGCM on the cause of the Younger Dryas. In: Oceanography 1988, ed. A. Ayala-Castanarez, W. Wooster, and A. Yanez-Arancibia, pp. 87–100. Mexico: UNAM.

Neftel, A., H. Oeschger, T. Staffelbach, and B. Stauffer. 1988. CO_2 record in the Byrd ice core 50,000–5000 years B.P.. *Nature* **331**:609–611.

Oeschger, H. 1991. Paleodata, paleoclimates and the greenhouse effect. In: Climate Change: Science, Impacts and Policy, ed. J. Jäger and H.L. Ferguson, pp. 211–224. Proc. 2nd World Climate Conference. Cambridge: Cambridge Univ. Press.

Raynaud, D., and J.M. Barnola. 1985. CO_2 and climate; information from Antarctic ice core studies. In: Current Issues in Climate Research, ed. A. Ghazi and R. Fantecchi, pp. 240–246. Dordrecht: Reidel.

Raynaud, D., J. Chappellaz, J.M. Barnola, Y.S. Korotkevich, and C. Lorius. 1988. Climatic and CH_4 cycle implications of glacial–interglacial CH_4 change in the Vostok ice core. *Nature* **333**:655–657.

Shackleton, N.J., M.A. Hall, J. Line, and C. Shuxi. 1983. Carbon isotope data in core V19–30 confirms reduced carbon dioxide concentration of the ice age atmosphere. *Nature* **306**:319–322.

Shackleton, N.J., and N.G. Pisias. 1985. Atmospheric carbon dioxide, orbital forcing and climate. In: The Carbon Cycle and Atmospheric CO_2: Natural Variations Archaean to Present, ed. E.T. Sundquist and W.S. Broecker. *Geophys. Monog.* **32**:303–317. Washington, D.C.: American Geophysical Union.

Sowers, T., M. Bender, D. Raynaud, Y.S. Korotkevich, and J. Orchardo. 1991. The $\delta^{18}O$ of atmospheric O_2 from air inclusions in the Vostok ice core: Timing of CO_2 and ice volume changes during the penultimate deglaciation. *Paleoceanog.* **6(6)**:679–696.

Staffelbach, T., B. Stauffer, and H. Oeschger. 1988. A detailed analysis of the rapid changes in ice-core parameters during the last ice age. *Ann. Glaciol.* **10**:167–170.

Staffelbach, T., B. Stauffer, A. Sigg, and H. Oeschger. 1991. CO_2 measurements from polar ice cores: More data from different sites. *Tellus* **43H**:91–96.

Stauffer, B., H. Hofer, A. Neftel, J. Schwander, and U. Siegenthaler. 1984. Atmospheric CO_2 concentration during the last glaciation. *Ann. Glaciol.* **5**:160–164.

Stauffer, B., E. Lochbronner, H. Oeschger, and J. Schwander. 1988. Methane concentration in the glacial atmosphere was only half that of the preindustrial Holocene. *Nature* **332**:812–814.

Wenk, T., and U. Siegenthaler. 1985. In: The Carbon Cycle and Atmospheric CO_2: Natural Variations Archaean to Present, ed. E.T. Sundquist and W.S. Broecker. *Geophys. Monog.* **32**:185–194. Washington, D.C.: American Geophysical Union.

13

Role of Ice Sheets and Sea Level

K. HERTERICH
Fachbereich 5–Geowissenschaften, Postfach 330440,
2800 Bremen 33, F.R. Germany

ABSTRACT

Climate modeling is a general tool that permits insight into the role of ice sheets and sea-level changes in generating observed climatic variability. However, the climate-related proxy data derived from the geologic record cannot always be directly compared with the output of presently available ice-sheet climate models. Future model developments should provide this link. Results obtained with a hierarchy of ice-sheet climate models are in reasonable agreement with paleoclimatic data, concerning some first order patterns in space and time. In addition, these modeling activities have shown that a considerable part of the climatic variance can be explained by linear models. Nonlinear mechanisms are nevertheless at work within the climate system. They are most probably responsible for the major climatic fluctuations which occurred during the Quaternary: the 100,000 year ice-age cycle and the Younger Dryas cold interval which interrupted the last deglaciation phase.

INTRODUCTION

How can we use paleoclimatic data to learn about the role of ice sheets and sea level in generating the observed variability within the climate system? One tool is modeling the climate in its general sense. Climate models are defined by a set of equations (physical, chemical, and biological). They range from very simple models (visual inspection of the data and simple reasoning) to complex 3-D models designed to simulate the coupled climate system (atmosphere–ocean–ice–continent–biosphere). If the output of a specific model fits the data, the processes incorporated in the model are expected to be reasonable, and in this way "explain" the observations.

Obviously we cannot infer all mechanisms related to ice sheets and sea-level changes simply by fitting models to the data on ice quantities and sea level alone. However, modeling experiments have shown that it is possible to learn about first order processes by using an ice-sheet model as the major prognostic component and considering only a subset of further climatic variables. For example, atmospheric

Global Changes in the Perspective of the Past

energy-balance models, defining the mass balance at the surface of the model ice sheet (Pollard 1982), were quite successful in explaining at least part of the observed variance in the global ice-volume fluctuations during the last several 100,000 years. In these models, oceanic climate variables are completely left out.

The ultimate test of our understanding of how the climate system works will be a simulation of all the existing data, using 3-D circulation-type models of the coupled climate system. These models must be tested under paleoclimatic conditions very different from those existing at present before their predictions concerning future climate change can be trusted.

DATA FOR MODEL VERIFICATION

The basic quantities which are calculated by a 3-D ice-sheet model are the evolution of the ice-sheet thickness distribution, ice velocity, and ice temperature. The latter two quantities are generally not available from paleodata. The ice-sheet thickness distribution is only partially documented in terms of the global ice volume and the ice-sheet margins at glacial maximum (possibly also during the following retreat phase of the ice-age ice sheets; Boulton et al. 1985). Part of the global ice volume (ice shelves and some fraction of marine-based ice sheets) does not contribute to sea-level changes.

The data taken from the geologic record, such as isotopic ratios ($\delta^{18}O/\delta^{13}C$) and faunal abundances, cannot be directly compared with these model output variables. For example, the global ice-volume changes are assumed to be approximately proportional to the changes in the mean $\delta^{18}O$ of ocean waters documented in the ocean sediment (Shackleton and Opdyke 1973), and sea-level changes are derived from observed positions of fossil coast lines. At present, most modelers using the simpler ice-sheet climate models are attempting to fit these proxy data, mainly the SPECMAP-stack $\delta^{18}O$ record (Imbrie et al. 1984), rather than the corresponding climatic quantities.

Future model developments should also incorporate the transport of $\delta^{18}O$ within the atmosphere–ocean–ice system. The $\delta^{18}O$ of ocean waters will then be a model output variable which can be compared more directly with the $\delta^{18}O$ data. In order to predict the position of coast lines, a model of bedrock dynamics is required, from which the height of the continent with respect to sea level can be calculated (depending on ice load, sea level, and tectonic effects).

Fairbanks (1989) has reconstructed a sea-level record for the last 17,000 years from Barbados coral reefs (assuming a constant rate of tectonic uplift of 34 cm per 1000 years). By calculating the rate of sea-level change, two melt-water peaks—before and after the Younger-Dryas cold interval—could be identified (Figure 13.1). This record will serve as a valuable data set (together with data on atmospheric temperature and oceanic state (see Figure 13.1) to test coupled atmosphere/ocean/ice-sheet models.

For the Northern Hemisphere ice-age ice sheets, the following boundary conditions (not provided by the geological record) have to be modeled: the mass balance at the

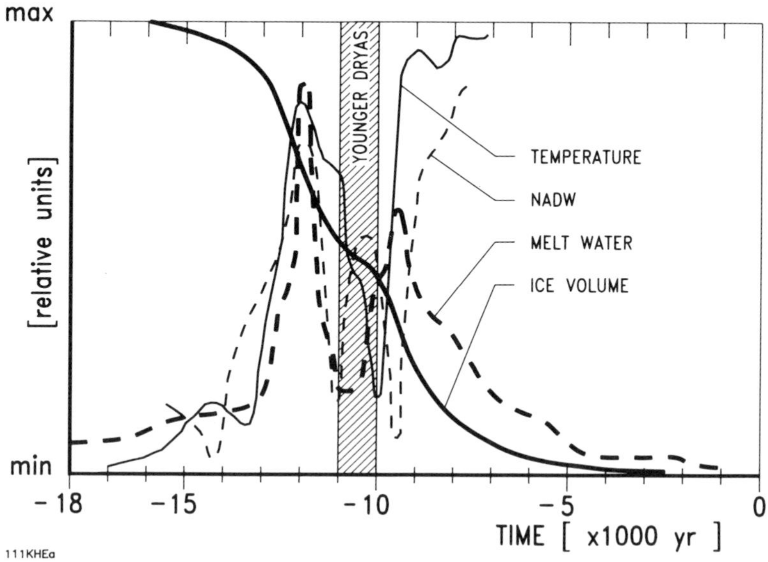

Figure 13.1 Qualitative changes in some climate variables during the last deglaciation: global ice volume and melt water (Fairbanks 1989); atmospheric temperature (Atkinson et al. 1987) and measure of North Atlantic deep-water (NADW) formation (Boyle and Keigwin 1987), as presented in Berger (1990). Note that uncertainties in the dating may introduce errors in the phases between different climate variables.

ice-sheet surface, the surface temperature, the sliding velocity at the ice-sheet bottom and the geothermal heat flux entering the ice sheet from below. However, in the case of the Antarctic ice sheet, some of these boundary conditions are provided by data. The surface temperature has been derived from the Vostok ice core for the last 160,000 years (Jouzel et al. 1987). Assuming presently observed lapse rates and measured correlations between surface temperature and mass balance, the Antarctic mass balance can also be determined during the last glacial cycle. Estimates of past sea-level changes (mainly caused by variations in the volume of the Northern Hemisphere ice-age ice sheets) can be used to force the marine-based part of the Antarctic ice sheet.

In addition to those paleodata from which ice-sheet quantities and sea level can be derived, paleodata on atmospheric and oceanic quantities are used to understand the impact of ice sheets on the climate system. In an atmospheric general circulation model (AGCM), the ice-sheet thickness distribution is a lower boundary condition (additional boundary conditions are insolation and sea-surface temperature). Available data for testing AGCM performance are temperatures and precipitation on land, and wind strength at a limited number of locations. Again, these climatic variables are estimates from proxy data (pollen, grain-size distributions) and involve rather complicated physical and biological sub-grid scale processes. In presently available AGCMs, with a typical horizontal resolution larger than 100 km, these sub-grid scale processes are not well represented.

Three-dimensional ocean general circulation models (OGCMs) have been developed (Maier-Reimer et al. 1991) which are able to calculate the stationary oceanic circulation (including salt and temperature distribution) under prescribed boundary conditions (fluxes of heat, momentum, and fresh water). With such models, the oceanic response to ice-sheet buildup and decay can be examined. Furthermore, the output from OGCMs can be used to drive bio-geochemical models predicting the distribution of tracers (such as $\delta^{13}C$, ^{14}C, and nutrients) which can be directly compared with corresponding data extracted from deep-sea sediments.

MODEL HIERARCHY AND MODELING RESULTS

Most of the models including an ice sheet as the major prognostic component view the ice-volume changes as a result of varying insolation (Milankovitch forcing), modified by internal dynamics within the climate system. The insolation changes during the last several million years that are used to force these models are generally those calculated by Berger (1978). The simplest nontrivial model of this sort is a linear first order differential equation relating ice-volume changes and insolation variations (Imbrie and Imbrie 1980). Such a linear relation was suggested by the data, showing that considerable variance of the $\delta^{18}O$ record is contained in the frequency bands of insolation variations (Hays et al. 1976). Fitting a linear model to the $\delta^{18}O$ record yields the time constant of the linear system and an amplification factor relating insolation to the mass balance of the ice sheet.

Further model developments added variables characterizing the ice-sheet shape and defined the chain of processes which transform the insolation signal into a mass balance at the ice-sheet surface in more detail (Pollard 1982). Most of these models were able to reproduce (at least qualitatively) the smaller fluctuations in the global ice volume during ice-sheet buildup (about 100,000 years), but experienced problems in modeling rapid ice-sheet decay (in less than 10,000 years) without substantial model tuning. At present, neither the chain of processes which link insolation and ice-sheet mass balance, nor the seemingly nonlinear mechanisms responsible for the observed asymmetry in the so-called 100,000 year ice-age cycle have been convincingly identified.

With the development of 3-D ice-sheet models, the changes in the ice-sheet margin of individual ice sheets as a function of time is a further model output. The main problem with the 3-D ice-sheet models is to define the mass balance at the ice-sheet surface as a function of longitude, latitude, and time. In the case of 2-D ice-sheet models (for a North–South cross section of the ice sheet), the atmosphere is generally modeled by a zonally symmetric energy-balance model, whereas an AGCM is required in order to provide an adequate 3-D mass balance. Given presently available computer resources, a typical AGCM cannot be integrated over an entire ice-age cycle. Thus, simpler atmospheric models will have to be developed (probably 2-layer models with a simplified hydrological cycle).

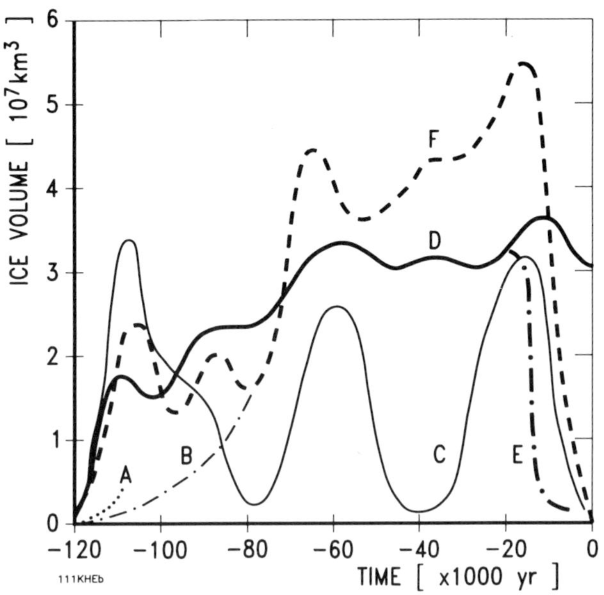

Figure 13.2 Variations in the global ice volume during the last glacial cycle as simulated with 3-D ice-sheet models by a number of authors, assuming different mass balances at the ice-sheet surface: (a) presently observed mass balances (Andrews and Mahaffy 1976); (b) snow line changes derived from observations (Oerlemans 1981); (c) assumptions on changes in the mass balance with insolation (Budd and Smith 1979); (d) energy-balance model with parameterized hydrological cycle (Deblonde and Peltier 1990); (e) pCO_2 influence on the mass balance (Lindstrom and MacAyeal 1989); (f) SPECMAP-stack $\delta^{18}O$ record, normalized to a glacial ice volume of 5.6×10^7 km^3.

At present, there is no 3-D model available which can account for all aspects of the observed ice-volume changes during the last ice-age cycle (see Figure 13.2). DeBlonde and Peltier (1990) succeeded in building up the Northern Hemisphere ice sheets during the last ice-age cycle by using an atmospheric global energy-balance model with a parameterized hydrological cycle. However, the model ice sheets did not decay as anticipated. Conversely, in the model of Lindstrom and MacAyeal (1989), ice-sheet decay occurred solely by increasing the atmospheric partial pressure of carbon dioxide (pCO_2) by 80 ppm. The impact of pCO_2 on the mass balance was extracted from AGCM sensitivity experiments. The rapid initial ice-sheet buildup within 5000 years (glacial inception) also seems to be difficult to explain (Andrews and Mahaffy 1976; Oerlemans 1981). Budd and Smith (1979) were able to simulate both initial ice-sheet buildup and final decay but overestimated ice-volume fluctuations induced by the Milankovitch insolation changes.

The role of sea-level change in causing climate variability has been studied by Huybrechts (1990) by simulating the Antarctic ice sheet during the last glacial cycle with a 3-D ice-sheet model. The results indicate that the global sea-level changes, caused by variations in the volume of the Northern Hemisphere ice sheets were responsible for the formation of Antarctic ice shelves in an interglacial and for their progressive replacement by the Antarctic ice sheet during glacial times (due to equatorward movement of the grounding line). However, the ice-sheet model of Böhmer and Herterich (1990) is rather insensitive to changes in sea level but depends strongly on the value of the bottom-friction coefficient in the area of ice streams. Both models do not show an unstable behavior of the West Antarctic ice sheet.

Atmospheric and oceanic circulation models have been successfully used to study the impact of ice sheets on the climate system by assuming that the atmosphere and the ocean respond rapidly to the boundary conditions imposed by the ice sheets. A comparison of paleodata with the results from a number of AGCM simulations of the last glacial maximum demonstrate that these models are able to simulate major features of the glacial atmosphere (Lautenschlager and Herterich 1990). A numerical experiment, related to ice sheets and using an OGCM, has also been performed. Estimated freshwater fluxes from the decay of the Laurentide ice sheet have been used to force the model ocean. When temperature and wind stress at the ocean surface are kept at their present values, these meltwater fluxes are able to stop deep-water production in the North Atlantic (Bryan 1986; Maier-Reimer and Mikolajewicz 1989).

LINEAR AND NONLINEAR CLIMATIC PROCESSES RELATED TO ICE SHEETS

Modeling experiments with simple models have shown that some 10-50% of the variance in the global ice volume can be explained as a linear response of the ice sheets to variations in insolation. This linearity guarantees that linear time-series analysis of the data provides insight into at least part of climate dynamics on ice-age time scales (Imbrie et al. 1989). Linear models may incorporate both negative and positive feedbacks (the latter only as long as they do not severely perturb the system). However, there is a considerable amount of climatic variance which seems to be a result of nonlinear dynamics within the climate system.

The ice sheets introduce delays in the transformation of the external solar forcing into variations of the climate on a number of different time scales (Hindmarsch et al. 1989). In a linear ice-sheet model only one time constant would be involved. In the initial phase of ice-sheet buildup, where the ice-thickness changes due only to snow accumulation without ice flow, the areal coverage of snow/ice fields follows the insolation with almost zero lag. The snow-albedo feedback is therefore active from the beginning of ice-sheet buildup. A further positive feedback is the so-called ice thickness/mass balance feedback (thicker ice implies colder surface temperature and thus higher positive mass balances) which is even stronger than the ice-albedo feedback. Due to the nonlinear dependence of the ice flow on ice thickness, extended flow starts only after the ice thickness exceeds some threshold. Climatic effects connected to the ice-sheet shape are therefore delayed by several thousand years. The longest delay derives from the temperature evolution within the ice sheet, with time scales of up to 100,000 years. Deformational properties within the ice (which strongly depend on ice temperature) may change appreciably, with resulting changes in the ice-sheet shape which are unrelated to the state of the rest of the climate system. If polythermal ice existed in ice-age ice sheets, with cold ice lying above temperate ice (a mixture of ice and water at the pressure melting point) this would add further nonlinear processes in determining ice-sheet evolution (Hutter et al. 1988).

In addition to the nonlinear response of the ice sheet to the forcing, the ice sheet is also driving the two other major components of the climate system (the atmosphere and the ocean) in a nonlinear way. Higher resolution AGCMs show that the Laurentide ice sheet caused a jet-splitting which forced the atmospheric flow around the ice sheet (Kutzbach and Guetter 1986; Lautenschlager and Herterich 1990). The response of the ocean to melt water from the ice-age ice sheets is possibly also highly nonlinear. The model ocean reacts to the melt water by stopping the formation of deep water in the North Atlantic. If this meltwater flux is switched off, however, the deep water formation in the model is not necessarily turned on again (Maier-Reimer, pers. comm.).

CONCLUSIONS

The ice sheets introduce a number of different time constants in the response of the climate to external insolation forcing. Climate models, to be compared with paleodata, should therefore contain an ice-sheet part (including thermomechanics of ice) as an active component. In the case of the smaller ice-volume fluctuations in the Milankovitch band, the ice sheets are responding almost linearly with some atmospheric and oceanic variables in turn following the boundary conditions (set by the ice sheets) in a quasi-stationary way. This gives support to using linear time-series analysis as a tool to extract physically meaningful phases and amplitudes from the deep-sea geological record at the Milankovitch frequencies. However, the major climatic fluctuations (such as glacial inception, deglaciation, and the Younger-Dryas cold interval) are most probably a result of nonlinear mechanisms within the coupled climate system (atmosphere/ocean/ice/continent/biosphere) which have not yet been uniquely identified.

ACKNOWLEDGEMENTS

Paleoclimatic modeling at the Max-Planck-Institut für Meteorologie is funded by the Bundesministerium für Forschung und Technologie (contract 07 KF012/7) and by the European Community (contract EPOC–0004–C).

REFERENCES

Andrews, J.T., and M.A.W. Mahaffy. 1976. Growth rate of the Laurentide ice sheet and sea level lowering (with emphasis on the 115,000 B.P. sea level low). *Quat. Res.* **6**:167–183.

Atkinson, T.C., K.C. Briffa, and G.R. Coope. 1987. Seasonal temperatures in Britain during the past 22,000 years, reconstructed using beetle remains. *Nature* **325**:587–592.

Berger, A.L. 1978. Long-term variations of daily insolation and Quaternary climatic changes. *J. Atmos. Sci.* **35**:2362–2367.

Berger, W.H. 1990. The Younger Dryas cold spell: A quest for causes. *Palaeogeog. Palaeoclim. Palaeoecol. (GPCS)* **89**:219–237.

Böhmer, W., and K. Herterich. 1990. A simplified 3-D ice sheet model including ice shelves. *Ann. Glaciol.* **14**:17–19.

Boulton, G.S., G.D. Smith, A.S. Jones, and J. Newsome. 1985. Glacial geology and glaciology of the last mid-latitude ice sheets. *J. Geol. Soc. (Lon.)* **142**:447–474.

Boyle, E.A., and L.D. Keigwin. 1987. North Atlantic thermohaline circulation during the past 20,000 years linked to high latitude surface temperature. *Nature* **330**:35–40.

Bryan, F. 1986. High latitude salinity effects and interhemispheric thermohaline circulations. *Nature* **305**:301–304.

Budd, W.F., and I.N. Smith. 1979. The growth and retreat of ice sheets in response to orbital radiation changes. In: Sea level, Ice and Climate Change. IAHS Publ. 131, pp. 369–409.

DeBlonde, G., and W.R. Peltier. 1990. A model of late Pleistocene ice sheet growth with realistic geography and simplified cryodynamics. *Clim. Dynam.* **5**:103–110.

Fairbanks, R.G. 1989. A 17,000 year glacio-eustatic sea-level record: Influence of glacial melting rates on the Younger Dryas event and deep ocean circulation. *Nature* **342**:637–642.

Hays, J.D., J. Imbrie, and N.J. Shackleton. 1976. Variations in the Earth's orbit: Pace maker of the ice ages. *Science* **194**:1121–1132.

Hindmarsh, R.C.A., G.S. Boulton, and K. Hutter. 1989. Modes of operation of thermo-mechanically coupled ice sheets. *Ann. Glaciol.* **12**:57–69.

Hutter, K., H. Blatter, and M. Funk. 1988. A model computation of moisture content in polythermal glaciers. *J. Geophys. Res.* **93(B10)**:12205–12214.

Huybrechts, P. 1990. A 3-D model for the Antarctic ice sheet: A sensitivity study on the glacial–interglacial contrast. *Clim. Dynam.* **5**:79–92.

Imbrie, J., J.D. Hays, D.G. Martinson, A. McIntyre, A.C. Mix, J.J. Morley, N.G. Pisias, W.L. Prell, and N.J. Shackleton. 1984. The orbital theory of Pleistocene climate: Support from a revised chronology of the marine δ^{18}O record. In: Milankovitch and Climate, ed. A.L. Berger, et al., part 1, pp. 269–305. Dordrecht: Reidel.

Imbrie, J., and J.Z. Imbrie. 1980. Modeling the climatic response to orbital variations. *Science* **207**:943–953.

Imbrie, J., A. McIntyre, and A. Mix. 1989. Oceanic response to orbital forcing in the late Quaternary: Observational and experimental strategies. In: Climate and Geo-Sciences. ed. A. Berger, et al., pp. 121–164. Dordrecht: Kluwer.

Jouzel, J., C. Lorius, J.R. Petit, C. Gethon, N.I. Barkov, V.M. Kotlyakov, and V.M. Petrov. 1987. Vostok ice core: A continuous isotope temperature record over the last climatic cycle (160,000 years). *Nature* **329**:403–408.

Kutzbach, J.E., and P.J. Guetter. 1986. The influence of changing orbital parameters and surface boundary conditions on climate simulations for the past 18,000 years. *J. Atmos. Sci.* **43**:1726–1759.

Lautenschlager, M., and K. Herterich. 1990. Atmospheric response to ice-age conditions: Climatology near the Earth's surface. *J. Geophys. Res.* **95(D13)**:22,547–22,557.

Lindstrom, D.R., and D.R. MacAyeal. 1989. Scandinavian, Siberian and Arctic Ocean glaciation: Effect of Holocene atmospheric CO_2 variations. *Science* **245**:628–631.

Maier-Reimer, E., and U. Mikolajewicz. 1989. Experiments with an OGCM on the cause of the Younger Dryas. Report No. 39. Hamburg: Max-Planck-Institut für Meteorologie.

Maier-Reimer, E., U. Mikolajewicz, and K. Hasselmann. 1991. On the sensitivity of the global ocean circulation to changes in the surface heat flux forcing. Report No. 68. Hamburg: Max-Planck-Institut für Meteorologie.

Oerlemans, J. 1981. Modeling of Pleistocene European ice sheets: Some experiments with simple mass balance parameterizations. *Quat. Res.* **15**:77–85.

Pollard, D. 1982. A simple ice sheet model yields realistic 100 Kyr glacial cycles. *Nature* **269**:334–338.

Shackleton, N.J., and N.D. Opdyke. 1973. Oxygen isotope and paleomagnetic stratigraphy of Equatorial Pacific core V28–238: Oxygen isotope temperatures and ice volumes on a 10^5 and 10^6 year scale. *Quat. Res.* **3**:39–55.

14

Global Ocean Circulation Changes

J.-C. DUPLESSY[1] and E. MAIER-REIMER[2]
[1]Centre des Faibles Radioactivités, Laboratoire Mixte CNRS–CEA,
91198 Gif-Sur-Yvette Cedex, France
[2]Max-Planck-Institut für Meteorologie, Bundesstrasse 55,
2000 Hamburg 13, F.R. Germany

INTRODUCTION

Increasing evidence suggests that the Earth's climate is paced by insolation variations due to orbital cycles (Hays et al. 1976; Imbrie et al. 1984). However, this forcing alone is not sufficient to account for the observed temperature and ice-extent changes (Rind et al. 1989). Moreover, abrupt climatic changes have been described which are far too rapid to be explained by the astronomical mechanism. As the global ocean circulation is an important component of the climatic system, Broecker and Denton (1989) proposed that Quaternary glacial–interglacial cycles were dominated by abrupt reorganizations of the ocean–atmosphere system, driven by changes in freshwater transports which impact salt structure in the sea and produce dramatic changes of the global oceanic circulation. These reorganizations may either be driven by astronomical insolation changes or be part of a self-sustained internal oscillation that would operate in the absence of changes in the Earth's orbital parameters.

It is therefore important to reconstruct the main trends of global ocean circulation during the last climatic cycles. The main methods developed to achieve this goal are reported in Table 14.1. The CLIMAP program built upon advances in marine micropaleontology (Imbrie and Kipp 1971) and isotopic geochemistry (Emiliani 1955; Shackleton and Opdyke 1973) mapped in detail the variability of surface ocean temperature and inferred surface circulation changes associated with the last glacial maximum (LGM) (CLIMAP 1976). Sea surface salinity estimates for the North Atlantic have been obtained only recently (Duplessy et al. 1991).

Geochemical methods enable us to reconstruct the flow lines within the deep-water realm. Numerous deep-sea cores have been analyzed for both faunal and geochemical tracers (Cd, $\delta^{13}C$, $\delta^{18}O$, ^{14}C), but only selected events of the last 150,000 years have been studied with enough detail to permit a reconstruction of the deep-water circulation on an almost global scale (Boyle 1988, 1992; Boyle and Keigwin 1982, 1987;

Global Changes in the Perspective of the Past
Edited by J.A. Eddy and H. Oeschger © 1993 John Wiley & Sons Ltd.

Table 14.1 Information derived from the study of deep-sea sediments.

Time Scale
- AMS ^{14}C
- U/Th
- Isotope stratigraphy
- Paleomagnetism record

Surface Water Realm
- Summer and winter sea surface temperature (global)
- Sea surface salinity (possibly global)
- Depth of the nutricline (tropical and subtropical ocean)
- Dominant winds (local)

Intermediate and deep-water realm
- Benthic foraminiferal $\delta^{18}O$ and deep-water temperature
- Cd and PO_4 content
- ^{13}C of total dissolved CO_2

Biological Parameters
- Concentration and flux of organic carbon to the sediment and paleoproductivity (from an empirical formula)
- Concentration and flux of biogenic carbonate to the sediment and carbonate paleoproductivity

CO_2 and Carbonate Chemistry Parameters
- Depth of the CCD and lysocline
- $\Delta\delta^{13}C$ between surface and deep waters

Broecker et al. 1984, 1988; Broecker, Peng, Trumbore et al. 1990; Curry et al. 1988; Duplessy et al. 1980, 1984, 1988; Duplessy and Shackleton 1985; Kallel et al. 1988; Labeyrie et al. 1987; Oppo and Fairbanks 1987; Shackleton et al. 1988). However the conditions prevailing in some important oceanic basins (in particular the Southern Ocean and the South Pacific) are still poorly documented, and a coherent picture of the global ocean circulation can only be obtained by ocean modeling. In this chapter, we will compare information derived from both sediment core data and ocean circulation models to analyze conditions prevailing during the LGM.

MAIN CHARACTERISTICS OF THE OCEAN CIRCULATION DURING LGM

The Atlantic Ocean

Duplessy et al. (1988) reconstructed a north–south section between 60°N and 45°S for the last glacial maximum (LGM). The $\delta^{13}C$ value of the total dissolved CO_2 has been estimated as the mean $\delta^{13}C$ value of the foraminifera genus *Cibicides* analyzed in the LGM level. Results have been plotted in Figure 14.1 as a depth-latitude section in the

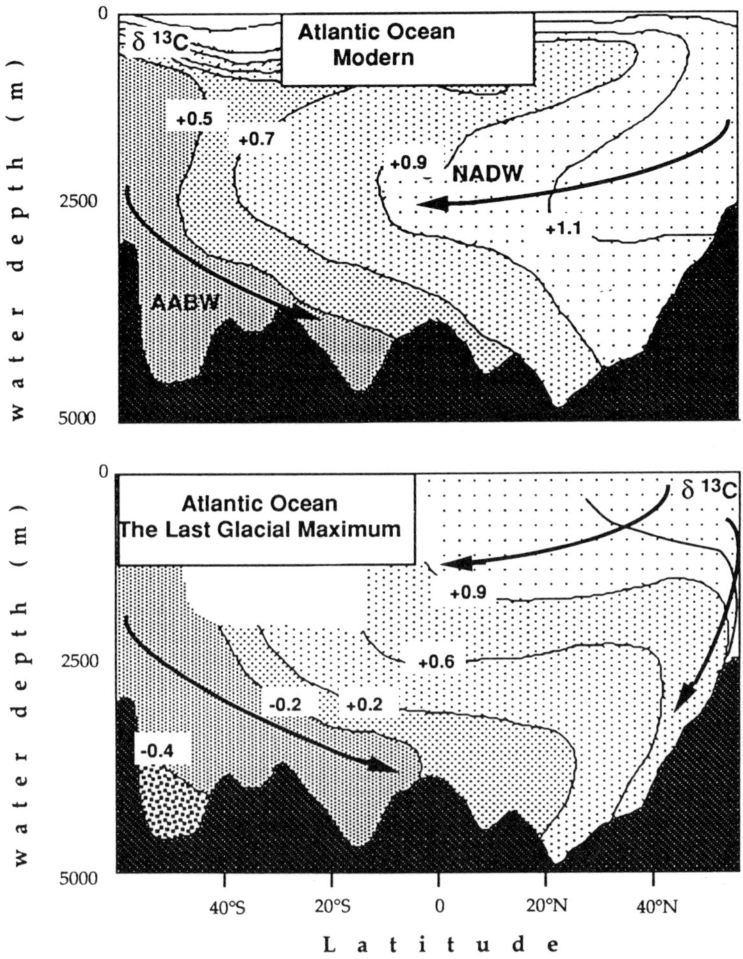

Figure 14.1 Reconstruction of the $\delta^{13}C$ changes across the Atlantic Ocean for the modern period (redrawn from Kroopnick 1984) and the LGM (adapted from the benthic foraminifera $\delta^{13}C$ distribution determined by Duplessy et al. 1988 and taking into account new measurements by Shackleton).

Atlantic Ocean, without taking into account the longitude of the analyzed cores. The original figure was modified to take into account recent measurements at 50°N by N.J. Shackleton, which indicate the presence of well-ventilated waters at about 1000 m water depth (Labeyrie et al. 1992). In the very high latitudes of the North Atlantic, a Glacial North Atlantic Deep Water mass, with high $\delta^{13}C$ values is found. This water mass invades the whole North Atlantic north of 45°N and extends as intermediate waters (above 2500 m) southward to 15°S. Below 3000 m, most of the deep eastern

Atlantic was filled with bottom water originating from the Southern Ocean so that the relative contribution of water from the Southern Hemisphere as compared to that from the Northern Hemisphere in the deep Atlantic was much greater than today.

Boyle and Keigwin (1982, 1985) and Boyle (1992) showed that the Cd content of the deep waters was about 50% higher than in the modern ocean, whereas the intermediate waters were Cd-depleted compared to deeper waters. The resulting hydrologic pattern agrees with the distribution of benthic foraminifera described by Streeter and Shackleton (1979) and Caralp (1987), and the changes of the carbonate compensation depth (Gardner and Hays 1976).

The Indian Ocean

Kallel et al. (1988) evidenced a very strong and sharp discontinuity separating the high $\delta^{13}C$ values of Intermediate Water to a depth of about 2000 m from negative $\delta^{13}C$ values deeper than 2300 m during LGM (Figure 14.2). This is in sharp contrast with the modern situation where the vertical $\delta^{13}C$ gradient in the North Indian Ocean is lower than 0.1 per mil in this depth range. This change in the vertical profile originates from a decrease of the $\delta^{13}C$ values of Bengal Gulf Deep Water by about 1 ‰ during LGM. At this time, the mean carbon isotopic composition of the total dissolved CO_2 in the global ocean was 0.32 ‰ smaller than the modern value (Duplessy et al. 1988) as a result of the reduced mass of the continental biosphere associated with glacial climatic conditions (Shackleton 1977). The smaller glacial to interglacial $\delta^{13}C$ change for Intermediate Water, compared with this global $\delta^{13}C$ signal, indicates an enhanced glacial ventilation of this water mass, while the larger $\delta^{13}C$ change for Deep Indian Water indicates a very weak ventilation of the deep Indian Ocean during the last glacial maximum.

The profile obtained by plotting the $\delta^{18}O$ values of benthic foraminifera against depth also shows a sharp discontinuity at the same level as the $\delta^{13}C$ profile in both the Bay of Bengal and Arabian Sea. This discontinuity indicates that intermediate and deep waters had much more distinct characteristics in glacial times than under modern conditions.

The Pacific Ocean

Under present conditions, most areas of the Pacific Ocean lie below the carbonate compensation depth, and foraminifera are absent in the sediment. However, it has long been observed that, in contrast with the deep north Atlantic Ocean, dissolution was less intense in most of the Pacific Ocean during LGM, with a deepening of a few hundred meters of the carbonate compensation depth (Broecker 1982). This indicates a significant change in the chemistry of the deep waters, with a drop in the water total CO_2 content which may possibly be related to an increase in ventilation at intermediate depth. By contrast with modern conditions, at the LGM, benthic foraminifera are often present in north Pacific sediments shallower than 4000 m. Duplessy et al. (1988) have

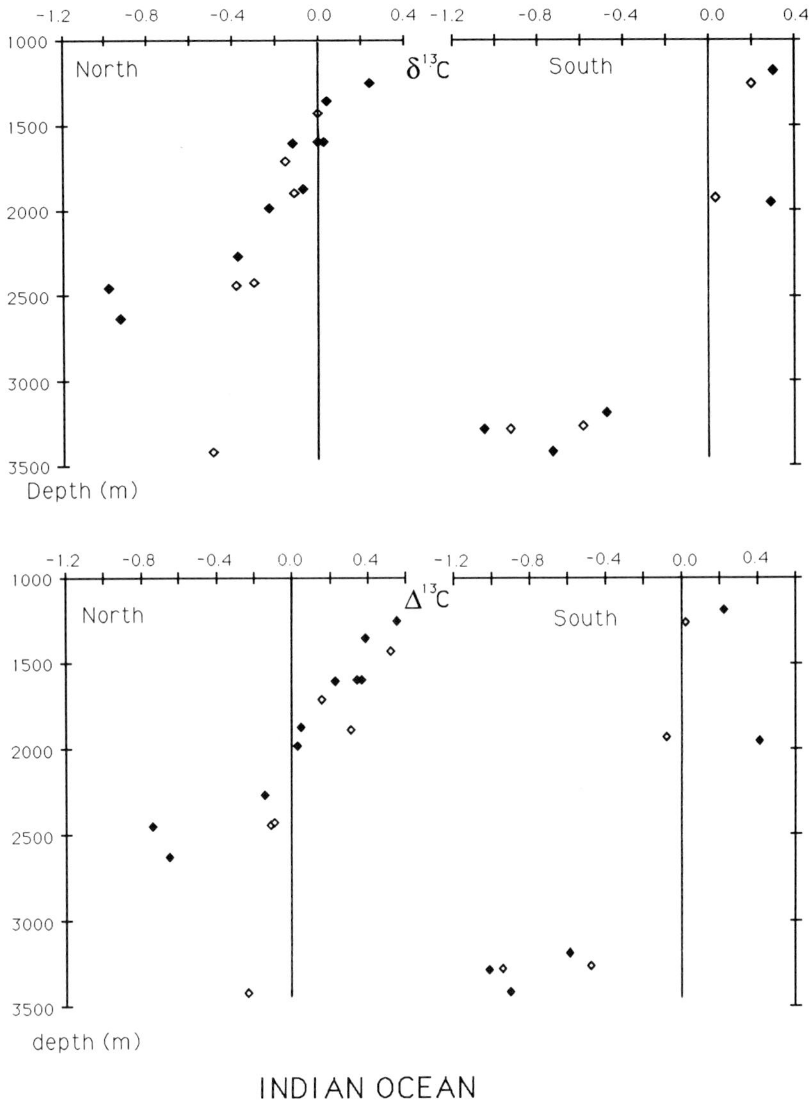

Figure 14.2 $\delta^{13}C$ profile reconstructed in the northern and southern Indian Ocean from the benthic foraminifera $\delta^{13}C$ distribution determined by Kallel et al. (1988), with additional measurements by Labeyrie. The $\Delta^{13}C$ value is calculated as $\delta^{13}C(LGM) - \delta^{13}C(modern) - 0.32$ to take into account the mean global oceanic $\delta^{13}C$ change (– 0.32‰) associated with the LGM conditions. White diamonds: western basin; black diamonds: eastern basin.

therefore generated a one-dimensional profile reconstructing the mean variations with depth of the $\delta^{13}C$ of the total dissolved CO_2 in the north and equatorial Pacific (Figure 14.3). A similar profile was generated for the present conditions (Figure 14.3) by interpolating from the GEOSECS data (Kroopnick 1974, 1985). The shape of the two profiles is markedly different, as the $\delta^{13}C$ minimum that is present today between 1 and 2.6 km water depth is absent during the LGM. The glacial profile for the Pacific Ocean exhibits $\delta^{13}C$ values which are identical to the modern values for the 1000–2600 m depth range. Deeper than 3000 m, $\delta^{13}C$ values are lower than modern ones by 0.3 ‰ in agreement with the data of Keigwin (1987).

Figure 14.3 also shows a depth profile of the difference in $\delta^{13}C$ between the LGM and today's values. A vertical line indicates the global mean shift of – 0.32 ‰. All the $\delta^{13}C$ values from water depths greater than 2600 m show a $\delta^{13}C$ shift indistinguishable from this global mean. By contrast, the mean $\delta^{13}C$ shift for the 1000–2600m depth range is significantly smaller than the global signal. This implies a relatively larger contribution of ventilated water to this water mass during LGM than in the modern ocean. The origin of the ventilated water is difficult to trace with the present data set. The contribution of low-nutrient, oxygenated surface water to the Pacific intermediate waters and the great thickness of this water mass is consistent with two features of glacial conditions in the Pacific Ocean: (a) the increase in abundance of *Cibicides* in the north Pacific, since this genus prefers to live in oxygenated waters (Streeter and Shackleton 1979) and (b) the decrease in carbonate dissolution and the deepening of the carbonate lysocline (Berger 1973; Thompson and Saito 1974).

The Southern Ocean

Under present conditions, the water which has upwelled from the deep circumpolar current in the vicinity of the Antarctic coast is further cooled by heat exchange with the atmosphere and shelf ice and is rendered more saline by sea-ice formation, so that it sinks rapidly. Its residence time at the surface is so short that neither phytoplankton utilization nor gas exchange with the atmosphere significantly reduce its CO_2 content (Weiss et al. 1979). As a result, the newly formed Antarctic Bottom Water (AABW) is strongly undersaturated with respect to carbonate, and we observed that calcareous foraminifera are entirely absent from the Weddell Sea. Planktonic and benthic foraminifera are only preserved in the Southern Ocean, far from the Antarctic continent, in sediment shallower than 3.5 km, slightly above the main body of the AABW. However, mixing is strong and the whole Southern Ocean deep-water mass below 1500 m has fairly homogeneous $\delta^{13}C$ values south of the polar front (Kroopnick 1985).

Only few $\delta^{13}C$ data from benthic foraminifera are available (Curry et al. 1988; Kallel 1988; Duplessy and Labeyrie 1989). All are characterized by very negative values during the glacial stages 2–4. These low $\delta^{13}C$ values in the Southern Ocean correspond with the values reported by Curry and Lohmann (1982) for the Vema

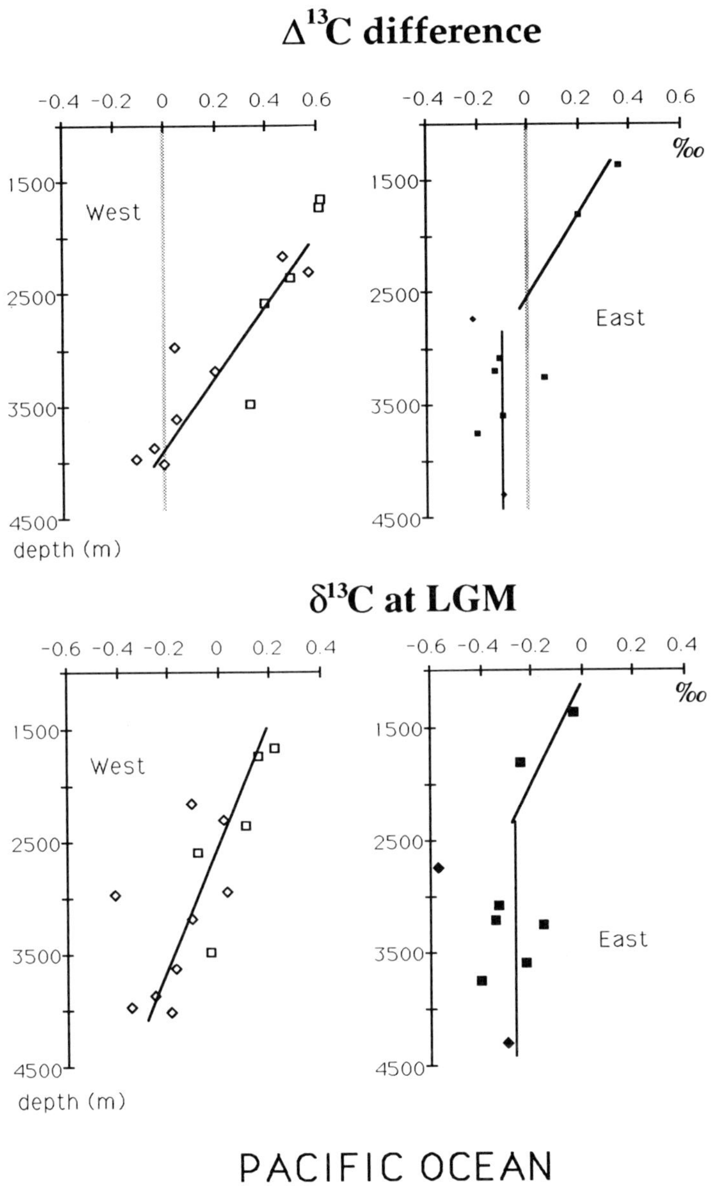

Figure 14.3 $\delta^{13}C$ profiles reconstructed in the western and eastern Pacific Ocean from the benthic foraminifera $\delta^{13}C$ distribution determined by Duplessy et al.(1988), with additional measurements by Labeyrie. The $\Delta^{13}C$ value is calculated as $\delta^{13}C(LGM) - \delta^{13}C(modern) - 0.32$ to take into account the mean global oceanic $\delta^{13}C$ change (– 0.32‰) associated with the LGM conditions.

Channel during the last glaciation. Such low $\delta^{13}C$ values suggest that the water which sank at high latitude was much less oxygenated than today, probably because the sea-ice cover extended about 1,000 km further north than under modern climatic conditions (CLIMAP 1976) and prevented gas exchange with the atmosphere when deep water upwelled close to the surface. The formation of deep water in the Southern Ocean would thus probably result from the activity of polynya, which developed within the ice pack.

UNSOLVED QUESTIONS

What Was the Origin of the Intermediate Water Mass with High Values in the North Atlantic Ocean?

Downcore analysis of $\delta^{13}C$ and Cd in benthic foraminifera from North Atlantic sediment cores shows the presence of deep waters (below 2500 m) enriched in nutrients relative to today and intermediate waters rich in $\delta^{13}C$ and significantly depleted in nutrients compared to the underlying waters. Whereas the reconstruction of the $\delta^{13}C$ profile in the Atlantic Ocean (Figure 14.1) shows that the low-$\delta^{13}C$, high-Cd deep waters formed in the Southern Ocean probably by a mechanism similar to the present one, the origin of the Intermediate waters is explained either by the sinking of well-ventilated nutrient-depleted surface water (Boyle and Keigwin 1987) or by the enhanced contribution of the Mediterranean outflow water (Oppo and Fairbanks 1987) observed off Gibraltar by Zahn et al.(1987).

A strong contribution of Mediterranean outflow water in the Atlantic intermediate water appears ruled out by the low Ba content of the Atlantic water (Lea and Boyle 1990). If we accept an Atlantic origin, the location of the sources of intermediate waters are not known and may be sought either north of the polar front or at the edge of the front itself. Jansen has even suggested that convection may have been significant in the Norwegian Sea, the salinity of surface water increasing during winter conditions by brine formation.

What Was the Origin of the Intermediate Water Mass with High $\delta^{13}C$ Values in the Indian Ocean?

Intermediate waters in the glacial Indian Ocean were enriched in $\delta^{13}C$ (Kallel et al. 1988) and Cd-depleted (Boyle 1992), while deep waters were strongly depleted in $\delta^{13}C$ and rich in Cd. This feature may be explained by either intermediate water formation in the Northern Arabian Sea or by expansion of Antarctic Intermediate Water north of 10°S. Increased flow of Intermediate water from the Pacific Ocean through the Indonesian Archipelago is ruled out on the basis of the low Cd values in the Intermediate Indian Ocean opposed to the high Cd values in the Pacific (Boyle 1992).

What Circulation Pattern Allowed Glacial $\delta^{13}C$ Values in the Deep Pacific Ocean to be Higher than Those of the Southern Ocean?

Very low $\delta^{13}C$ values (lower than – 0.40‰ vs. PDB) were measured at latitudes higher than 40°S in the deep Southern Ocean whereas the mean value for the deep Pacific Ocean is close to – 0.30. $\delta^{13}C$ values from North Pacific cores were also low (Keigwin 1987; Duplessy et al. 1987) suggesting that no deep water formed in the North Pacific during the LGM. Consequently, the mechanism by which the $\delta^{13}C$ value of Pacific deep water, which originates in the Southern Ocean, became progressively less negative is still unknown.

Why Did Cd and $\delta^{13}C$ Provide Discordant Evidence in the Southern and Pacific Oceans?

While $\delta^{13}C$ data were low and indicate that Antarctic (and Indian Ocean) deep waters were slightly more nutrient enriched than in the deep Pacific, Cd data indicate only minor changes in the nutrient content of glacial Antarctic deep waters compared with the modern ocean (Boyle 1992).

In the intermediate Pacific waters (above 2400 m), $\delta^{13}C$ data suggest increased ventilation and therefore nutrient levels depleted relative to the glacial deep waters and to the modern intermediate waters, Cd data indicate only a slight reduction of nutrient values in the glacial intermediate Pacific compared to modern values, but the glacial nutrient levels are higher in the intermediate waters than in the deep waters (Boyle 1992).

Was the Redfield Ratio Similar to the Modern One During LGM?

Shackleton et al. (1983) showed that the $\delta^{13}C$ difference between planktonic and benthic foraminifera increased during the last glaciation. Following a model developed by Broecker (1982), he showed that this $\Delta\delta^{13}C$ curve may be related to the low atmospheric CO_2 content during the last glaciation. However, Broecker (1982) credits P. Weyl with an alternative explanation, a change in the Redfield ratio, i.e., the ratio of organic carbon to phosphorus and nitrogen in organic matter formed in surface ocean waters. Shackleton and Pisias (1985) showed that the $\Delta\delta^{13}C$ curve obtained in Pacific core V 19–30 may be explained by an increase of the Redfield ratio from 106 under modern conditions to about 140 during LGM. No geological data provide a means to check this hypothesis.

What Was the Salinity Difference between Water Sinking to the Abyss in the Northern and Southern Hemispheres?

Under modern conditions, the newly formed NADW has a $\delta^{18}O$ value which is about 0.3% higher than the newly formed AABW, because the salinity of surface waters in the North Atlantic Ocean is greater than that of the Southern Ocean.

Under glacial conditions, the $\delta^{18}O$ value of the newly formed intermediate and deep waters is still unknown. This lack of knowledge is a major source of uncertainty for estimating deep-water temperature from the benthic foraminiferal $\delta^{18}O$ data from the various oceanic basins.

THE ROLE OF THE OCEANS: IMPLICATIONS FROM OCEAN MODELS

Up to now, coupled ocean–atmosphere models have suffered from the fact that the separately developed models do not balance each other with respect to the errors in the fluxes across the air–sea interface. Consequently, the climate of the coupled model exhibits a drift even in attempts to simulate the present climate. A systematic tuning of free parameters to remove such drifts would require so many experiments that even with today's computer facilities it is not feasible. It is convenient to introduce flux corrections (Sausen et al. 1988) in which mismatches of the fluxes of the present climate are added to the fluxes in the coupled system. By this technique the submodels can be coupled to a rather stationary model of the present climate which allows, at least, the investigation of the effect of small perturbations. For the simulation of a completely different climate state, however, with potentially increased impact of nonlinearities, the flux correction is conceptually not applicable.

Although substantial progress has been made in modeling the atmospheric response to LGM boundary conditions (e.g., Kutzbach and Guetter 1986; Broccoli and Manabe 1987; Rind 1987; Lautenschlager and Herterich 1990), relatively little has been done on the glacial ocean circulation. Aside from the somewhat inferential studies of Miller and Russell (1989) and Keffer et al. (1988), there has been no attempt to model the three-dimensional glacial response of the ocean and couple a global carbon model to a three-dimensional ocean simulation 18,000 years before present (18 Ka). For a critical appraisal of the modeling efforts, however, the circulation and sediment distribution of the deep ocean—which is the least determined part of the system by the available boundary conditions—are the most sensitive dynamic variables; the model sediment allows in addition a direct comparison with data without any empirical transfer functions.

As stated above, there are up to now only crude approximations of the contrast in surface salinity of the major oceans at the LGM as compared with the modern ocean. On the other hand, it is widely accepted that even small changes of salinity in the regions of deep-water formation crucially affect the deep circulation. The pronounced difference between the modern Atlantic and Pacific is attributed exclusively to the very different freshwater budgets in the northern parts, characterized by the high precipitation in the northern Pacific (Warren 1983) which is fed primarily by the evaporation excess in the Atlantic (Broecker, Peng, Jouzel et al. 1990). The modern oceanic conveyor belt is driven primarily by the cooling of salty water in the northern Atlantic (Gordon 1986). Even for constant fluxes the Atlantic is believed to have the potential for at least two different modes of operation (Broecker et al. 1985; Manabe and Stouffer 1988). This conjecture, which was anticipated by the analysis of an extremely simplified heat-salt oscillator (Stommel 1962), has recently been strongly supported by the results of numerous experiments with OGCMs (Bryan 1986). Since the atmosphere provides no direct feedback for salinity anomalies, the ocean is capable of remembering for a long time any anomalies in freshwater fluxes. Figure 14.4 shows how, by the internal dynamics of the ocean, a "white noise" behavior of freshwater

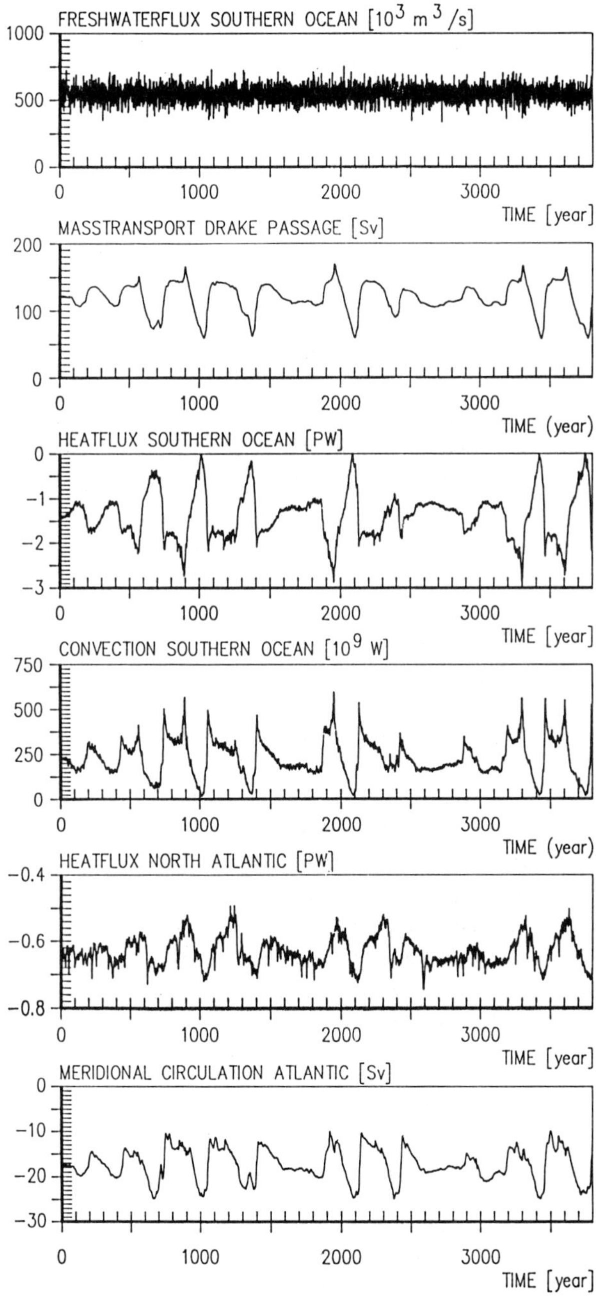

Figure 14.4 Time series of several circulation indices of the LSG circulation model in response to the precribed white noise freshwater flux (uppermost panel).

fluxes can be transformed into slow oscillations on a time scale of centuries of substantial amplitude (Mikolajewicz and Maier-Reimer 1990).

Model Chain

In a first attempt to simulate all the components of the LGM climate, we have performed at MPI consecutive runs of the individual component models: the AGCM used is the T21 model developed at the European Centre for Medium-Range Weather Forecasts (ECMWF) in Reading (Louis 1984) and modified for climate simulations at the Meteorological Institute of Hamburg University (Dümenil and Schlese 1987).

Two T21 experiments were performed over six annual cycles, a present-day control and the ice-age response (Lautenschlager and Herterich 1990). For the LGM, the lower boundary conditions were obtained from CLIMAP (1981). The last five model years were used to calculate the climate means and anomalies (ice age minus present). These five-year mean anomalies were taken as driving forces for the ocean experiments. The annual mean freshwater flux anomaly is shown in Figure 14.5.

The OGCM used is the large-scale geostrophic global model developed at the Max-Planck-Institut für Meteorologie (Maier-Reimer and Hasselmann 1987).

The carbon cycle model (cf. Heinze et al. 1990; Bacastow and Maier-Reimer 1990; and Maier-Reimer and Bacastow 1990) computes the distribution of ΣCO_2 of the three isotopes as the result of nutrient-limited new production in a given current field. A distinction is made between calcite production and the formation of soft tissue. Sinking particles of organic matter (POC) are remineralized, provided there is enough oxygen. The calcite pool interacts with a bioturbated sediment layer. The variables of the model are alkalinity, phosphate as a limiting nutrient, oxygen, dissolved CO_2,

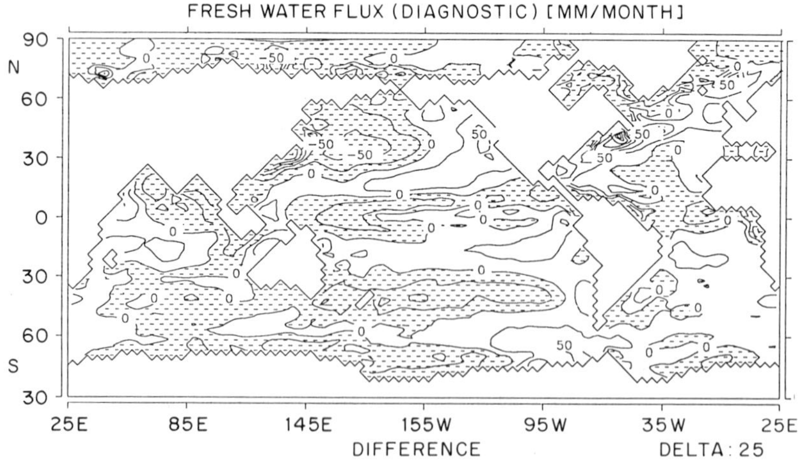

Figure 14.5 Change in the effective freshwater flux into the ocean including the smoothed response of river inputs resulting from the switch from modern SST to the CLIMAP reconstruction of SST at LGM.

POC, and calcite. In all carbon tracers the three isotopes are treated separately. In the atmosphere, CO_2 concentrations are supposed to be zonally mixed with a meridional diffusion of one hemisphere per year. In the gas exchange between ocean and atmosphere, the ^{13}C is fractionated according to

$$F^{13} = (\ 1.02389 - 9.483/(T + 273.16))\ F^{12}\ ,$$

where T is the local SST and F denotes the fugacity. For the implementation into soft tissue material we assume a constant fractionation of –20‰ . For radiocarbon we assume the square of these fractionation factors. Details of the model are given in Bacastow and Maier-Reimer (1990) and in Maier-Reimer and Bacastow (1990). The inventories of the tracers were carefully tuned to produce a stationary preindustrial CO_2 concentration of 275 ppm for the control run. Figure 14.6 shows the modern distribution of $\delta^{13}C$ on a section of the eastern Atlantic. The agreement with Kroopnick's (1985) evaluation of the GEOSECS data on the corresponding sections is quite obvious.

The changes in the zonally averaged meridional circulation are shown in Figure 14.7 for the Atlantic and the Pacific. Compared to the control experiment, the

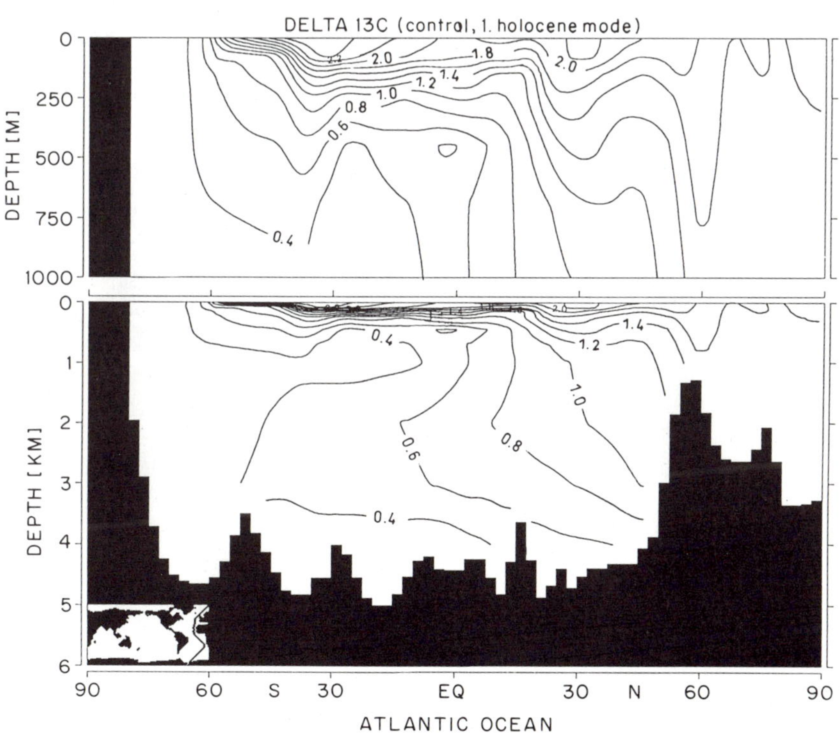

Figure 14.6 $\delta^{13}C$ as simulated with the carbon cycle model for the modern ocean.

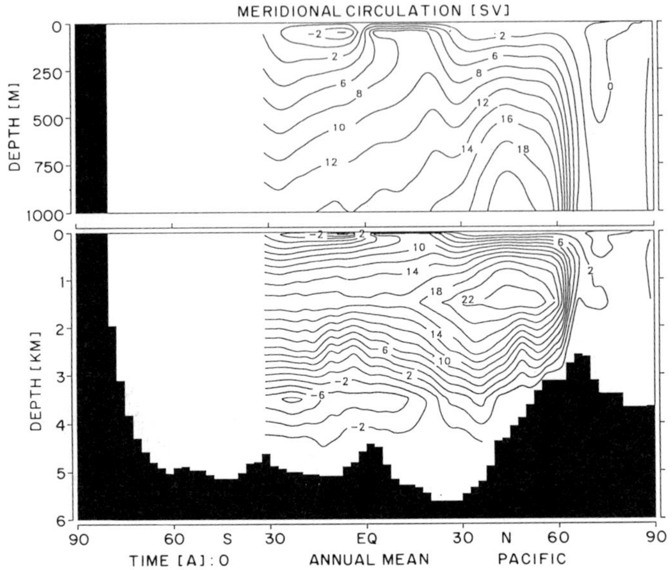

Figure 14.7a Meridional overturning as simulated from the LSG for the modern Atlantic. Units are 10^6 m^3/s (Sv). Positive numbers denote "clockwise" rotation.

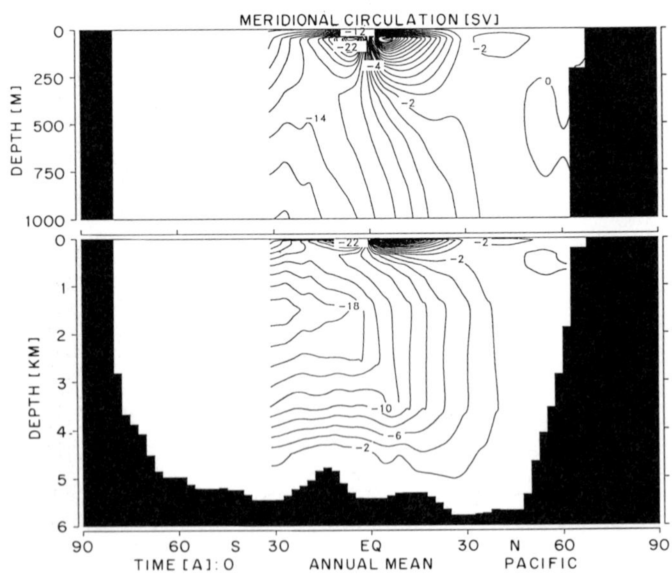

Figure 14.7b Meridional overturning as simulated from the LSG for the modern Pacific. Units are 10^6 m^3/s (Sv). Positive numbers denote "clockwise" rotation.

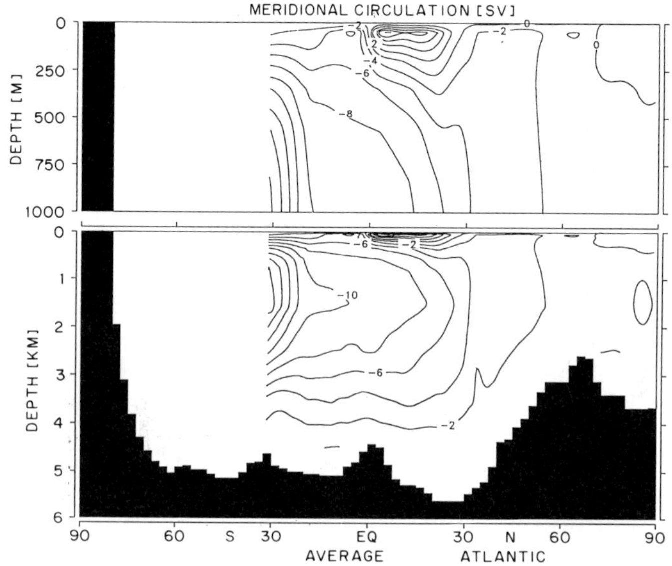

Figure 14.7c Meridional overturning as simulated from the LGM experiment for the modern Atlantic. Units are 10^6 m^3/s (Sv). Positive numbers denote "clockwise" rotation.

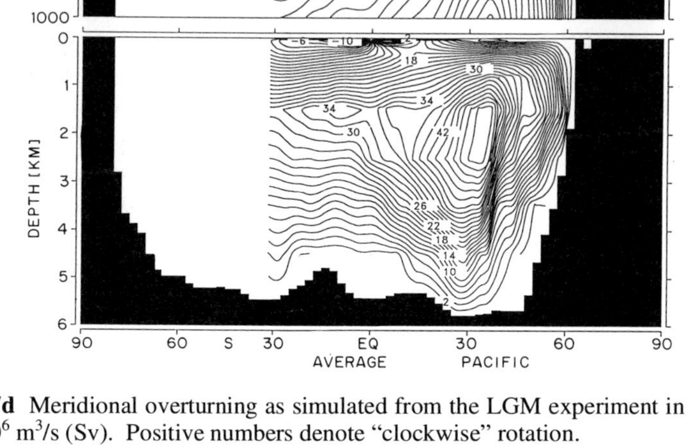

Figure 14.7d Meridional overturning as simulated from the LGM experiment in the Pacific. Units are 10^6 m^3/s (Sv). Positive numbers denote "clockwise" rotation.

simulated glacial contributions to deep-water production of the Atlantic and the Pacific were exchanged. In the North Atlantic, deep-water production stopped completely and in the North Pacific, thermohaline circulation started intensively. (This reversed pattern has a remarkable similarity to the first attempts to simulate the present climate with coupled OAGCMs (Bryan et al. 1975)). Consequently, in the ice-age response, the bottom water was younger in the equatorial Pacific (~ 300 years, C^{14} age) and older in the equatorial Atlantic (~ 300 years, C^{14} age).

The simulated temperature reduction of 1–3°C in the deep ocean fits into the suggested, but not generally accepted, temperature decrease of 1–2 K (Chappell and Shackleton 1986; Labeyrie et al. 1987).

Major discrepancies between data and model appear in the glacial deep ocean response. The intermediate waters in the North Pacific and North Atlantic should be more strongly ventilated (Boyle and Keigwin 1987; Curry et al. 1988). The North Atlantic thermohaline circulation seems to be substantially reduced. The glacial source of global deep water should be around Antarctica with a strength comparable to that observed today (Oppo and Fairbanks 1987; Duplessy et al. 1988). This sketch yields a globally reduced deep-water overturn at 18 Ka. The meridional stream function in the Atlantic of our glacial run seems to be roughly compatible with this outline, while the meridional stream function in the Pacific shows incompatibility. A comparison of our carbon isotopes simulation based on the ice-age response reveals a more pronounced discrepancy with the data than the foregoing qualitative arguments. Whereas Duplessy describes a $\delta^{13}C$ distribution similar in structure to the modern one, but with a much shallower spreading of the North Atlantic deep water (rather an intermediate water than a deep water), our model produces a pronounced minimum of $\delta^{13}C$ (Figure 14.8). In the northern Pacific, the situation is similarly inconsistent: the meridional stream function shows a pronounced overturning which would be compatible with ideas of a stronger production of intermediate water; the age difference between pelagic and benthic foraminifera (Broecker et al. 1988; Shackleton et al. 1988), however, is in clear contrast to our ice-age response on radiocarbon distribution. At this stage of model verification, we have to state clearly that our model results are wrong.

The freshwater flux taken from the glacial AGCM response was identified as the most probable reason for the "dissimulation." It would be very useful to get more information about glacial ocean surface salinity, especially in the North Pacific, to verify the regional T21 ice-age response in the freshwater flux. The glacial Aleuten pressure system was simulated 20° farther eastwards than in the present-day control run. Therefore, the simulated freshwater anomaly was positive in the eastern North Pacific and negative in the western part, where the glacial deep-water production was simulated by the OGCM. The glacial location of the Aleuten pressure system could not be obtained from geological data in Alaska (Kutzbach and Wright 1985). However, the T21 pattern of freshwater flux anomaly is also simulated by other AGCMs using 18 Ka CLIMAP boundary conditions (Joussaume, pers. comm.). The local driving of the AGCM should be connected with the prescription of the large sea-ice

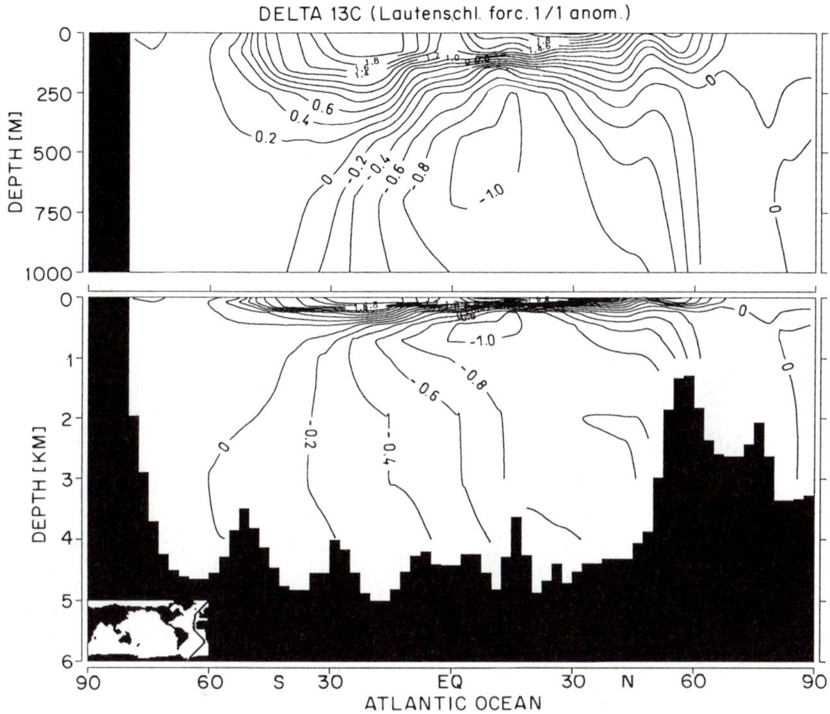

Figure 14.8 $\delta^{13}C$ for the LGM experiment in the Atlantic. A better agreement with data is achieved with the second circulation mode, which can stably develop from restoring SST with the freshwater fluxes from the control run. Figure 14.9 shows the distribution of δC^{13}.

extent in the northern Pacific. A recent reexamination of the primary geological data in that region indicated that the CLIMAP reconstruction of ice-coverage may have been overestimated (Ruddiman, pers. comm.). If so, we can expect that with a revised sea-ice reconstruction, our model results would exhibit a smaller departure from the modern ocean simulation than the ice-age response discussed in this chapter.

The CLIMAP SST field has also been challenged because of disagreements between AGCM results and observations over land (Webster and Streeten 1978; Rind and Peteet 1985). Although initial attempts at independently validating the CLIMAP SSTs appear to have been supportive (Prell 1985; Broecker 1986), a more recent reexamination of the transfer functions in the Pacific does suggest that the SST field may need revision (Chen et al. 1991).

Thus, the critical dependency of climate model results on the CLIMAP SSTs is not just based on atmospheric models and tropical land areas, where in some cases the observations themselves are sparse. Our results indicate that, when ocean

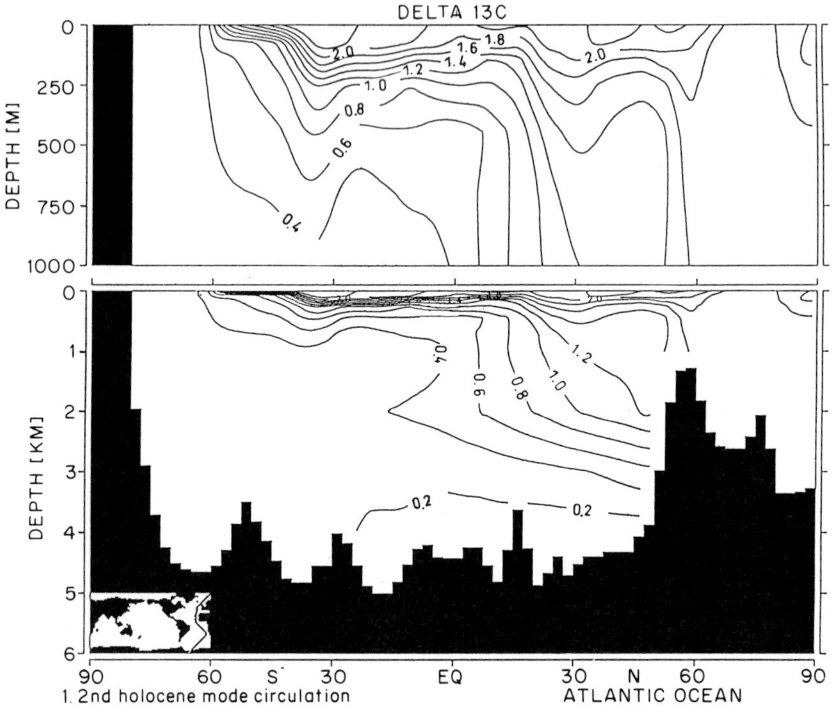

Figure 14.9 $\delta^{13}C$ in the Atlantic for the second stable circulation under modern boundary conditions.

models are considered, model data discrepancy may also extend to the validation of results with deep-sea sediments. Although more work clearly is required on OGCM studies of ice-age circulation, our results suggest that a satisfactory reconciliation with the sediment record may require reevaluation of CLIMAP SST reconstruction.

ACKNOWLEDGEMENTS

We are greatly indebted to M. Lautenschlager, U. Mikolajewicz, and C. Heinze who performed the numerical experiments of the atmospheric circulation (M.L.), the oceanic circulation (U.M.) and the carbon cycle (C.H.). Thanks are also due to T. Crowley, K. Hasselmann, and K. Herterich for valuable discussions and comments. Part of this work has been funded by the German Ministry of Research and Technology (BMFT) under grant 07KF0127.

REFERENCES

Bacastow, R., and E. Maier-Reimer. 1990. Ocean-circulation model of the carbon cycle. *Clim. Dynam.* **4**:95–126.

Berger, W.H. 1973. Deep-sea carbonates: Pleistocene dissolution cycles. *J. Foram. Res.* **3**:187–195.

Boyle, E. 1988. Cadmium: Chemical tracer of deep-water paleoceanography. *Paleoceanog.* **3**:471–489.

Boyle, E.A. 1992. Cadmium and $\delta^{13}C$ paleochemical ocean distributions during the stage 2 glacial maximum. *Ann. Rev. Earth Planet. Sci.*, in press.

Boyle, E.A., and L.D. Keigwin. 1982. Deep circulation of the North Atlantic over the last 200,000 years: Geochemical evidence. *Science* **218**:784–787.

Boyle, E.A., and L.D. Keigwin. 1985. Comparison of Atlantic and Pacific paleo-chemical records for the last 215,000 years: Changes in deep ocean circulation and chemical inventories. *Earth Planet. Sci. Lett.* **76**:135–150.

Boyle, E.A., and L.D. Keigwin. 1987. North Atlantic thermohaline circulation during the last 20,000 years: Link to high latitude surface temperature. *Nature* **330**:35–40.

Broccoli, A.J., and S. Manabe. 1987. The influence of continental ice, atmospheric CO_2, and land albedo on the climate of the last glacial maximum. *Clim. Dynam.* **1**:87–99.

Broecker, W.S. 1982. Glacial to interglacial changes in ocean chemistry. *Prog. Oceanog.* **11**:151–197.

Broecker, W.S., M. Andrée, G. Bonani, W. Wolfli, H. Oeschger, M. Klas, A. Mix, and W. Curry. 1988. Preliminary estimates for the radiocarbon age of deep water in the glacial ocean. *Paleoceanog.* **3**:659–669.

Broecker, W.S., and G.H. Denton. 1989. The role of ocean–atmosphere reorganizations in glacial cycles. *Geochim. Cosmo. Acta* **53**:2465–2501.

Broecker, W.S., A. Mix, M. Andrée, and H. Oeschger. 1984. Radiocarbon measurements on coexisting benthic and planktonic foraminifera shells: Potential for reconstructing ocean ventilation times over the past 20,000 years. *Nucl. Inst. Meth.* **B5**:331–339.

Broecker, W.S., T.-H. Peng, J. Jouzel, and G. Russel. 1990. The magnitude of global freshwater transports of importance to ocean circulation. *Clim. Dynam.* **4**:73–79.

Broecker, W.S., T.-H. Peng, S. Trumbore, G. Bonani, and W. Wolfli. 1990. The distribution of radiocarbon in the glacial ocean. *Glob. Biogeochem. Cyc.* **4**:103–117.

Broecker, W.S., D.M. Peteet, and D. Rind. 1985. Does the ocean–atmosphere system have more than one stable mode of operation? *Nature* **315**:21–26.

Bryan, F. 1986. High-latitude salinity effects, and interhemispheric thermohaline circulations. *Nature* **323**:301–304.

Caralp, M. 1987. Deep-sea circulation in the northeastern Atlantic over the past 30,000 years: The benthic foraminiferal record. *Oceanol. Acta* **10**:27–40.

Chappell, J., and N.J. Shackleton. 1986. Oxygen isotopes, and sea level. *Nature* **324**:137–140.

Chen, M.-T., J. Imbrie, and W.L. Prell. 1991. Estimating sea-surface temperatures from tropical Pacific planktonic foraminifers. Comparison of Imbrie-Kipp method, and modern analog technique. *EOS Trans.* **72(17)**:157.

CLIMAP. 1976. The surface of the Ice Age Earth. *Science* **191**:1131–1137.

CLIMAP. 1981. Seasonal reconstruction of the Earth's surface at the last glacial maximum. *Geol. Soc. Am. Map Chart. Ser.* MC–36.

Curry W.B., J.C. Duplessy, L.D. Labeyrie, and N.J. Shackleton. 1988. Changes in the distribution of $\delta^{13}C$ of deep water ‰ CO_2 between the last glaciation and Holocene. *Paleoceanog.* **3**:317–341.

Curry, W.B., and G.P. Lohmann. 1982. Carbon isotopic changes in benthic foraminifera from the western south Atlantic: Reconstruction of glacial abyssal circulation patterns. *Quat. Res.* **18**:218–235.

Dümenil, L., and U. Schlese. 1987. Description of the general circulation model. In: Climate Simulations with the T21–Model in Hamburg, ed. G. Fischer. Rep. No. 1, Meteor. Inst. Large–Scale Atmosph. Model.

Duplessy, J.C., and L.D. Labeyrie. 1989. Global ocean circulation during the last ice age: Enhanced intermediate water formation and reduced deep-water circulation. In: Oceanography 1988, ed. A. Ayala Castanares, A.W. Wooster, and A. Yanez Arancibia. Conacyt–UNESCO.

Duplessy, J.C., J. Moyes, and C. Pujol. 1980. Deep water formation in the North Atlantic Ocean during the last ice age. *Nature* **286**:476–482.

Duplessy, J.C., and N.J. Shackleton. 1985. Response of global deep-water circulation to the Earth's climatic change 135,000–107,000 years ago. *Nature* **316**:500–507.

Duplessy, J.C., N.J. Shackleton, R.G. Fairbanks, L.D. Labeyrie, D. Oppo, and N. Kallel. 1988. Deep water source variations during the last climatic cycle and their impact on the global deep-water circulation. *Paleoceanog.* **3**:343–360.

Duplessy, J.C., N.J. Shackleton, R.K. Matthews, W. Prell, W.F. Ruddiman, M. Caralp, and C.H. Hendy. 1984. ^{13}C record of benthic foraminifera in the last interglacial ocean: Implications for the carbon cycle and the global deep-water circulation. *Quat. Res.* **21**:225–243.

Emiliani, C. 1955. Pleistocene temperatures. *J. Geol.* **63**:538–578.

Gardner, J.V., and J.D. Hays. 1976. Responses of sea-surface temperature and circulation to global climatic change during the past 200,000 years in the Eastern Equatorial Atlantic Ocean. *Mem. Geol. Soc. Am.* **145**:221–246.

Gordon, A.L. 1986. Interocean exchange of thermocline water. *J. Geophys. Res.* **91(C4)**:5037–5046.

Hays, J.D., J. Imbrie, and N.J. Shackleton. 1976. Variations in the Earth's orbit: Pacemaker of the ice ages. *Science* **194**:112–132.

Heinze, C., E. Maier-Reimer, and K. Winn. 1990. Glacial pCO_2 reduction by the world ocean: Experiments with the Hamburg carbon cycle model. *Paleoclim.* **6**:395–430.

Imbrie, J., J.D. Hays, D.G. Martinson, A. McIntyre, A.C. Mix, J.J. Morley, N.G. Pisias, W. Prell, and N.J. Shackleton. 1984. The orbital theory of Pleistocene climate: Support from a revised chronology of the marine ^{18}O record. In: Milankovitch and Climate, ed. A. Berger, J. Imbrie, J.D. Hays, G. Kukla, and B. Saltzman, pp. 269–305. Hingham, MA: Reidel.

Imbrie, J., and N.G. Kipp. 1971. A new micropaleontological method for quantitative paleoclimatology: Application to a late Pleistocene Caribbean core. In: The Late Cenozoic Glacial Ages, ed. K.K. Turekian, pp. 77–181. New Haven: Yale Univ. Press.

Kallel, N. 1988. Variations de la composition isotopique de l'oxygène et du carbone des foraminifères: Traceur de la circulation océanique pendant le dernier maximum glaciaire et lors de la dernière déglaciation. Thesis, Université Paris XI.

Kallel, N., L.D. Labeyrie, A. Juillet-Leclerc, and J.C. Duplessy. 1988. A deep hydrological front between intermediate and deep-water masses in the Glacial Indian Ocean. *Nature* **333**:651–655.

Keffer, T., D.G. Martinson, and B.H. Corliss. 1988. The position of the Gulf Stream during Quaternary glaciations. *Science* **241**:440–442.

Keigwin, L.D. 1987. North Pacific deep-water formation during the latest glaciation. *Nature* **330**:362–364.

Kroopnick, P. 1974. The dissolved O_2–CO_2–^{13}C system in the eastern equatorial Pacific. *Deep Sea Res.* **21**:211–222.

Kroopnick, P. 1985. The distribution of ^{13}C of CO_2 in the world oceans. *Deep Sea Res.* **32**:57–84.

Kutzbach, J.E., and P.J. Guetter. 1986. The influence of changing orbital parameters, and surface boundary conditions on climate simulations for the past 18,000 years. *J. Atmos. Sci.* **43**:1726–1759.

Kutzbach, J.E., and H.E. Wright. 1985. Simulation of the climate 18,000 years B.P.: Results for the North American/North Atlantic/European sector, and comparison with the geological record of North America. *Quat. Sci. Rev.* **4**:147–187.

Labeyrie, L.D., J.C. Duplessy, and P.L. Blanc. 1987. Variations in mode of formation and temperature of oceanic deep waters over the past 125,000 years. *Nature* **327**:477–482.

Labeyrie, L.D., J.C. Duplessy, J. Duprat, A. Juillet-Leclerc, J. Moyes, N. Kallel, and N.J. Shackleton. 1992. Changes in the vertical structure of the North Atlantic Ocean between glacial and modern times. *Quat. Sci. Rev.*, in press.

Lautenschlager, M., and K. Herterich. 1990. Atmospheric response to ice-age conditions—climatology near the Earth's surface. *J. Geophys. Res.* **95**:22547–22557.

Lea, D.W., and E.A. Boyle. 1990. Foraminiferal reconstruction of barium distributions in water masses of the glacial oceans. *Paleoceanog.* **5**:719–742.

Louis, J.-F. 1984. ECMWF Forecast Model: Physical Parameterization. ECMWF Research Dept., Research manual 3.

Maier-Reimer, E., and R. Bacastow. 1990. Modeling of geochemical tracers in the ocean. In: Climate-Ocean Interaction, ed. M. Schlesinger, pp. 233–267. Dordrecht: Kluwer.

Maier-Reimer, E., and K. Hasselmann. 1987. Transport, and storage of CO_2 in the ocean: An inorganic ocean-circulation carbon cycle model. *Clim. Dynam.* **2**:63–90.

Manabe, S., and R.J. Stouffer. 1988. Two stable equilibria of a coupled ocean–atmosphere model. *J. Climate* **1**:841–866.

Mikolajewicz, U., and E. Maier-Reimer. 1990. Internal secular variability in an oceangeneral circulation model. *Clim. Dynam.* **4**:145–156.

Miller, J.R., and G.L. Russell. 1989. Ocean heat transport during the last glaciation. *Paleoceanog.* **4**:141–155.

Oppo, D.W., and R.G. Fairbanks. 1987. Variability in the deep and intermediate water circulation of the Atlantic during the past 25,000 years: Northern Hemisphere modulation of the Southern Ocean. *Earth Planet. Sci. Lett.* **86**:1–15.

Östlund, A., and M. Stuiver. 1980. GEOSECS Pacific radiocarbon. *Radiocarbon* **22**:25–53.

Prell, W.L. 1985. The stability of low-latitude sea–surface temperatures: An evaluation of the CLIMAP reconstruction with emphasis on the positive SST anomalies. U.S. DOE Cont. Report DE–ACO2–83ER60167. Washington, D.C.: U.S. Dept. of Energy.

Rind, D. 1987. Components of the Ice Age circulation. *J. Geophys. Res.* **92(D4)**:4241–4281.

Rind, D., and D. Peteet. 1985. Terrestrial conditions at the last glacial maximum, and CLIMAP sea-surface temperature estimates: Are they consistent? *Quat. Res.* **24**:1–22.

Rind, D., D. Peteet, and G. Kukla. 1989. Can Milankovitch orbital variations initiate the growth of ice sheets in a general circulation model? *J. Geophys. Res.* **94**:12851–12871.

Sausen, R., K. Barthel, and K. Hasselmann. 1988. Coupled ocean-atmosphere models with flux correction. *Clim. Dynam.* **2**:145–163.

Shackleton, N.J. 1977. Tropical rainforest history and the equatorial Pacific dissolution cycles. In: The Fate of Fossil Fuel CO_2 in the Oceans, ed. N.R. Anderson and N.A. Malahoff, pp. 401–428. New York: Plenum.

Shackleton, N.J., J.C. Duplessy, M. Arnold, P. Maurice, M. Hall, and J. Cartlidge. 1988. Radiocarbon age of last glacial Pacific deep water. *Nature* **335**:708–711.

Shackleton, N.J., M.A. Hall, J.Line, and C. Shuxi. 1983. Carbon isotope data in core V19–30 confirmed reduced carbon dioxide concentration of the ice-age atmosphere. *Nature* **306**:319–322.

Shackleton, N.J., and N.D. Opdyke. 1973. Oxygen isotope and paleomagnetic stratigraphy of equatorial Pacific core V28–238: Oxygen isotope temperatures and ice volumes on a 10^5 year and 10^6 year scale. *Quat. Res.* **3**:39–55.

Shackleton, N.J., and N.G. Pisias. 1985. Atmospheric carbon dioxide, orbital forcing, and climate. In: The Carbon Cycle and Atmospheric CO_2: Natural Variations Archean to Present, ed. E.T. Sundquist and W.S. Broecker. *Geophys. Monog.* **32**:303–317. Washington, D.C.: American Geophysical Union.

Stommel, H. 1962. Thermohaline convection with two stable regimes of flow. *Tellus* **13**:224–230.

Streeter, S.S., and N.J. Shackleton. 1979. Paleocirculation of the deep North Atlantic: 150,000-year record of benthic foraminifera and oxygen-18. *Science* **203**:168–171.

Thompson, P.R., and T. Saito. 1974. Pacific Pleistocene sediments planktonic foraminifera: Dissolution cycles and geochronology. *Geology* **2**:333–335.

Warren, B.A. 1983. Why is no deep water formed in the North Pacific? *J. Marine Res.* **41**:327–347

Weiss, R.F., H.G. Östlund, and H. Craig. 1979. Geochemical studies of the Weddell Sea. *Deep Sea Res.* **26**:1093–1120.

Zahn, R., M. Sarnthein, and H. Erlenkeuser. 1987. Benthic isotope evidence for changes of the Mediterranean Outflow during the Late Quaternary. *Paleoceanog.* **2**:543–559.

15

The Role and Response of Continental Vegetation in the Global Climate System

J.T. OVERPECK

Paleoclimatology Program, National Geophysical Data Center, National Oceanic and Atmospheric Administration, 325 Broadway, E/GC, Boulder, CO 80303 U.S.A. and

Institute of Arctic and Alpine Research, University of Colorado at Boulder, 1560 30th Street, Boulder, CO 80309 U.S.A.

ABSTRACT

Future climate change is likely to be much greater than that of the past two centuries, making the late Quaternary record of climate and vegetation change a crucial resource for understanding both the response and role of vegetation within the climate system. Predictions of future change will have to be model-based, because this change will not have past analogs. Simulation of observed past change, however, provides needed insight for simulating future change. Current models can simulate many aspects of past climate-induced vegetation change and suggest that future change could be as large as that which took place over the last glacial to interglacial transition, and more rapid. This possibility highlights the need to build better models for simulating vegetation change and for including realistic influences of vegetation in climate and trace gas models. Climate models will not be able to simulate regional climates or hydrologic change correctly without the incorporation of realistic vegetation feedbacks. The record of global change for the late Quaternary must be reconstructed more thoroughly before it will be possible to gain a complete understanding of the response and role of vegetation in a climate system subjected to rapidly changing boundary conditions.

INTRODUCTION

The terrestrial biosphere, and vegetation in particular, plays an important role in the behavior of the climate system. Most importantly, changes in vegetation can exert a strong influence on the dynamics of terrestrial ecosystems, soils, atmospheric trace gas concentrations, and climatic change. Efforts to understand and model these

Global Changes in the Perspective of the Past
Edited by J.A. Eddy and H. Oeschger

dynamics thus call increasingly for a better understanding of how vegetation responds to climatic change and how this response can be modeled in a realistic fashion. The observational record of the past 150–200 years is too short to understand the response of vegetation to climatic change and the influence of the terrestrial biosphere on the long-term dynamics of the climate system. This chapter reviews what the late Quaternary record of climate and vegetation change reveals about the role and response of vegetation in the climate system, and how this long-term record can be used to test hypotheses and models regarding climate–vegetation interactions.

THE POSSIBLE RESPONSE OF VEGETATION TO CLIMATE CHANGE: PAST VERSUS FUTURE

Climate Change: Past Versus Future

Future climate and vegetation change is likely to be significantly larger and faster than any change of the past several centuries. Hemispheric to global warming of 0.3 to 0.6°C appears to have characterized the last century, as the Earth emerged from the "Little Ice Age" (Houghton et al. 1990). Moreover, Little Ice Age cooling during the 17th and 19th centuries may have been concentrated in the Northern Hemisphere (Figure 15.1; Cook et al. 1991; Bradley and Jones 1991). The cause(s) of this change is not understood, but changing atmospheric trace gases, solar output, volcanic aerosols, and/or processes internal to the climate system may have played important roles (Bradley and Jones 1991; Overpeck 1990). The response of natural vegetation to this small climate change is also difficult to decipher because of the large human impact on natural vegetation in many parts of the world during this time.

In addition to providing valuable networks of long paleoclimatic time series, including ones that may aid in the detection of recent greenhouse warming (Figure 15.1), tree ring records from selected tree species enable the study of how small variations in climate influence the growth of these different tree species. This information may help improve time-dependent models of vegetation change (Cook and Cole 1991), even though simulations of future vegetation change will probably require extrapolation beyond the range of variability recorded in the last few centuries.

In sharp contrast to the small range of climatic variability over the last several centuries, doubling of atmospheric trace gas concentrations may produce a 1.5° to 4.5°C global mean warming by the end of the next century (Houghton et al. 1990). Unless dramatic policy initiatives are enacted, this global warming could continue beyond A.D. 2100. Regional temperature increases, particularly at mid- to high-latitudes, may exceed 10°C in winter and accompany significant changes in summer moisture availability (Houghton et al. 1990). This change clearly exceeds any of the past few centuries and requires that we turn to the geological record of global change for a complete understanding of how the climate system (including the biosphere) responds to large changes in climatic boundary conditions.

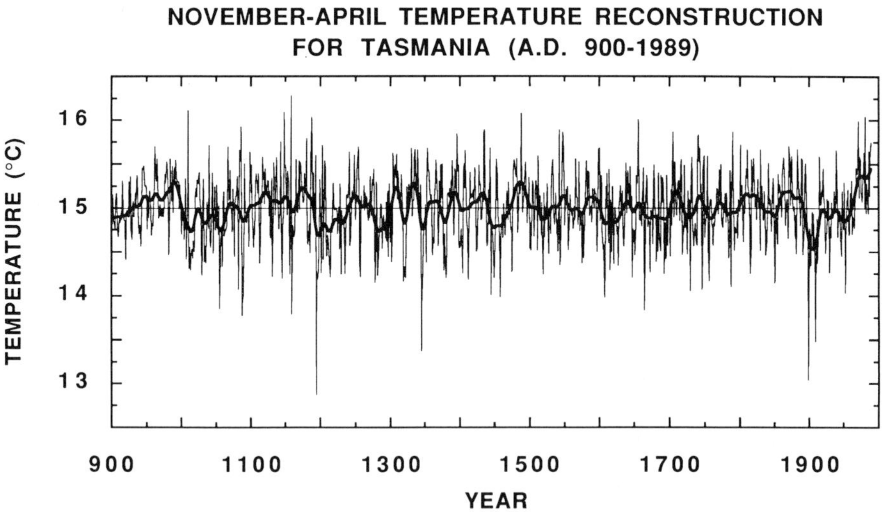

Figure 15.1 November–April average surface air temperatures reconstructed from Tasmanian tree rings. The heavy solid line is the series smoothed with a 25-year low-pass filter. These data show no significant long-term "Little Ice Age" cooling between A.D. 1450 and 1850, but do suggest that recent growing season temperatures may have been among the warmest of the last millennium. Tree ring data offer important opportunities to study the growth responses of selected tree taxa to climatic change. Figure from Cook et al. 1991.

The primary cause of the climate changes expected for the near future is the ongoing steep increase in atmospheric trace gas concentrations. In contrast, climatic change of the late Quaternary was forced by a more complex array of changing boundary conditions, including variations in the distribution of solar energy reaching the Earth, ocean circulation, atmospheric trace gas concentrations, and global ice extent and height (Hansen et al. 1984; COHMAP 1988). Future and past forcing mechanisms are different, and the climatic responses to these forcings will also differ. This means that future global change will not have exact geological analogs (Webb and Wigley 1985; Crowley 1990). Therefore, assessments of future change cannot be based on direct comparisons with the past, but instead have to be model-based.

Perhaps the best period for study of how vegetation responds to large changes in climatic forcing is the glacial–interglacial transition spanning the last 18,000 years. Many well-dated paleoclimate and paleovegetation records are available for this period, and macro-evolutionary processes are not important over such short time scales. Furthermore, modeling and observational studies suggest that the mean global surface temperatures warmed by about 4°–5°C over this period of time (COHMAP 1988), about the same as the warming that is predicted to occur in the next 100–200 years. Regional changes were, and will often be, larger or smaller than the mean global response.

Vegetation Change: Past Versus Future

The response of vegetation to climatic change over the past 18,000 years has been studied using paleoecological data from sites worldwide, and several conclusions seem fundamental (Delcourt and Delcourt 1987; Huntley and Webb 1989; Davis 1990; Prentice et al. 1991). First, climatic change of this magnitude results in large changes in vegetation. Second, plant taxa respond individualistically to climatic change. Vegetation regions, or biomes, commonly appear, change, and disappear through time as each taxon tracks its favored climatic conditions. Climate conditions without modern analogs are likely to produce vegetation regions without modern analogs. These rules have operated in the past and are likely to operate in the future.

Fossil pollen data and the quantitative method of modern analogs (Overpeck et al. 1992; Webb et al. 1991) have been used to produce paleovegetation maps for eastern North America illustrating the typical magnitude of vegetation change in this region over the past 18,000 years (Figure 15.2). The map for 18,000 yr B.P. (Figure 15.2a) is marked by a low density of sites, but still defines regional vegetation patterns that differ greatly from modern-day distributions (Figure 15.2d). Most of the ice-free landscape was covered by conifer-dominated woodlands and forests, while tundra vegetation grew between woodlands and the ice sheet in some areas. By 12,000 yr B.P. (Figure 15.2b), data coverage improves. To the east of the NE–SW trending Appalachian Mountains, a modern-like latitudinal zonation of vegetation regions developed, but to the west, a large region of vegetation without any modern analog existed. This no-analog region that was essentially a spruce-dominated woodland with high amounts of sedge in the southwest and significant amounts of sedge, ash, elm, and ironwood/hornbeam in the north and east (Wright 1967; Webb 1988; Overpeck et al. 1992). After about 10,000 yr B.P., the modern vegetation patterns of eastern North America began to emerge (Webb 1988), and only small areas of no-analog vegetation existed at any one time (Figure 15.2c).

In an effort to put possible future climate-induced vegetation change into a perspective of past change, Overpeck et al. (1991) used climate-pollen response surfaces coupled with output from three atmospheric general circulation models (GCMs) to simulate the vegetation patterns that would exist in eastern North America under climatic conditions in equilibrium with doubled atmospheric CO_2 concentrations (Figure 15.3). They then compared this change to that observed in the past. This approach, although preliminary and ignorant of potentially important transient and nonclimatic effects (Solomon 1986; Davis 1989; Overpeck et al. 1990), suggests that future vegetation change may be quite significant. Just as they did over the past 18,000 years, plant ranges and abundance maxima could shift by as much 500 to 1000 km during the next 200 to 500 years (Webb 1988; Overpeck et al. 1991). These changes simulated using the static response surface approach (Figure 15.3) do agree qualitatively with those simulated using a time-dependent dynamic forest-stand simulation model (Solomon 1986; Overpeck and Bartlein 1989; Overpeck et al. 1992). Vegetation

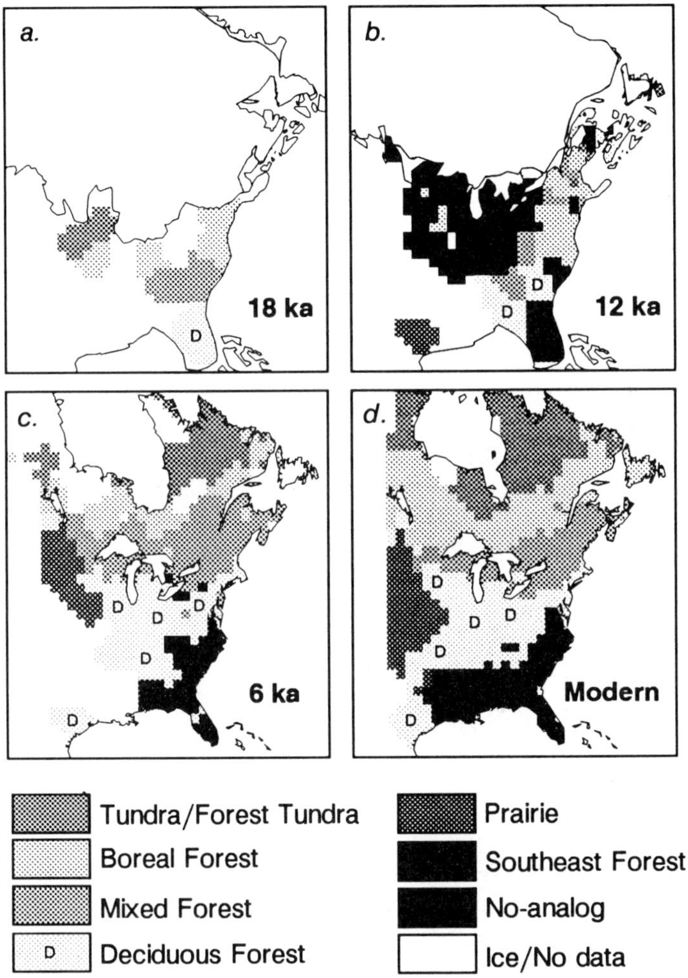

Figure 15.2 Selected paleovegetation maps reconstructed using the method of modern analogs and over 13,000 samples of fossil and modern pollen (Overpeck et al. 1992, Webb et al. 1991). **A**) 18,000 yr B.P., **B**) 12,000 yr B.P., **C**) 6,000 yr B.P., and **D**) modern. No vegetation is mapped in areas without fossil pollen data, and "no-analog" refers to vegetation without any modern analog. Figure after Overpeck et al. 1992.

biomes may change in composition and, when certain transient factors are considered, in structure as well (Overpeck et al. 1990).

The response surfaces used to produce the doubled CO_2 maps (Figure 15.3) have also been used to simulate vegetation patterns of the past 18,000 years (Figure 15.4; Webb et al. 1987; Overpeck and Bartlein 1989). The comparison of the simulated vegetation maps with maps generated using actual fossil pollen data illustrates the

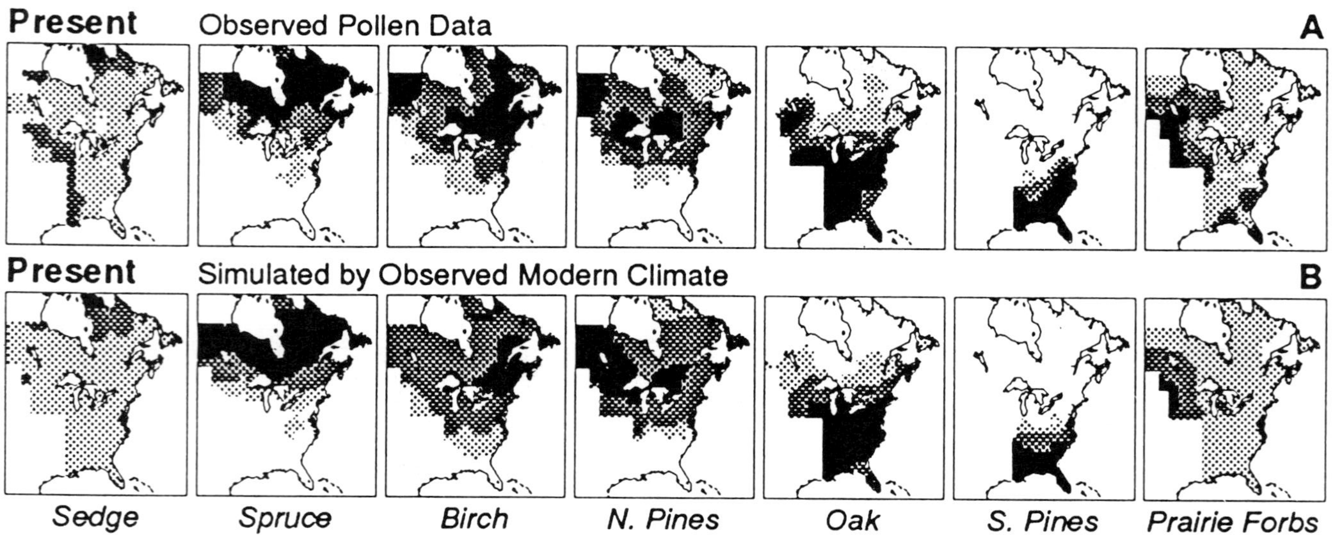

Figure 15.3 Maps of pollen abundance: **A**) observed modern pollen abundance; **B**) equilibrium modern pollen abundances estimated using pollen response surfaces and modern climatic data;

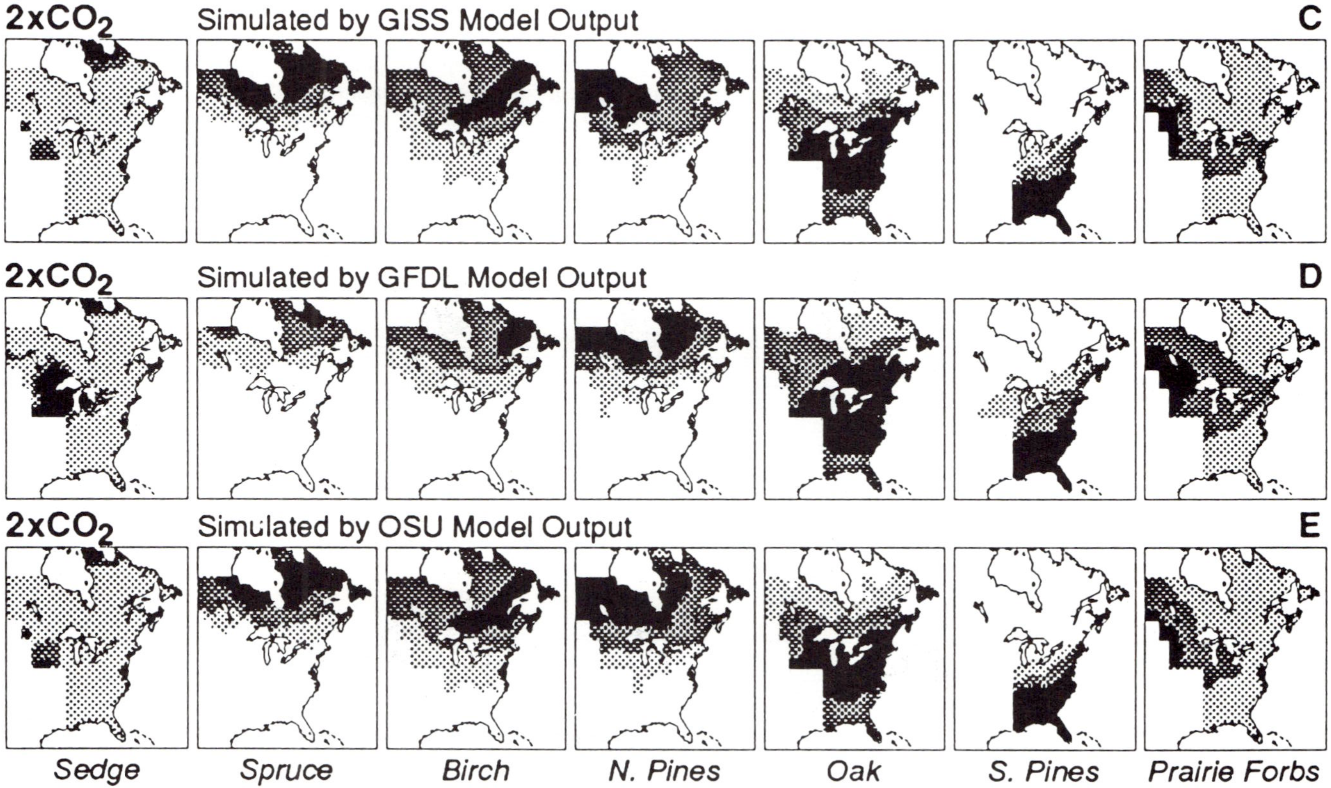

Figure 15.3 *continued* **C–E**) doubled CO_2 equilibrium pollen abundances estimated using doubled CO_2 scenarios from each of the three climate models as input to the response surfaces. The shading represents three different levels of abundance: between 0 and 5% (lightest shading), 5 to 20% (medium shading), and greater than 20% (black). The subcontinental patterns observed in modern observed and simulated isopoll maps correspond closely with those in the potential vegetation of the region (Webb 1988). Figure from Overpeck et al. (1991).

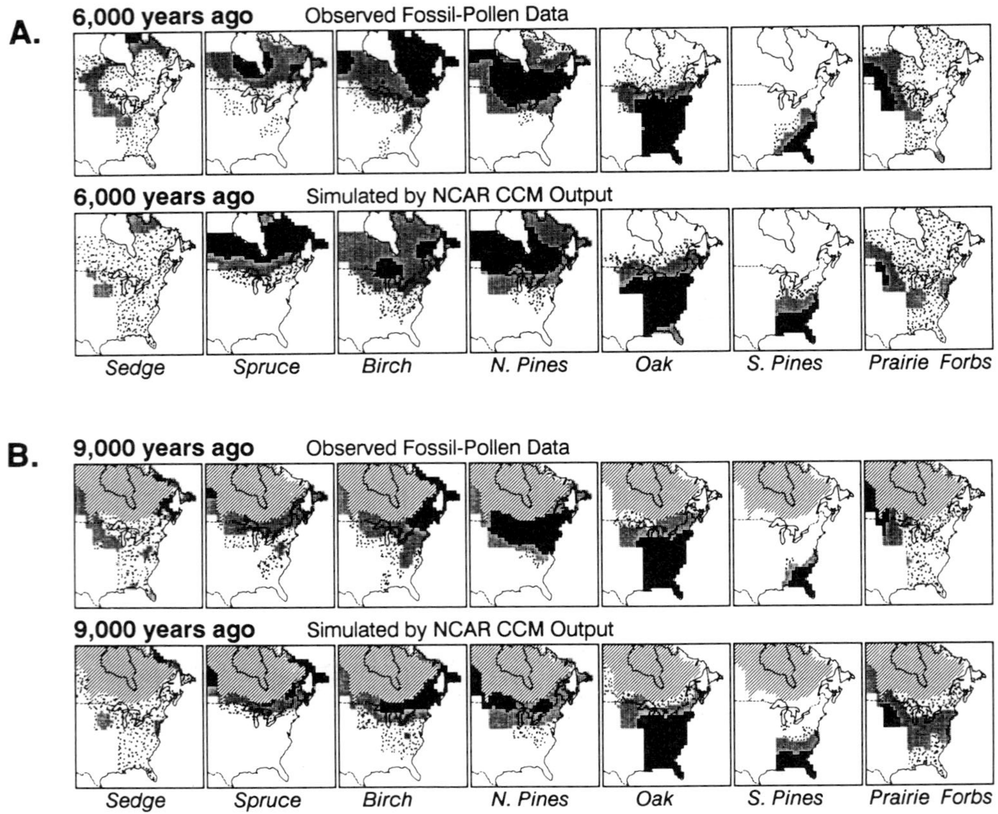

Figure 15.4 Observed and simulated pollen data for selected time intervals: **A**) 6,000 yr B.P. and **B**) 9,000 yr B.P. The observed maps were generated using fossil pollen data from over 250 lakes and mires in eastern North America, whereas the simulated maps were generated using pollen response surfaces (same as in Figure 15.3) and climates simulated using the NCAR CCM General Circulation Model. Three levels of shading indicate pollen percentages greater than 1% (lightest), 5%, and 20% (darkest). The extent of the Laurentide Ice Sheet and associated glacial lakes is indicated by diagonal shading. Portions of the map area without data were not mapped. Figures from Overpeck and Bartlein 1989.

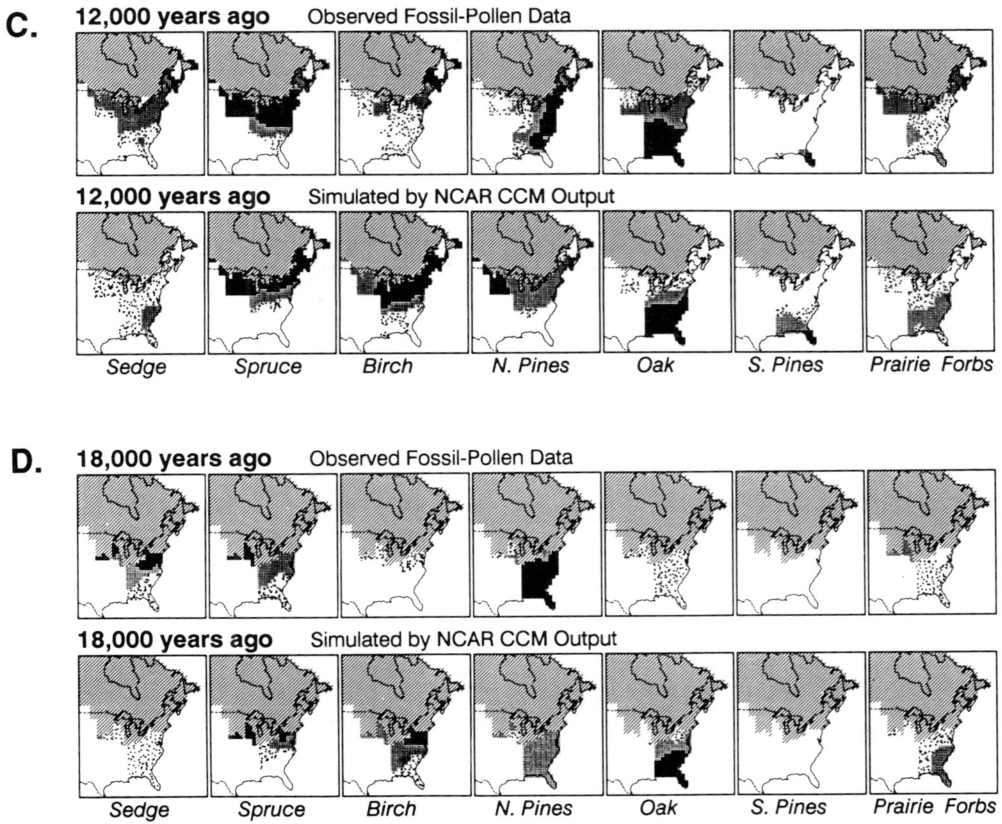

Figure 15.4 *continued* Observed and simulated pollen data for selected time intervals: **C**) 12,000 yr B.P. and **D**) 18,000 yr B.P.

power of the response surface methodology. The degree of fit between the simulated and observed maps is striking for the period 12,000 yr B.P. to present (Figure 15.4a–c; Overpeck and Bartlein 1989). Lack of fit at 18,000 yr B.P. (Figure 15.4d) is most likely the result of poor boundary condition specification or GCM performance (Webb et al. 1987; Overpeck and Bartlein 1989). This conclusion is supported by recent successful simulations of pre-12,000 yr B.P. vegetation patterns using climate inferred from independent paleoclimatic data (Prentice et al. 1991). The successful simulation of past vegetation change provides added confidence in our simulations of future change.

Overpeck et al. (1991) used a multivariate dissimilarity measure to compare rates and magnitudes of observed past vegetation change with those simulated for the future using doubled CO_2 climate scenarios generated by the GISS, OSU, and GFDL

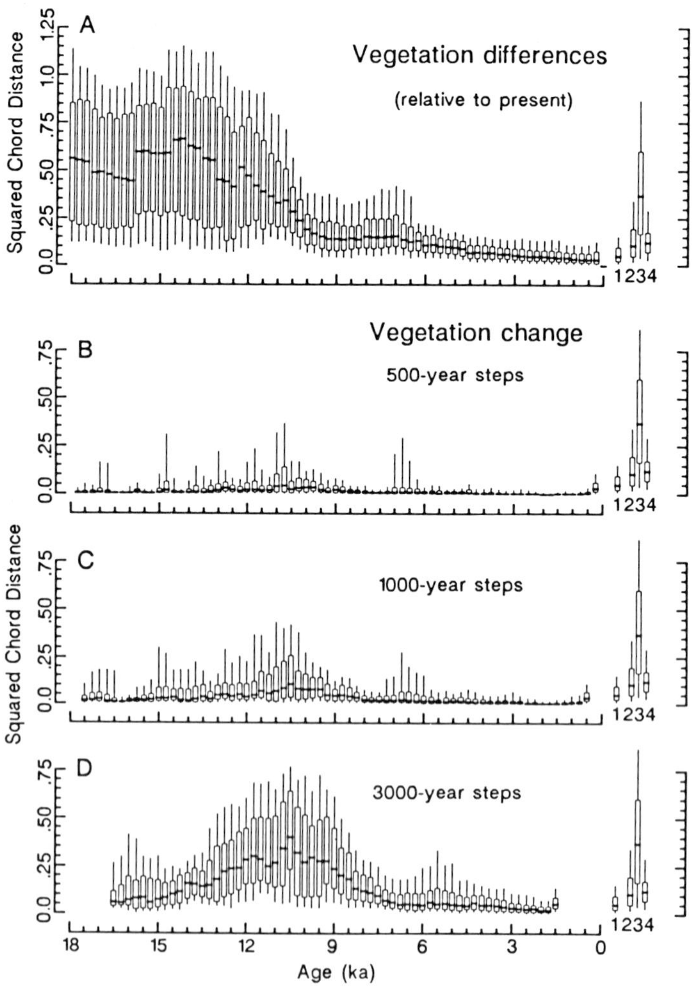

GCMs. The results of these comparisons (Figure 15.5a) suggest that the differences between the vegetation distribution for simulated doubled CO_2 and present are as large as those between 7000 yr B.P. and present (GISS and OSU simulations) or 11,000 yr B.P. and present (GFDL). The potential change in eastern North American vegetation within the next 200 to 500 years may therefore exceed the total amount of change that occurred during the past 10,000 years. Averaged over eastern North America, the maximum rate of change in the past occurred between 12,000 and 9000 yr B.P. during deglaciation (Figure 5b–d). This was a period of large-scale vegetational reorganization across eastern North America (Jacobson et al. 1987; Webb 1988). Although some of the changes in pollen distribution that have occurred over 500-year intervals in the past rival the magnitude of changes simulated by the three doubled CO_2 experiments (Figure 5b), the median magnitude of the future changes simulated by the OSU and GISS models is comparable to that which occurred over a 1000-year interval of maximum vegetation change during the last glacial–interglacial transition (Figure 15.5c). The GFDL-simulated future change is equivalent in magnitude to the that which required 3000 years over this transition (Figure 15.5d).

Although the possibility exists that the rates of future climate change may be slowed by the response of oceans, the above comparison suggests that the response of vegetation to future climate change will be substantial, and perhaps even unprecedented during the past 20,000 years. Climate-induced vegetation change in other parts of the world was also large over the past 18,000 years (e.g., Huntley and Webb

---

Figure 15.5 A) Vegetation differences diagram comparing the potential future vegetation change with that of the past 18,000 years. Square chord distances (Overpeck et al. 1985) were calculated to measure the differences between: (i) observed modern pollen distributions and those of 250 years ago, 500 years ago, etc. back to 18,000 yr B.P. (from geological data, left graph), (ii) observed modern pollen distributions and the simulated modern distributions (distribution #1 on right), and (iii) the simulated modern pollen distributions and each of the sets of distributions simulated by doubled CO_2 climate model experiments (distribution #2 for the GISS GCM, #3 for the GFDL GCM, and #4 for the OSU GCM on right). The squared chord distances for each comparison are shown as box plots. The center line in each box is the median dissimilarity between two pollen distributions, the edges of the box are the 25th and 75th percentiles, and the lines extend to the 10th and 90th percentiles. Distribution #1 (right) indicates the kinds of dissimilarity that arise from the lack-of-fit of the response surfaces and from the omission of other potentially useful predictor variables. Distributions #2 to #4 show that the differences between the simulated modern maps and those simulated by the doubled CO_2 experiments are large. **B–D**) Vegetation change diagrams showing comparisons between fossil pollen distributions separated by 500 years **B**) 1000 years **C**), and 3000 years **D**) over the last 18,000 years. Fossil sample comparisons are plotted at the midpoint of a sliding window (with a width of 500, 1000, 3000 years) centered on 250-year intervals. For example, the rightmost box on the 500-year time step diagram shows the comparison of 0 yr B.P. with 500 yr B.P. and is plotted at 250 yr B.P., the next box to the left is for the comparison of 250 yr B.P. and 750 yr B.P., and so on. Distributions 1–4 are repeated for comparison. Figure from Overpeck et al. (1991).

1988), implying that the future response of vegetation to climate change could be significant over much of the Earth. The possibility of significant future vegetation change is supported by recent efforts to model the global vegetation response to include an interactive biosphere in GCM simulations of future climate change (Henderson-Sellers 1990).

THE ROLE OF VEGETATION IN GLOBAL CHANGE

The response of vegetation to climate change has been characterized both by model and observational studies. Less understood, but equally important, is the role that vegetation played, and will play, as an interactive component of the climate system. Future vegetation change will likely impact the climate system both indirectly, via the effects on ecosystem dynamics, soils, and atmospheric trace gases, and directly, via changes in land surface characteristics (Houghton et al. 1990).

Soils

Rapid climate and vegetation change, particularly when coupled with current land-use practices, could lead to an accelerated rate of soil erosion in many areas. Vegetation also plays a role in the long-term dynamics of natural soil development. Although soils hold more carbon than living vegetation, the slowness of soil processes means that soil dynamics and carbon storage in soils are not likely to change much within the next century (Schlesinger 1990). Over longer time scales, however, soil dynamics are more susceptible to climate and vegetation change, and are likely to affect trace gas fluxes, vegetation distribution, and surface hydrology. An improved understanding of vegetation's impact on changing climate could be gained by mapping, analysis, and modeling of the large changes in soils that occurred during the late Quaternary, particularly in areas formerly covered by ice sheets.

Trace gases

Vegetation also plays important roles in the regulation of atmospheric greenhouse gases, such as water vapor, CO_2, CH_4, N_2O (via soil influences), nonmethane hydrocarbons (NMHCs), and via the influences of NMHCs, on tropospheric O_3 (Cicerone and Oremland 1988; Lashof 1989; Prentice and Fung 1990; Tans et al. 1990). A sharp deglacial increase in atmospheric CH_4 concentrations (from 350 to 650 ppbv) is currently attributed to large changes in the terrestrial biosphere between 14,000 and 11,000 yr B.P. (Chappellaz et al. 1990), but may also relate to changes in atmospheric CH_4 sinks (Peltier et al., this volume). A similar large deglacial increase in CO_2 (ca. 70 ppmv; Barnola et al. 1987) was probably dominated by abrupt changes in ocean circulation (Broecker and Peng 1989; Boyle 1991), but the effects on the global carbon cycle of large post-12,000 yr B.P. shifts in terrestrial vegetation and peatland development are potentially significant. The exact role of the biosphere in the changing global carbon cycle remains to be worked out.

Vegetation-driven changes in trace gases lead to changes in the planetary radiation balance. The large glacial–interglacial changes in atmospheric CO_2 and CH_4 have been linked to a direct radiative forcing of about 2 W/m^2 or a 0.7°C increase in global surface air temperatures when feedbacks are ignored (Lorius et al. 1990). This trace gas forcing, some of which can be attributed to vegetation change, translates to a warming of 2° to 2.5°C (40 to 50% of the total glacial–interglacial change) when feedbacks such as water vapor, clouds, sea ice, and snow cover are factored in (Lorius et al. 1990). Deglacial trace gas changes may have been partly responsible for the apparent global synchroneity of large abrupt climatic change at about 13,000 yr B.P. A complete understanding of how vegetation can influence trace gases, and thus climate, will require a careful study of past global change.

Direct Climate Feedbacks

Realistic representations of vegetation and soils are required in climate models in order to yield accurate simulations of temperature and moisture fields, and thus past and future climatic change. This fact has been demonstrated in global scale model sensitivity tests (e.g., Rind 1984; Henderson-Sellers 1990), as well as detailed examinations of the effects of vegetation change on regional climates (e.g., Dickinson and Henderson-Sellers 1988).

Vegetation change interacts with the climate system primarily through changes in surface albedo and the transport of moisture from the land surface into the atmosphere, as well as via effects on land surface roughness, snow masking, and surface runoff (Rind 1984; Dickinson and Henderson-Sellers 1988). Vegetation change has apparently exerted a significant direct impact on past climates (Hansen et al. 1984; Street-Perrott et al. 1990) and is likely to do so in the future (Henderson-Sellers 1990). Vegetation change is also likely to have an indirect effect on climate in its role as a land-surface stabilizer. The establishment of vegetation in desert and proglacial areas sharply slowed tropospheric dust loading during the last deglaciation (Royer et al. 1983), thereby warming the air at the surface and allowing greater convective activity (Figure 15.6). Future desertification could lead to climatically important changes in surface albedo, and evapotranspiration, and the frequency of dust events.

CONCLUSIONS: WHAT NEEDS TO BE DONE

The magnitude of future climate-induced vegetation change will likely equal that observed during the last deglaciation, and the rate of this change may be 2 to 5 times more rapid. Because vegetation change could play a significant role affecting many other components of the climate system, efforts must be accelerated to produce models capable of simulating realistic climate-induced vegetation change and feedbacks of

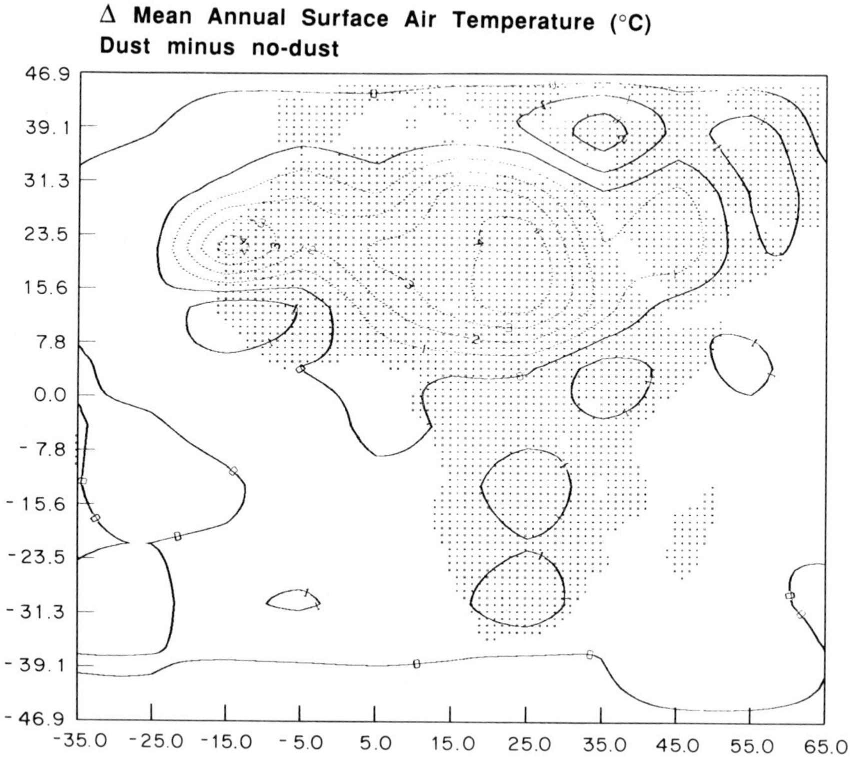

Figure 15.6 Anomaly map illustrating the potential effect of glacial-age tropospheric dust on the surface air temperature of Africa prior to ca. 12,500 yr B.P. After this time, small increases in available moisture and in vegetative cover apparently reduced dust entrainment. Anomalies plotted are those between two identical NASA GISS GCM experiments, except that one ("dust" experiment) included levels of desert dust estimated for peak late-Pleistocene atmospheric dust loading (optical depth = 4X modern = 3.0; Royer et al. 1983), and the other ("no-dust" experiment) had no desert dust in the troposphere. Results from Overpeck and Rind (in prep.).

this change within the climate system. The geological record of past climate and vegetation change is the only observational record of large change and thus a necessary guide in building realistic models. Ultimately, vegetation should be modeled in a way that acknowledges the individualistic response of plant taxa to climatic change, and the time-varying nature of plant associations. The possible rapidity of future change highlights the need for models that can simulate time-dependent change. Although the importance of direct effects of CO_2 on natural plant communities should not be ignored (Houghton et al. 1990), there is still no evidence that direct effects altered climate-plant relationships in any significant way in the past (e.g., Prentice et al. 1991).

In addition to guiding interpretations, the record of past climate, vegetation, and trace gas change should serve as a primary resource for testing hypotheses and models.

If we think we know what caused the large changes in atmospheric CO_2 and CH_4 in the past, we should be able to model these changes. Similarly, if we think we understand the feedbacks of vegetation on climatic change, we should be able to test this understanding against the many large climate and vegetation changes of the past. Ice core records already provide a remarkable record of past trace gas change, but need to be dated more accurately. Higher resolution records of CO_2 are also needed to determine if CO_2 fluctuations played a role in climate events, such as the Younger Dryas, and to improve the observational baseline against which models of CO_2 change are tested. Efforts to map global climate and vegetation change of the past need to be accelerated and focused on selected time periods, such as the last glacial maximum, to examine the equilibrium response and role of vegetation, and on periods of most rapid change (e.g., between 14,000 and 9000 yr B.P., and perhaps 5000 and 3000 yr B.P. in the Tropics), to study the time-dependent response and role of vegetation. Successful hypothesis and model testing will require more accurate temporal resolution and intra-site correlations than are usually obtained (Overpeck 1990).

This review emphasizes past change at mid-latitudes primarily because far more data are available for these latitudes than for elsewhere. A complete understanding of the response and role of vegetation within the climate system requires that special efforts be made to produce detailed subcontinental-scale reconstructions from all latitudes. Without a well-dated record of past global change, simulations of future global change can not adequately be tested. The ability of the current generation of climate and vegetation models to simulate important aspects of past change demonstrates important progress.

ACKNOWLEDGEMENTS

I thank P. Bartlein, E. Cook, B. Frenzel, R. Goldberg, I.C. Prentice, D. Rind, R. Webb, and T. Webb III for discussions of climate and vegetation change. Thanks also to K. Anderson, P. Bartlein, B. Lipsitz, R. Goldberg, and R. Healy for help with graphics, and to J. Cole for a valuable review of the manuscript. The support of facilities and colleagues at the NASA Goddard Institute for Space Studies and NOAA National Geophysical Data Center is gratefully acknowledged, as is the help from the Dahlem staff.

REFERENCES

Barnola, J.M., D. Raynaud, Y.S. Korotkevich, and C. Lorius. 1987. Vostok ice core provides 160,000-year record of atmospheric CO_2. *Nature* **329**:408–414.

Boyle, E.A. 1991. Quaternary ocean paleochemistry. *Rev. Geophys. Supp.*, pp. 634–638.

Bradley, R.S. and P.D. Jones. 1991. Climate since A.D. 1500. London: Routledge.

Broecker, W.S. and T.-H. Peng. 1989. The cause of the glacial–interglacial atmospheric CO_2 change: A polar alkalinity hypothesis. *Glob. Biogeochem. Cyc.* **3**:215–239.

Chappellaz, J., J.M. Barnola, D. Raynaud, Y.S. Korotkevich, and C. Lorius. 1990. Ice-core record of atmospheric methane over the past 160,000 years. *Nature* **354**:127–131.

Cicerone, R.J. and R.S. Oremland. 1988. Biogeochemical aspects of atmospheric methane. *Glob. Biogeochem. Cyc.* **2**:299–327.

COHMAP. 1988. Climatic changes of the last 18,000 years: Observations and model simulations. *Science* **241**:1043–1052.

Cook, E., T. Bird, M. Peterson, M. Barbetti, B. Buckley, R. D'Arrigo, R. Francey, and P. Tans. 1991. Climatic change in Tasmania inerred from a 1089-year tree-ring chronology of Huon pine. *Science* **253**:1266–1268.

Cook, E., T. Bird, M. Peterson, M. Barbetti, and R. Francey. 1992. The Little Ice Age in Tasmanian tree rings. In: Proc. International Symposium on the Little Ice Age Climate. Tokyo: Tokyo Metropolitan Univ. Press.

Cook, E.R. and J. Cole. 1991. On predicting the response of forests in eastern North America to future climatic change. *Clim. Change* **19**:271–282.

Crowley, T.J. 1990. Are there any satisfactory geologic analogs for a future greenhouse warming? *J. Climate* **3**:1282–1992.

Davis, M.B. 1989. Lags in vegetation response to greenhouse warming. *Clim. Change* **15**:75–82.

Davis, M.B. 1990. Research questions posed by the paleoecological record of global change. In: Global Changes of the Past, ed. R. Bradley, pp. 385–395. Boulder, CO: UCAR/Office for Interdisciplinary Earth Studies.

Delcourt, P.A. and H.R. Delcourt. 1987. Long-term Forest Dynamics of the Temperate Zone. New York: Springer.

Dickinson, R.E. and A. Henderson-Sellers. 1988. Modeling tropical deforestation: A study of GCM land-surface parameterizations. *Q. J. R. Meteorol. Soc.* **114**:439–462.

Hansen, J., A. Lacis, D. Rind, G. Russell, P. Stone, I. Fung., R. Ruedy, and J. Lerner. 1984. Climate sensitivity: Analysis of feedback mechanisms. In: Climate Processes and Climate Sensitivity, ed. J.E. Hansen and T. Takahashi. *Geophys. Monog.* **29**:130–163. Washington, D.C.: American Geophysical Union.

Henderson-Sellers, A. 1990. Predicting generalized ecosystem groups with the NCAR CCM: First steps towards an interactive biosphere. *J. Climate* **3**:917–940.

Houghton, J.T., G.J. Jenkins, and J.J Ephraums, eds. 1990. Climatic Change: The IPCC Scientific Assessment. Cambridge: Cambridge Univ. Press.

Huntley, B. and T. Webb III. 1988. Vegetation History. Dordrecht: Kluwer.

Jacobson, G.L., Jr., T. Webb III, and E.C. Grimm. 1987. Patterns and rates of vegetation change during the deglaciation of eastern North America. In: North America and Adjacent Oceans during the Last Deglaciation, ed. W.F. Ruddiman and H.E. Wright Jr., pp. 277–288. Vol. K–3 of The Geology of North America. Boulder, CO: Geological Society of America.

Lashof, D.A. 1989. The dynamic greenhouse: Feedback processes that may influence future concentrations of atmospheric trace gases and climatic change. *Clim. Change* **14**:213–242.

Lorius, C., J. Jouzel, D. Raynaud, J. Hansen, and H. Le Treut. 1990. The ice-core record: Climate sensitivity and future greenhouse warming. *Nature* **347**:139–145.

Overpeck, J.T. 1990. Century- to millennium-scale climatic variability during the late Quaternary. In: Global Changes of the Past, ed. R. Bradley, pp. 139–173. Boulder CO: UCAR/Office for Interdisciplinary Earth Studies.

Overpeck, J.T. and P.J. Bartlein. 1989. Assessing the response of vegetation to future climate change: Ecological response surfaces and paleoecological model validation. In: U.S. Environmental Protection Agency Report EPA–230–05–89–54. chap. 1. Washington, D.C.: U.S. EPA.

Overpeck, J.T., P.J. Bartlein, and T. Webb III. 1991. Potential magnitude of future vegetation change in eastern North America: Comparisons with the past. *Science* **254**:692–695.

Overpeck, J.T., D. Rind, and R. Goldberg. 1990. Climate-induced changes in forest disturbance and vegetation. *Nature* **343**:51–53.

Overpeck, J.T., R.S. Webb, and T. Webb III. 1992. Mapping eastern North American vegetation change over the past 18,000 years: No-analogs and the future. *Geology* **20:**1071–1074.

Overpeck, J.T., T. Webb III, and I.C. Prentice. 1985. Quantitative interpretation of fossil pollen spectra: Dissimilarity coefficients and the method of modern analogs. *Quat. Res.* **23**:87–108.

Prentice, I.C., P.J. Bartlein, and T. Webb III. 1991. Vegetation and climate change in eastern North America since the last glacial maximum. *Ecology* **72**:2038–2056.

Prentice, K.C. and I.Y. Fung. 1990. The sensitivity of terrestrial carbon storage to climate change. *Nature* **346**:48–51.

Rind, D. 1984. The influence of vegetation on the hydrological cycle in a global climate model. In: Climate Processes and Climate Sensitivity, ed. J.E. Hansen and T. Takahashi. *Geophys. Monog.* **29**:73–91. Washington, D.C.: American Geophysical Union.

Royer, A., M. De Angelis, and J.R. Petit. 1983. A 30,000 year record of physical and optical properties of microparticles from an East Antarctic ice core and implications for paleoclimate reconstruction models. *Clim. Change* **5**:381–412.

Schlesinger, W.H. 1990. Evidence from chronosequence studies for a low carbon-storage potential of soils. *Nature* **348**:232–234.

Solomon, A.M. 1986. Transient response of forests to CO_2-induced climate change: Simulation modeling experiments in eastern North America. *Oecologia* **68**:567–579.

Street-Perrott, F.A., J.F.B. Mitchell, D.C. Marchand, and J.S. Brunner. 1990. Milankovitch and albedo forcing of the tropical monsoons: A comparison of geological evidence and numerical simulations for 9000 y BP. *Trans. R. Soc. Edin. Earth Sci.* **81**:407–427.

Tans, P.P., I.Y. Fung, and T. Takahashi. 1990. Observational constraints on the global atmospheric CO_2 budget. *Science* **247**:1431–1438.

Webb, R.S., J.T. Overpeck, and T. Webb III. 1991. Reconstructed vegetation patterns for eastern North America: 18,000 yr B.P. to present (in prep.)

Webb, T., III. 1988. Eastern North America. In: Vegetation History, ed. B. Huntley and T. Webb III, pp. 385–414. Dordrecht: Kluwer.

Webb, T., III, P.J. Bartlein, and J.E. Kutzbach. 1987. Post-glacial climatic and vegetational changes in eastern North America since 18 ka: Comparison of the pollen record and climate model simulations. In: North America and Adjacent Oceans during the Last Deglaciation, ed. W.F. Ruddiman and H.E. Wright Jr., pp. 447–462. Vol. K–3 of The Geology of North America. Boulder, CO: Geological Society of America.

Webb, T., III and T.M.L. Wigley. 1985. What past climates can indicate about a warmer world. In: Projecting the Climatic Effects of Increasing Carbon Dioxide, ed. M.C. MacCracken and F.M. Luther, pp. 237–258. Report DOE/ER–0237. Washington, D.C.: U.S. Dept of Energy.

Wright, H.E., Jr. 1967. The use of surface samples in Quaternary pollen analysis. *Rev. Palaeobot. Palyn.* **2**:321–330.

Standing, left to right:
U. Siegenthaler, I. Levin, M. McElroy, D. Raynaud, E. Maier-Raimer, J.C. Duplessy

Seated, left to right:
W.R. Peltier, C.A. Burga, J.T. Overpeck, K. Herterich

16

Group Report: How Can We Use Paleodata to Evaluate the Internal Variability and Feedbacks in the Climate System?

W.R. PELTIER, Rapporteur

C.A. BURGA, J.-C. DUPLESSY,
K. HERTERICH, I. LEVIN, E. MAIER-REIMER,
M. MCELROY, J.T. OVERPECK, D. RAYNAUD,
U. SIEGENTHALER

INTRODUCTION

Our knowledge of the modern climate system is encoded in conventional atmosphere–ocean general circulation models (AOGCMs), and we expect them to be useful as aids in predicting the way in which global climate will evolve in response to changes in internal or external forcing. In spite of their considerable numerical complexity, however, the finite resolution at which we can afford to integrate them ensures that many crucial physical processes can be represented only in a heavily parameterized form (e.g., tropical cumulous convection, clouds, boundary layer processes, radiative transfer, etc.). Although we may adjust the variables that control the impact of these parameterized processes to obtain a high quality fit to modern climate, we have no reason to believe that the model "tuned" in this way will behave realistically when differently forced. There is, therefore, good reason to attempt to test these models further by comparing their predictions to observed states of the climate system that differ significantly from that which obtains at present. If the AOGCM is able to pass such additional tests, then we will be more justified in placing credence in its prediction of the response to the ongoing anthropogenic increase of the radiatively active trace gases carbon dioxide (CO_2) and methane (CH_4). Paleodata provide the only source of the information required to perform such additional tests.

Global Changes in the Perspective of the Past
Edited by J.A. Eddy and H. Oeschger

In the following discussion, we will focus especially on the way in which such data on past states of the climate system may be invoked to constrain variations through time of the behavior of several important subcomponents of the system. Such information is vital in helping us to appreciate the nature of internal climate system variability and the physical–chemical–biological feedbacks that underlie this variability. Since the response to external forcing associated with variations of the Earth's orbit around the Sun is apparently a crucial factor in triggering such variability throughout late Quaternary time, and since the dating of significant climatic events is far better in this most recent epoch of Earth history than for earlier times, we will focus most of the ensuing discussion on this period. Within this period, the dominant climatic signal is that associated with the 100,000-year cycle of the ice ages, which began around 900,000 years ago. This cycle is now known to involve dramatic changes in the thermohaline circulation of the oceans (Shackleton et al. 1983; Boyle and Keigwin 1985), in the terrestrial and marine biosphere, and in atmospheric trace gases as well as in continental ice cover. We will explore the interactions between these subelements of the Earth system that are involved in the glacial cycle. As we will suggest, the inferences that follow from this exploration provide useful insights into climate system function and sensitivity.

GREENHOUSE GASES AND CLIMATIC DETERMINISM

Before entering our discussion of climate system variability and specific feedbacks, it will be useful to review a number of general concepts concerning the behaviour of complex nonlinear systems. One of the more important concepts is connected to the question of the existence of multiple-equilibria. If a system is "transitive," for example, it is one which has a unique (perhaps statistical) equilibrium state for each set of boundary conditions, regardless of the initial conditions from which this end state is approached. To be specific, if the Earth's climate system were transitive and the atmospheric concentration of CO_2 were 300 ppm, then the same steady-state climate would obtain, whether this state was reached by reducing CO_2 from 450 ppm to 300 ppm or by increasing CO_2 from 150 ppm to 300 ppm. On the other hand, if the system is intransitive (or "nonergodic"), then it may posses more than a single equilibrium state for the same set of boundary conditions. In some climate models, for example, if the Earth is initially covered by ice, it will remain ice-covered even with the incident solar insolation fixed at its present-day value. On the other hand, if the Earth is initially ice-free and the solar intensity is similarly fixed, then the same model will reproduce the present-day climate. In other words, the model (or the Earth itself if its climate system is also intransitive) remembers its history, such that different equilibrium states may be realized even when the boundary conditions are fixed. Whether the actual Earth has a climate system that is transitive or intransitive is still being actively debated.

A specific current challenge to climatologists is to provide some insight into the changes of climate that might be expected in the future, in response to increasing levels of CO_2 and other greenhouse gases. In attempting to reduce the issue to one that is unrelated to the details of the actual greenhouse gas increase, it has become standard practice to ask what equilibrium climate might ensue in response to a doubling of atmospheric CO_2 concentration, all other features of the climate system remaining fixed. In particular, would this new climate state differ significantly from that which obtains today? In this circumstance, we would expect that the increased levels of CO_2 would induce an increase in the emission of infrared radiation from the stratosphere to the troposphere–surface system of about 4 W/m^2. This serves as a useful estimate of the magnitude of the associated climatic forcing, neglecting feedbacks associated with changes in stratospheric O_3, stratospheric temperature, tropopause height, tropospheric H_2O, clouds and other effects due to enhanced CO_2. The first-order issue that we must address is, then, whether a change in climatic forcing of 4 W/m^2 should be considered significant. In particular, can we provide insight into this problem based on information from the past?

On the basis of all the information at our disposal, it is clear that the climate system displays exceptional variability in both space and time on all scales. Over the past two million years, it is clear, in particular, that the changes in radiative forcing have been extremely modest. (For example, the June insolation variations at a period of 10^5 years associated with the change of orbital eccentricity have fallen from about 8 W/m^2 to about 4 W/m^2; annually averaged, of course, the variation in radiative forcing through time is vanishingly small). Nevertheless, the variations in global climate over this time have been extremely large. The most recent million years of climate history, for example, have been characterized by major episodic advances of continental ice sheets punctuated by relatively brief epochs of deglaciation, during which climate was relatively benign.

As we will discuss in much greater detail below, prior to about one million years ago the dominant period of variability in the climate system was the 41-Ka period of the obliquity cycle. Beginning about 900,000 years ago, a shift to a dominant period of 100,000 years occurred, an event to which many have referred to as the mid-Pleistocene climate transition. It is in this interval of time that the massive expansions and contractions of northern hemisphere land ice occurred, individual pulses of which were accompanied by rises and falls of global sea level in the range of 100–150 m.

In earlier periods (3–100 Ma ago), climate was apparently much warmer than at present. This is evidenced, as are the glacial cycles discussed above, by isotopic data from deep-sea sediments which indicate temperatures for the deep oceans as high as 15°C (e.g., Crowley and North 1991) during the Cretaceous (Brass et al. 1982). Even a cursory examination of the climate record, therefore, suggests that change, even profound change, has been the norm rather than the exception. As we will further elaborate below, the weight of the paleo evidence suggests that a change in the atmospheric concentration of CO_2 by a factor of 2 should be considered significant. However, the details of the climatic response to such a change of trace gas loading will depend critically upon a complex network of interacting positive and negative feedback loops. Circumstantial evidence from the past suggests, in particular, that the equable climates of the Eocene and Cretaceous periods were associated with levels of CO_2 that were between 2 and 3 times present levels (e.g., Crowley and North 1991).

THE 100,000-YR ICE-AGE CYCLE AND THE MID-PLEISTOCENE CLIMATE TRANSITION

No single observation of climate system behavior illustrates the importance of internal feedbacks in the determination of climate state better than the 100,000-year cycle of late Pleistocene ice volume, which has been so clearly revealed by oxygen isotope data from deep-sea sedimentary cores (Broecker and Van Donk 1970). Detailed analyses of the power spectra of $\delta^{18}O$ time series from such archives for the late Pleistocene (Hays et al. 1976; Imbrie et al. 1984) demonstrate the indelible imprint of the orbital insolation signal upon them. Since $\delta^{18}O$ is understood to be dominated by the signal associated with variations of continental ice volume (Shackleton 1967; approximately two-thirds of this variation is currently believed to be due to ice-volume change and one-third due to temperature change during the last half of Pleistocene time), and since the spectra reveal 95% significant lines at periods of 100,000 years (the dominant period for the past 900,000 years), 41,000 years, 23,000, years and 19,000 years, it should not be too surprising that such data are now widely accepted as verifying the essential idea in Milankovitch's theory of ice ages. The 100-Ka period is the dominant period on which changes in orbital eccentricity occur; the 41-Ka period is that which dominates the variation of orbital obliquity; and the lines at 23-Ka and 19-Ka periods are those that dominate changes of the precision parameter (e.g., Berger 1978). That the climate system's response to the seasonal insolation anomalies associated with changes in the Earth's orbit involve critical feedback effects, however, is made clear by the fact that the power spectrum of the seasonal insolation time series that is driving the variations of ice volume differs profoundly from that of $\delta^{18}O$. Although the ice-volume time series is dominated by the spectral line corresponding to a period near 100,000 yr, the summer seasonal insolation anomaly time series has essentially zero power at this period. Only by isolating the anomaly associated with a single summer month (June, say) may one find any significant signal at 100-Ka period in the incoming insolation signal that is apparently driving the ice-volume response. To illustrate this fact we show in Figure 16.1 (Hyde and Peltier 1985, 1987) the time series and power spectrum for the summer seasonal insolation anomaly at 65°N latitude (based on the analysis of Berger 1978).

A variety of different mechanistic models of the climate system have been proposed in an attempt to understand how this system converts the stimulus to which it is subject into a response that differs so radically from the stimulus itself. These models all embody some degree of nonlinearity, since it is rather difficult to imagine how any linear system could replicate the disparity between input and output documented above.

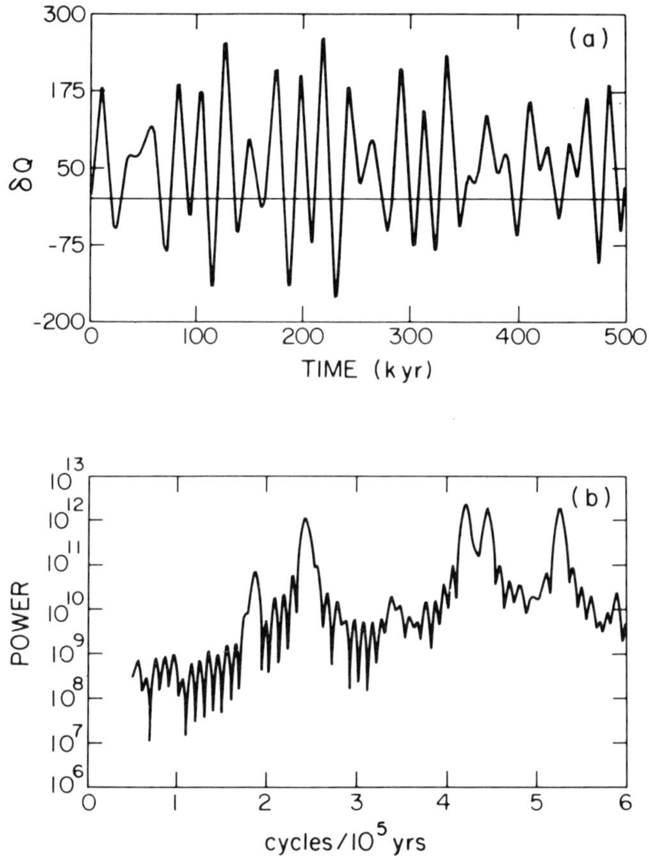

Figure 16.1 (a) The variation of caloric summer seasonal insolation at 65°N latitude over the past 500 Ka computed from Berger (1978a, 1978b). (b) The power spectrum of 1000 Ka of this insolation anomaly time series. Negligible power is observed at 100 Ka, while strong peaks are observed at 41 Ka, 23 Ka, and 19 Ka. From Hyde and Peltier (1985).

However, aside from explaining the origin of the 100-Ka component in the ice-volume variability, such models are also obliged to explain why this signal was essentially absent prior to about 900 Ka ago. This further fact is illustrated in Figure 16.2a (from Deblonde and Peltier 1991a), where the $\delta^{18}O$ record from ODP 677 (presented in Shackleton et al. 1990) is shown. The lower panels of this figure display a number of time-dependent statistics that have been employed to characterize the rather sharp transition in the record that occurs at mid-Pleistocene time. Even visually it is apparent that the 100-Ka cycle is essentially absent prior to 900 Ka ago. This fact is quantified in Plate I, in which we show the digital sonogram (time-dependent power spectrum) for ODP 667 on the time scale of Shackleton et al. (1990). Inspection of this figure

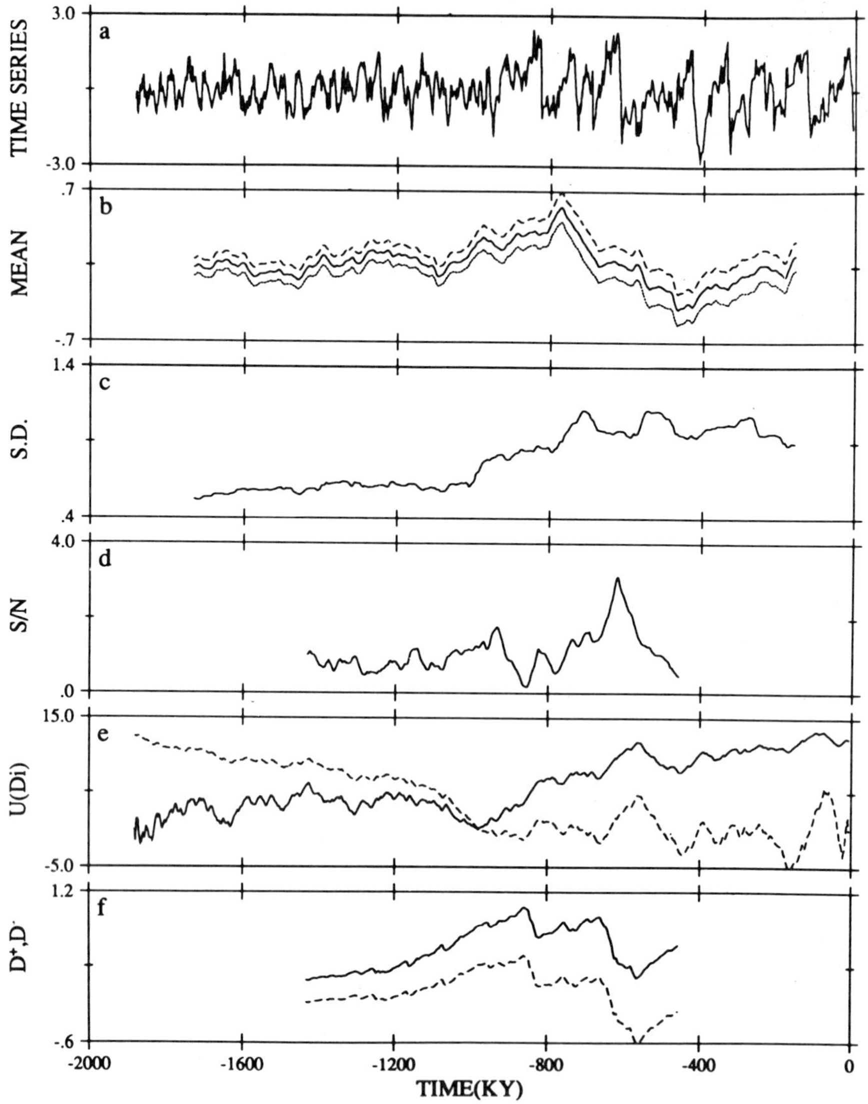

Figure 16.2 Statistical analysis of ODP site 677 $\delta^{18}O$ record with the depth-age model taken from Shackleton et al. (1990). (a) Time series with linear trend removed. (b) Running mean of the time series shown in (a) with segment length of 300 Ka. (c) Changing standard deviation of the time series shown in (a) computed for a segment length of 300 Ka. (d) Signal over noise ratio averaged over a range of segment lengths of 300 to 450 Ka (the confidence level is 99%). (e) Mann-Kendall rank statistic, $U(D_i)$, for forward (solid) and retrograde (dashed) time series. (f) Test for jump in variance, D^+ (solid line) and D^- (dashed line), averaged over the same segment length as in (d). From Deblonde and Peltier (1991a).

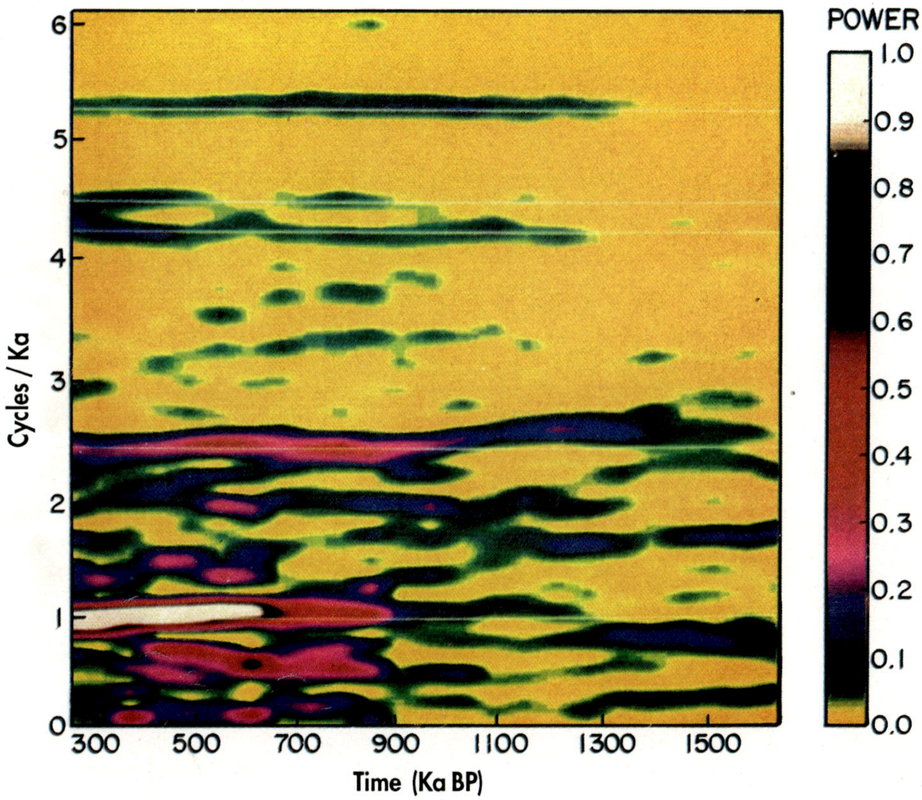

Plate I Time-dependent power spectrum (digital sonogram) of the ODP 677 $\delta^{18}O$ time series (part a of Figure 16.2) on the time scale of Shackleton et al. (1990). A segment length of 500 Ka has been employed for the purpose of computing the running spectrum.

reveals a number of further interesting aspects of the way in which the 100-Ka line rather suddenly appears as the dominant feature of the climate record, even though no substantial variation in the orbital insolation signal accompanies it.

There currently exist two relatively simple models that have been advanced to "explain" both the origin and the internal dynamics of the 100-Ka cycle. These two models, by Maasch and Saltzman (1990) and Deblonde and Peltier (1991a), "explain" the 100-Ka cycle as being due, respectively, to a free oscillation of the system that could exist even in the absence of the insolation stimuli and a forced oscillation that would die out if the stimulus were removed. The former model is of the so-called "dynamical systems" type, which is based upon a coupled set of nonlinear ordinary differential equations with only three degrees of freedom. The latter is based upon a set of coupled nonlinear partial differential equations that are derived from a two-dimension (zonally averaged) description of the interaction between an applied insolation anomaly time series, a large-scale ice sheet (whose nonlinear dynamics are described explicitly and whose mass balance is parameterized in terms of the applied insolation anomaly), and the underlying Earth, which is assumed capable of adjusting isostatically under the weight of the ice load (e.g., Peltier 1982). For the former model, the 100-Ka cycle may be induced to appear by hypothesizing the action on the climate system of a long time scale additional forcing, which causes the sudden appearance of the 100-Ka cycle at 900 Ka B.P.. In the latter model, the appearance of the 100-Ka cycle is associated with the activation of a feedback loop that impacts the mass balance only when the time rate of change of ice volume exceeds some threshold. Such a feedback loop was first hypothesized by Pollard (1982) as being associated with amplification of meltback due to calving into proglacial lakes. In the more detailed model of Deblonde and Peltier (1991b), the feedback is left generic (unspecified in detail as to its precise cause). It should be understood that in both the model of Maasch and Saltzman (1990) and that of Deblonde and Peltier (1991b), some additional forcing must be invoked beyond that associated with orbital insolation variations to understand why the 100-Ka cycle appears. We find the slow decrease of atmospheric CO_2 concentration that apparently has been occurring over the past several tens of millions of years (Crowley and North 1991) to be a good candidate for the nonorbital contribution to the forcing that is required in the context of either of these models to stimulate the onset of the 100-Ka cycle. The ability of such simple models to mimic the $\delta^{18}O$ time series (and therefore the variations of Pleistocene ice volume) is demonstrated in Figure 16.3. This figure presents a synthetic ice history from the model of Deblonde and Peltier (1991a) that may be compared to the $\delta^{18}O$ time series from ODP 677 previously shown on Figure 16.2. On the lower plates of Figure 16.3, the same statistical tests are applied to the synthetic record as were previously applied to ODP 677. Except for the apparent 400-Ka component of the model response in the earliest Pleistocene, which is not evident in ODP 677, the synthetic time series fits the observed time series rather well.

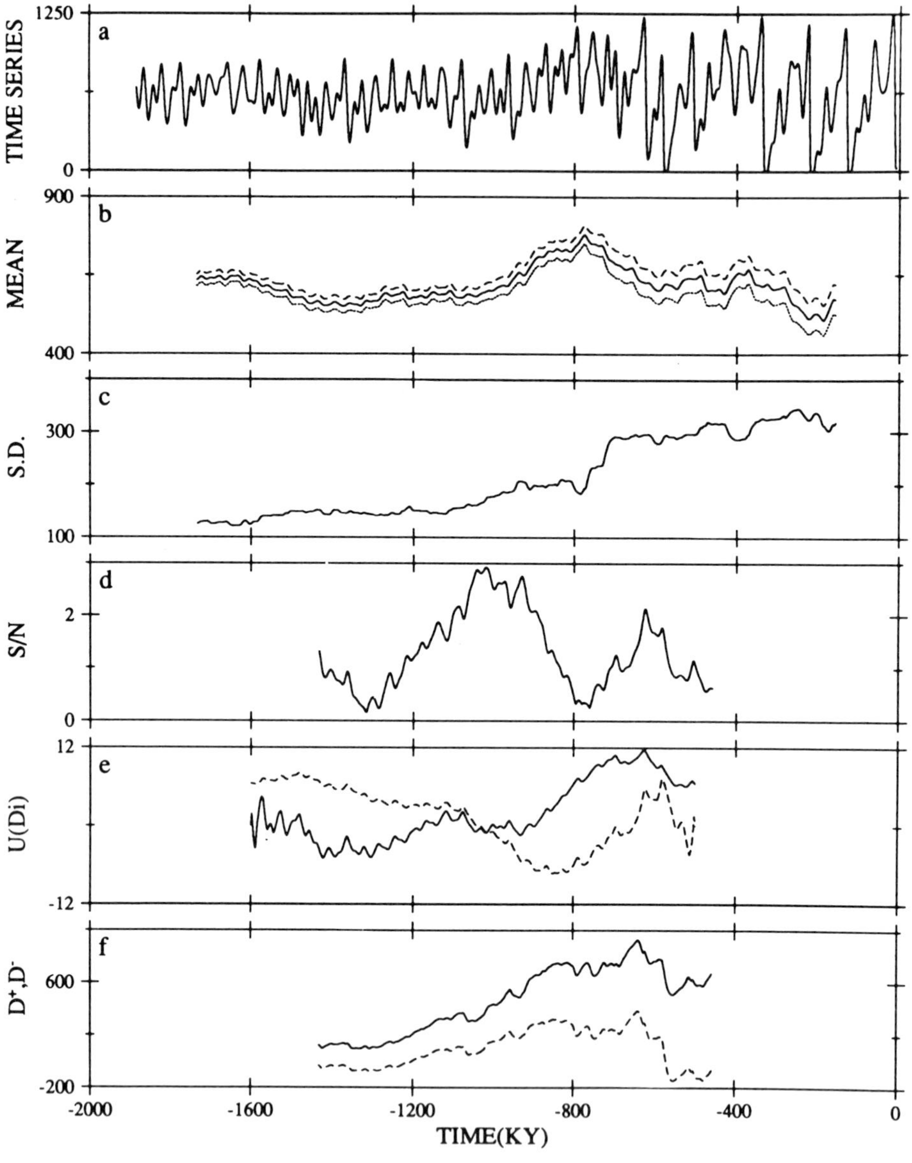

Figure 16.3 Same as Figure 16.2 but for a simulated ice volume time series (plate a) based upon the model of Deblonde and Peltier (1991a).

A number of more complex models of the 100-Ka cycle have also been proposed to describe more accurately the feedbacks that are operative within the system and which support the dominance of this cycle in late Pleistocene time. The model described by Gallée et al. (1991), for example, is based on a zonally averaged formulation in which, when it is subject to the Milankovitch variations of solar insolation, the predicted ice volume rises slowly and collapses rapidly in a way that is similar to that implied by the $\delta^{18}O$ data from deep-sea sedimentary cores (Labeyrie et al. 1987; Broecker and Van Donk 1970) and by the ice-core record from Vostok (e.g., Jouzel et al. 1987). In this model, the rapid retreats of continental ice that mark the terminations of each glacial cycle are driven by a sharp decrease of the albedo of the ice induced by the "aging" of the snow on its surface. The model of Deblonde and Peltier (1991b), on the other hand, is based on a two-dimensional global seasonal energy balance model due to North et al. (1983), in which separate subdomains are imbedded wherein the dynamics of the accumulation and flow of ice in response to applied orbital insolation anomalies are described separately. When it is initialized with ice-free (with the exception of Greenland) Northern Hemisphere conditions beginning 122,000 years ago (the last interglacial), this model successfully predicts the observed locations and the maximum thickness achieved by both the North American and the Northwest European ice complexes as implied by the geophysically inferred ICE–3G model (Tushingham and Peltier 1991, 1992). With the same minimal set of interacting physical ingredients as were employed in the simpler model of Deblonde and Peltier (1991a), however, the termination that commences near 21 Ka B.P. is not predicted. Just as in the simpler model, an additional feedback loop must be added. Recently completed additional tests have demonstrated that neither the observed variation of CO_2 (Barnola et al. 1987) nor the variation of high-latitude heating associated with the varying strength of the thermohaline circulation (see below) can by themselves induce the observed full deglaciations. One feedback, in the context of this model, has proven effective in the simulation of complete terminations is the modulation of ice-sheet albedo due to the intense rise in atmospheric dust concentration, which appears to be a generic feature of the paleo record prior to each major termination of ice volume (De Angelis et al. 1987). Although this influence is similar to that of the seasonally "snow aging" invoked by Gallée et al. (1991) to cause terminations, it is derivative of an entirely different cause. GCM experiments run with the GISS AGCM also suggest that high tropospheric dust loading at glacial maxima over the desert regions may have led to tropospheric warming at high latitudes and thus contributed to rapid terminations. How are we to decide which, if either, of these feedback loops is actually involved in the termination process?

We hope that this brief review of theories of the 100-Ka cycle will help to put in evidence the fundamental issues that are raised by investigations of climate system variability extracted from paleo records. In what follows we will "zoom-in" on the separate subcomponents of the global climate system, which paleo observations suggest were intimately involved in the most recent termination event of the present ice age. As we will attempt to make clear, fundamental issues remain concerning the relative timing of the imprint of termination upon these different subsystems, including the global thermohaline circulation of the oceans, the terrestrial biosphere, and the trace gas component of the global atmosphere. In each of these cases, the paleodata will be shown to shed considerable light on the functioning of the system, as well as upon the timing of variations in its level of "activity."

MODES OF CIRCULATION IN THE GLOBAL OCEAN

That there may be more than one single mode of deep circulation of the oceans is an idea first postulated by Stommel (1961) on the basis of a rather simple box model. Bryan (1986) later showed this same "intransitivity" to be characteristic of a very detailed oceanic general circulation model (OGCM) (see Duplessy and Maier-Reimer, this volume). The possible importance of this intransitivity to the understanding of late Pleistocene glacial history has been stressed by Broecker and Denton (1990), among others, especially in the context of the terminations mentioned above. This has spawned a small industry devoted to the construction of simplified models of oceanic behavior that might be suitable for application to the understanding of long time-scale climate variability (e.g., Marotzke et al. 1988, Marotzke 1989; Stocker and Wright 1991).

That much simpler hydrodynamic systems than oceans may exhibit such intransitive behavior has, of course, been well understood on the basis of the observed behavior of laboratory systems. The most famous of these is the differentially heated rotating annulus which, beginning with the very important early work of Fultz (1961), has continued to deliver extraordinarily useful insights into the behavior of nonlinear systems in general. With the outer cylindrical wall heated and the inner cylindrical wall cooled, the rotating system exhibits two principle regimes of behavior, which may be distinguished in the plane with the thermal Rossby number and the Taylor number as coordinate axes. At low rotation rates for all levels of heating, the flow driven by the differential heating takes the form of a zonally symmetric (Hadley) circulation. With the differential heating (intended in this laboratory system to mimic the differential heating between equator and poles to which the planetary atmosphere is subject) set at intermediate levels, and for sufficiently rapid rotation, the steady solution loses stability to a zonally wavy solution (via the hydrodynamic instability process), with the zonal wave number of the physically realized solution being higher, the lower the heating rate. When the heating rate is set to be close to the boundary separating two distinct wavy regimes, the flow is then observed to execute unpredictable excursions from the regime dominated by one wave number to the regime dominated by the adjacent wave number. This is an excellent example of a dynamical system evolving under the influence of a "strange attractor." Similar behavior might be anticipated for the dynamical behavior of the ocean as a whole in response to the buoyancy flux that drives its large-scale overturning.

The work of Bryan (1986) was the first to point out that large-scale numerical models of the oceans, which have traditionally been driven by the prescription of wind stress, temperature, and salinity at the surface, were capable of this behavior. He pointed out that such models were strictly intransitive in the sense that, when driven by freshwater fluxes, they may either remain in their existing state of circulation or flip into a completely different state when driven by the same freshwater flux, but initializing with a slight perturbation of the surface salinity. At the Max Planck Institute in Hamburg, the first experiments were conducted in which small amplitude fluctuating white noise was superimposed upon the main freshwater flux. These integrations demonstrated the existence of a relatively long period oscillation in the resulting response, especially in the Atlantic, which might possibly be connected to the fast $\delta^{18}O$ variations observed in Greenland ice cores. These experiments have also been reported by other groups and the same result derived.

The extent to which paleodata may be brought to bear on the issue of whether or not such variations in the mode of thermohaline circulation are actually observed is extremely interesting and constitutes a main thrust of modern research in the area of paleooceanography. In fact, two distinct modes of deep ocean circulation have been tentatively identified: that associated with the modern ocean (based upon both hydrological data and GEOSECS geochemical data) and that associated with the last glacial maximum (LGM). The latter mode of circulation is characterized by:

— reduced North Atlantic Deep Water (NADW) flow compared to Antarctic Bottom Water (AABW) flow (Curry et al. 1982; Streeter and Shackleton 1979; Boyle and Keigwin 1982);
— enhanced intermediate water flux in the three oceans (Duplessy et al. 1988; Duplessy and Labeyrie 1989);
— a sharp separation between intermediate waters more ventilated than today, and deep waters that are less ventilated than today (Kallel et al. 1988); and
— significantly lower deep water temperatures (Labeyrie et al. 1987). Among the major unknowns concerning the oceanic circulation at the LGM we must note especially the following: (1) the circulation of the Southern Ocean, especially concerning the areas of deepwater formation, deep convection, and surface-water sinking; (2) the location of the main zones of intermediate water formation in both hemispheres; and (3) the mode of intermediate water ventilation in the Pacific.

It is perhaps likely that modes of oceanic circulation other than the "full glacial" and "modern" modes, which have been recognized to date in the paleo record, also exist (Duplessy and Shackleton 1985; Boyle and Keigwin 1982; Boyle 1984), and it is important to realize that the mere recognition of differences between LGM and modern circulation patterns is insufficient evidence to prove bistability of the climate system. Paleooceanographic reconstructions have, however, shown that the modern mode of oceanic circulation also prevailed during the previous interglacial about

125 Ka ago (Duplessy et al. 1984). $\delta^{18}O$ and $\delta^{13}C$ analyses of benthic foraminifera demonstrate that both the temperature and dynamics of the circulation also changed at the end of the Eemian period about 115 Ka ago (CLIMAP Project Members 1984; Duplessy and Shackleton 1985; Boyle 1988a,b). However, the glacial mode of deep-water circulation prevailed only during the period from 65,000 to 15,000 years ago in the Atlantic ocean, i.e., during isotope stages 2, 3, and 4 (Boyle and Keigwin 1982; Duplessy et al. 1988). For the period of 115,000 to 65,000 years ago, the nature of the circulation is still poorly documented; however, available evidence shows that it exhibited characteristics intermediate between the modern and LGM modes (Duplessy and Shackleton 1985, Duplessy et al. 1988). We do not yet know, however, whether this corresponds to a well-defined and distinct circulation pattern or whether the ocean is in fact able to exhibit a continuum of dynamical states between the modern and LGM extremes. To model properly past circulation changes of the ocean using an OGCM, we require knowledge of the freshwater fluxes at the ocean–atmosphere interface (Broecker 1989b, 1990 Duplessy et al. 1991).

Concerning the circulation changes associated with the last glacial–interglacial transition, we note that over time scales of several centuries the atmospheric CO_2 concentration is slave to ocean circulation and chemistry (Broecker 1982, 1989a; Boyle 1986). Paleooceanographic data show that oceanic changes continued for several thousand years and that we are not yet able to describe accurately the complete set of changes that accompanied the last deglaciation. Nevertheless, based upon present evidence, we may safely assert that the first evidence of circulation change is provided by the observed difference in ^{14}C age between planktonic and benthic foraminifera from the same sediment level. This provides an estimate of the ventilation age of the deep water (Broecker et al. 1988, 1990; Duplessy et al. 1991). By this means, the ventilation age is observed to decrease in the deep and intermediate Pacific ocean shortly following the LGM, although we do not yet know the mechanism whereby the circulation rate is enhanced (Shackleton et al. 1988). Deep-water formation in the Norwegian Sea did not begin until much later, following the first phase of rapid deglaciation about 12,000 years ago (Sejrup et al. 1984). The modern thermal structure of the Pacific ocean was established only when the deglaciation event was essentially complete, namely about 9000 years ago (Labeyrie et al. 1990). A major effort should be undertaken in the future to understand more fully the processes by which the ocean shifted from its glacial to its interglacial mode of operation, by analyzing a set of high sedimentation rate deep-sea cores from the major ocean basins.

Another type of climate instability involves the small ice-cap instability (SICI), which is related to albedo discontinuities at the snow–ice transition. This phenomenon has been studied for over twenty years and a stability proof has been developed for some of the cases (e.g., Crowley and North 1989). It may be possible to test for this instability at certain intervals in the geologic record. For example, at the end of the last interglacial, ice volume started to increase while the North Atlantic was still in its present circulation mode. High resolution studies of unmixed $\delta^{18}O$ records might provide some evidence for abrupt transitions that implicate SICI or some other instability mechanism.

From a somewhat different perspective, we note that the role of the tropical oceans is frequently neglected in considering the influence of ocean circulation on climate, in spite of the very large surface area which they cover. This is clearly due to the fact, discussed above, that the polar oceans are the locations of the formation of deep water and thus the ultimate motors that drive the thermohaline circulation. Small changes in the tropical oceans may nevertheless be rather effective agents of climate change; for example, the climatic effect of the most important greenhouse gas (H_2O) increases exponentially with temperature, and the effect of phase transformation (a release of the heat of condensation) adds to the greenhouse effect.

During the recent past (the 41-year period, 1949–1989), the surface temperature in the Tropics has risen by 0.4°–0.5°C (see Flohn, this volume), which should have been accompanied by increases in evaporation, winds, the strength of the extratropical circulation, and the average temperature of the tropical troposphere (from 3–12 km altitude). Disregarding what triggered this process, this example demonstrates the important role that water vapor plays as an amplifier of climatic processes, one that could also have been effective during abrupt climatic change, as, for example, during the transition from full glacial to Holocene conditions. The Quaternary ice core described in Figure 19.11 (Flohn, this volume) provides information beginning from about 50 years prior to the most marked glacial advance of the Little Ice Age (following A.D. 1550) and documents a large increase of precipitation with an Atlantic moisture source. This moist period ends by about 1700, simultaneously with the arrival of the most intense period of the Little Ice Age in Europe. Later cold periods occurring until about 1850 were accompanied by a dry period in the Andes of Peru. This sequence of events might be taken to suggest that tropical ocean evaporation played a leading role in initiating the glacier response in the first half of the 16th century.

TERRESTRIAL BIOSPHERIC VARIATIONS THROUGH THE GLACIAL–INTERGLACIAL TRANSITION

Overpeck (this volume) summarizes recent research directed toward refining our understanding of the changes in continental vegetation that might be expected in a world characterized by double the present concentrations of atmospheric trace gases. This research builds upon a variety of prior activities directed towards an understanding of the changes in vegetation that accompanied the last glacial–interglacial transition. The point is made that possible future changes in surface vegetation might be of the same order as those that accompanied the glacial–interglacial transition itself. Significant variations in the terrestrial biosphere potentially involve significant changes in the quantity of carbon stored in this component of the Earth system and it is therefore important that we investigate whether and to what extent such changes may be associated with the change in CO_2 concentration that has been observed in the ice-core record (e.g., Raynaud et al. 1988) to have accompanied the glacial–interglacial transition. Because trace gas concentration is a determinant of climate state, terrestrial vegetation could be implicated in a significant climate feedback loop, making this an extremely important issue that we shall attempt to briefly review. It should also be recognized that the biosphere may influence climate through means unconnected with trace gases (see Overpeck, this volume), for example through changes in surface albedo and its role as a land surface stabilizer.

First, we note that there are large differences among the estimates of the change in terrestrial carbon storage (biomass, soil organic carbon, and peat) between the LGM and present (or more accurately, pre-forest clearance). K.C. Prentice and Fung (1990) estimated little change; Adams et al. (1990) estimated that C storage decreased by over 1000 Gt (i.e., a decrease of more than 30%), whereas the estimate based on an interpretation of the change in global ocean $\delta^{13}C$ is around 500 Gt (Duplessy et al. 1988; I.C. Prentice and Sarnthein, this volume). Any data-based estimate of LGM carbon storage (e.g., Adams et al. 1990) is difficult to constrain. For many areas, paleobotanical data are sparse and poorly dated. There is particular concern about Amazonia and Indonesia (including the exposed Sunda shelf), which are key areas in terms of possible changes in C storage. There are, fortunately, ongoing efforts to obtain ^{14}C-dated pollen records from these areas extending back to 18 Ka.

The estimate of Prentice and Fung (1990) uses a model approach (a GISS GCM experiment, with CLIMAP boundary conditions and a modified Holdridge climate–vegetation classification scheme) to predict the distribution of vegetation in regions at 18 Ka. Use of a model approach is probably unavoidable because we may never have enough data to obtain reliable estimates entirely from them alone. On the other hand, the Prentice and Fung results were not compared extensively with paleodata and may be inconsistent with some data (Adams et al. 1990).

There are also a number of general problems with all land-based estimates of 18 Ka C storage to date, namely:

1. They are sensitive to variations in the values assumed for the biomass and (especially) soil C storage in each vegetation type; these values themselves are not well constrained.
2. There is a major assumption involved in applying these values in the past, e.g., we know from pollen evidence that many vegetation types had a different species composition at 18 Ka, and, we might guess, different C fluxes and storages also.
3. Existing studies have not taken into account possible direct effects of the low CO_2 at 18 Ka. This would have meant generally lower net primary production and possibly also lower net ecosystem production and lower equilibrium C storage. However, it should be noted that there is no necessary relationship among these variables. Organic matter decomposition rates are also crucial and are limiting in cold climates.

4. It has been assumed that C storage is now and was also at 18 Ka and in pre-forest clearance time, in equilibrium with climate. This is adequate for biomass and *may* be a reasonable approximation for soil C storage (assuming residence time on the order of 1–2 Ka). However, it is not reasonable for peat, which has accumulated to the extent of perhaps 450 Pg during the Holocene.

The marine-based estimates of around 500 Pg C transfer cannot be definitively rejected, but these estimates may also be confounded, e.g., by a change in oceanic DOC (Siegenthaler's suggestion), an increase in continental plant $\delta^{13}C$ linked to the lower available pCO_2 (through possible changes in the ratio of C_3 to C_4 plants), and possibly other mechanisms (Prentice and Sarnthein, this volume). Despite all these uncertainties, it seems unlikely that the terrestrial biosphere could have been a net contributor to the CO_2 increase during the glacial–interglacial transition.

Group discussion emphasized the importance of real (i.e., pollen or plant macrofossil-based) and well-dated information of glacial-age vegetation in the Tropics. It is suspected that tropical forests may have been significantly changed at 18 Ka (see Overpeck, this volume). The evidence for this seems to be strong in Africa but less strong in Amazonia and Indonesia, where there is some evidence for aridity but not enough data to allow a quantitative assessment. The issue has been clouded by the acceptance of the "refugium hypothesis" to account for local centers of diversity in Amazonia; this hypothesis is questioned by many ecologists and paleoecologists. Discussion also emphasized the need for process-based modeling, not just of past climates but also of the ecological processes determining above- and below-ground C storage as a function of CO_2 and climate. Life is obviously not in equilibrium with climate. Major evolutionary events, such as the invention of carbon fixation and, later, of oxygen evolution, had enormous effects on the internal dynamics of the atmosphere–Earth system. Less tumultuous events continued as, for example, continents collided and terrestrial species from one invaded the other. It would therefore seem dangerous to assume that the biosphere will interact with the rest of the system in a manner which depends only on climate (CO_2), etc., and not on time, or evolutionary stage.

When marine fauna and flora are analyzed to infer past climate, a great deal of effort is expended in buffering the calibrations against "noise" in the abundance of individual species. This "noise" probably reflects complex microevolutionary changes in the competition between species which probably operate on all time scales and not only between individual species, but between major groups. This could well have a significant effect on climate and may represent a significant component of unforced low-frequency climatic variability.

FEEDBACK AND CONTROL OF THE CONCENTRATIONS OF CO_2 AND CH_4

Because the physical and chemical controls on the concentrations of CO_2 and CH_4 are distinct, though both trace gases are currently increasing in concentration and are strong contributors to potential greenhouse warming, it will be useful to address the issues concerning them separately. Once again we will look to the paleorecord for insights concerning the feedbacks in which these gases are involved.

Glacial–Interglacial Variations of CO_2

Polar ice studies have shown that atmospheric CO_2 was 80 to 100 ppm lower during the LGM than during the Holocene. There is a general consensus among carbon cycle modellers that oceanic processes must have been responsible for these glacial–interglacial CO_2 variations. Changes of the terrestrial biomass (perhaps excepting humus) went the wrong way, as discussed above. According to the majority of studies (reviewed briefly above), biomass most likely increased from glacial time to the Holocene and, therefore, CO_2 would have been withdrawn from the atmosphere–ocean system. This alone would have caused a drop of the atmospheric CO_2 level. The atmospheric CO_2 concentration is determined by the physical-chemical conditions in the ocean's surface waters (pCO_2, alkalinity, temperature, and salinity).

Temperature and salinity variations were too small to explain more than a 10 to 20 ppmv decrease in atmospheric CO_2. A main mechanism for changing the concentration of pCO_2 in surface water is the so-called biological pump, i.e., the downward transport of carbon by dead organic particles which then are oxidized at depth and returned to inorganic carbon (pCO_2). A stronger biological pump could therefore withdraw CO_2 from the surface waters and the atmosphere. The biological productivity is limited in the large oceanic regions by lack of nutrients (P, N) but not so in the high-latitude southern ocean; therefore, changes in that region have been suggested to be responsible for the CO_2 variations. For an extreme scenario in which all available surface nutrients would be used up by marine biology, carbon cycle model studies yield an atmospheric level of 150 to 170 ppmv (not considering land biomass changes). For the opposite extreme of a dead ocean (no biological carbon pump), the simulated CO_2 level is in the range of 450 to 550 ppmv. The observed LGM concentration is 180–200 ppmv, that is, near to the lower limit of the range that is possible if the strength of the biological carbon pump (but nothing else) is varied in carbon cycle models. If, based on a whole-ocean $\delta^{13}C$ change of ca. 0.3‰, the glacial-to-Holocene change in terrestrial biomass is assumed to be 500 Gt C, this alone would have decreased atmospheric CO_2 at LGM by roughly 30 ppm. To explain the LGM CO_2 level by the biological carbon pump would then mean that surface water nutrients would have to have been zero in the whole ocean. This would mean that $\delta^{13}C$ in surface water should have been higher ($\delta^{13}C$ being anticorrelated with the pCO_2 concentration) and the Cd/Ca ratio (a proxy for the nutrient phosphate), smaller. These predictions are, however, not confirmed by deep-sea sediment studies on planktonic foraminifera. Furthermore such an active biological pump would have created extensive zones with zero dissolved oxygen in disagreement with data from the sedimentary record. Therefore, the biological pump discussed above seems not to have been solely responsible for the observed change of CO_2. Some other mechanism must also be invoked to explain the observed low glacial CO_2 concentration. One possible mechanism might involve a change in the oceanic alkalinity budget, since lower CO_2 concentration can be caused by higher surface-water alkalinity.

The required alkalinity changes might themselves be caused by a stronger dissolution of carbonate sediments at depth, for instance, due to a different depth distribution of pCO_2 in the glacial ocean (more aggressive deep waters, as in the "bottom heavy" hypothesis of E. Boyle). Another conceivable way to change oceanic alkalinity would involve changes of the river input of organic and inorganic carbon to the oceans; this has not been studied in detail to date. A problem with this mechanism is that sediment-water interaction could change total-ocean alkalinity within a period of only several Ka, while ice-core data show that the CO_2 increase occurred early during the most recent two deglaciations, approximately simultaneously with the southern hemisphere warming and several Ka prior to the most recent change of ice volume (see Raynaud and Siegenthaler, this volume). To constrain oceanic alkalinity changes, more information on variations of the lysocline depth in different ocean basins is required. On the other hand, detailed model studies on the river input sedimentation cycle must also be performed.

According to current understanding, a combination of different mechanisms seems to have been responsible for the observed glacial–interglacial CO_2 variations. To identify these mechanisms, a number of questions will have to be addressed:

1. What were the global mean changes of ^{13}C in surface water and in the ocean as a whole in time? How did ^{13}C in atmospheric CO_2 change?
2. Is it possible that the Redfield ratios C/P and C/N in marine biomass were different during the ice age than today? What controls these Redfield ratios during the formation of biomass, and is there a fractionation of nutrients versus carbon during the remineralization?
3. Shifts between plankton communities producing $CaCO_3$ shells (e.g., coccoliths) and those not producing $CaCO_3$ shells (e.g., diatoms) may have an impact on oceanic alkalinity. Were there such shifts on a large scale, and by what factors are they controlled?
4. Can the changes in oceanic alkalinity be better constrained from sediment data? Could changes in the riverine input of dissolved and particulate matter have affected atmospheric CO_2? Note that during low sea-level periods, accumulation of river sediments occurred at great depths, not on the (then not water-covered) continental shelf, as today.
5. A significant fraction of the global new production occurs in relatively small, fertile areas of the ocean (coastal reprocessing zones). There is good evidence for a sizably larger ice-age productivity in these low-latitude regions. Could this have significantly influenced atmospheric CO_2?
6. A new method to reconstruct pCO_2 levels is based on ^{13}C analyses on sedimentary organic carbon. The basis for this method should be studied further, and potential applications should be investigated.

On one hand, global reconstructions of ocean paleoproductivity by means of accumulation rates of organic matter (likewise by means of planktonic foraminifera assemblages) show that low- and mid-latitude productivity was increased by a factor of 2–5 in the high-productivity upwelling belts during the LGM, except for the high-productivity belt offshore of southern Arabia. On the other, productivity generally was slightly decreased, by an average of about 20%, in the low-productivity subtropical gyres and in ice-covered portions of the ocean (Sarnthein et al. 1988, see also Prentice and Sarnthein, this volume). If one summarizes the areas with increased and decreased export productivity during the LGM, one arrives at a LGM-to-Present decrease by about 2–4 Gt C yr^{-1}, depending on different values of modern productivity published by various authors. The value of this difference contradicts most recent model calculations and is mainly based on narrow belts of the ocean surface.

Increased glacial productivity values were clearly linked to an increased nutrient reservoir in the surface ocean, as deduced from $\delta^{13}C$ values of near-surface dwelling planktonic foraminifera (*Globigerinoides ruber*); lower productivity values paralleled lower to equal nutrient concentrations.

In a core from right below the equatorial upwelling belt in the East Atlantic (2°S, 10°W), i.e., in clearly pelagic sediments, glacial productivity maxima (stages 2, upper 3, 4, 6) parallel distinct decreases in the CO_2 concentration of the surface ocean (top 10 m) by 10–20%. These variations have, for the first time, been reconstructed using ^{13}C data from the organic carbon in the sediment, which were corrected for a uniform UK-sea-surface temperature and a uniform nutrient ($C = {}^{13}C$) content of the surface water, based on ^{13}C values of surface dwelling planktonic foraminifera (*G. ruber*).

The Glacial–Interglacial Variations of CH_4

The increase in the concentration of CH_4 over the last several centuries is even more spectacular than that of CO_2. Glacial levels of CH_4 were about 300 ppb: the interglacial value was about 600 ppb (Raynard et al. 1988, Stauffer et al. 1988, Chappellaz 1990), while the contemporary value is close to 1,800 ppb (Blake and Rowland 1986, 1988). Temporal changes in CH_4 can arise as a consequence of either a change in production or a change in loss. Previous discussions of the ice-core CH_4 record have stressed changes in sources (Chapellaz et al. 1990); however, our group felt that changes in sinks could be equally important. Production, according to present understanding, is dominated by microbially mediated processes on land (with contributions in the modern era from biomass burning and leakage of natural gas). Loss proceeds mainly in the atmosphere, by reaction with the hydroxyl radical, OH.

The concentration of OH itself depends on the column abundance of stratospheric O_3 (determining the flux of ultraviolet radiation penetrating to the troposphere), on the abundance of tropospheric O_3 (the source of O^1D), on the abundance of tropospheric H_2O (the source of OH through reaction with O^1D), on the abundance of CH_4 and CO (the primary loss reactions), and indirectly on species such as NO_2. Part of the rise of CH_4 in the contemporary atmosphere may be due to a temporal decline in OH, reflecting perhaps increases in CO due to enhanced incidence of biomass burning. Estimates of concentrations of OH, and consequently of production and loss processes for CH_4 during earlier times depend critically on assumptions made concerning the properties of O_3.

Measurements of CH_2O and formic acid (and if possible CO), found in air bubbles of ice cores in conjunction with CH_4, can provide an indirect estimate of the abundance of OH. It would be useful if such measurements could be carried out not only in ice from polar regions but also from mid-latitudes. The concentration of OH is expected to vary appreciably, both with latitude and season. The lifetime of CH_4 is set mainly by reaction with OH in the Tropics.

Studies of the isotopic composition of CH_4 can provide important clues that can be exploited to refine our understanding of the significance of the various possible sources. In particular, measurements of the isotopic composition of CH_4 at glacial time, and of its evolution from the last termination to present, would restrict the range of unknowns in present models.

The global budget of recent atmospheric methane, and particularly the relative importance of anthropogenic vs. natural sources, is still known only with great uncertainties. Most budget estimates have been constrained by the atmospheric methane sinks (oxidation by OH and uptake by soils) which seemed to be better known (Cicerone and Oremland 1988). However, just recently the OH sink has been revised by as much as 25% (Vaghijani and Ravishankara 1991; Crutzen 1991). Another constraint on the global methane budget has been derived from isotope (^{13}C, ^{14}C) observations on atmospheric methane and its sources. This, however, implies an exact knowledge of the kinetic isotope fractionation factor, which has been measured under laboratory conditions (Cantrell et al. 1990) but must be confirmed by isotopic measurements in the upper troposphere and lower stratosphere as well as in the Southern Hemisphere (Cassey et al. 1992) where the concentration variations are assumed to be mainly OH driven.

Concerning the interpretation of glacial–interglacial atmospheric methane concentration changes as derived from ice-core measurements, a much better knowledge of the recent methane budget, and particularly of its natural sources and sinks, is necessary. In this context, a closer look at sources and sinks that are of minor importance today but which might have been of relevance in the past (such as oceans and soils) should perhaps be emphasized. Bacterial methane production is strongly linked to anaerobic environments. Changes in these anaerobic conditions from the present-day situation to glacial times, with a different climate, may cause considerably different methane production rates as well as changes in the global distribution of the sources. Methane oxidation within source areas, already today, significantly influences the net methane flux into the atmosphere. It may have become even more important during glacial time. The processes involved in methane production, its final emission into the atmosphere, as well as the removal processes have to be studied in order to model the methane budget under modified climatic conditions.

The issue of microbial oxidation of methane in aerated soils as a sink for atmospheric methane has been investigated in several studies in recent years (Keller et al. 1983; Stendler et al. 1989; Born et al. 1990; Whalen and Reeburg 1990; Dörr et al. 1992). Dorr et al. (1992) found a strong correlation between soil permeability (soil texture class) and the methane uptake rate by soils, derived from parallel ^{222}Radon flux and ^{222}Radon and methane concentration measurements in the soil air at various sites in Central Europe. Methane decomposition in soils is basically controlled by the gas transport resistance within the soil, since soil temperature has been found to be of minor influence.The soils investigated in this study were classified with respect to their soil texture class; the mean observed methane uptake rates within each class were then used together with the NASA–GISS global digital data set of soil types (Staub et al. 1987) to estimate the global methane uptake rate by soils. With this parameterization, a global methane soil sink in the range of 9–56 Tg CH_4 yr^{-1} with a best estimate of 29 Tg CH_4 yr^{-1} was calculated. This corresponds to ca. 7% of the total recent global destruction rate by OH (Crutzen 1991). As the soil permeability strongly depends on soil humidity in a postulated drier climate during glacial periods, the total global uptake of atmospheric methane by soils may have been larger than in the present-day climate.

CONCLUSIONS

We have attempted in the preceding discussion to review the main issues upon which our group focused in attempting to come to grips with the ways in which proxy data records of past climates could be used to deepen our understanding of the main feedbacks operating within the climate system as a whole. The dominant signal in the climate record throughout the last 2 Ma of Earth history, especially in the time since 0.9 Ma, is that associated with the variation of continental ice volume, which may be accurately tracked using $\delta^{18}O$ data from deep-sea sedimentary cores. This signal is very strongly influenced by the variations of solar insolation associated with variations of the geometry of the Earth's orbit around the Sun. The dominant 100-Ka signal of the late Pleistocene, however, can apparently not be understood in terms of a linear model; nonlinear dynamics with strong coupling to parts of the climate system outside of the ice sheets themselves appear to be required to reconcile the observations. Many additional components of the Earth system appear to be deeply implicated in this oscillation of planetary climate. We have discussed three of the most important: the oceans, the biosphere, and the atmospheric load of radiatively active trace gases CO_2 and CH_4. Other components of the system, especially the atmospheric dust level, may have a very specific role, particularly in the physics of terminations. In providing a means to determine the relative phasing of the activity of these various subsystems during the glacial cycle, paleo records clearly provide a valuable key to the understanding of whole system function.

REFERENCES

Adams, J.M., H. Faure, L. Faure-Denard, J.M. McGlade, and F.I. Woodward. 1990. Increases in terrestrial carbon storage from the Last Glacial Maximum to the present. *Nature* **348:**711–714.

Barnola, J.M., D. Raynaud, Y.S. Korotkevitch, and C. Lorius 1987. Vostock ice core provides 160,000-year record of atmospheric CO_2. *Nature* **329**:408–414.

Berger, A.L. 1978. Long-term variations of daily insolation and Quaternary climatic changes. *J. Atmos. Sci.* **35**:2362–2367.

Blake, D.R., and F.S. Rowland 1986. Worldwide increase in tropospheric methane. *J. Atmos. Chem.* **4**:43–62.

Blake, D.R., and F.S. Rowland 1988. Continuing worldwide increase in tropospheric methane, 1978 to 1987. *Science* **239**:1129–1131.

Born, M., H. Dorr, and I. Levin 1990. Methane consumption in aerated soils of the temperate zone. *Tellus* **42B**:2–8.

Boyle, E.A. 1984. Cadmium in benthic foraminifera and abyssal hydrography: Evidence for a 41-kyr obliquity cycle. In: Climate Processes and Climate Sensitivity, ed. J. Hansen and T. Takahashi. *American Geophys. Union Monographs* **29**:360–368.

Boyle, E.A. 1986. Deep ocean circulation, preformed nutrients and atmospheric carbon dioxide: Theories and evidences from oceanic sediments. In: Mesozoic and Cenozoic Oceans, ed. K.H. Hsu, pp. 49–60, Geodynamics Series N15. American Geophy. Union.

Boyle, E.A. 1988a. The role of vertical chemical fractionation in controlling late Quaternary atmospheric carbon dioxide. *J. Geophys. Res.* **93**:15,701–15,714.

Boyle, E.A. 1988b. Vertical oceanic nutrient fractionation and glacial/interglacial CO_2 cycles. *Nature* **331**:55–56.

Boyle, E.A., and L.D. Keigwin 1982. Deep circulation of the North Atlantic over the last 200,000 years: Geochemical evidence. *Science* **218**:784–787.

Boyle, E.A., and L.D. Keigwin. 1985. Comparison of Atlantic and Pacific paleochemical records for the last 215,000 years: changes in the deep ocean circulation and chemical inventories. *Earth Planet. Sci. Lett.* **76**:135–150.

Brass, G.W., J.R. Southam, and W.H. Peterson 1982. Warm saline bottom water in the ancient ocean. *Nature* **296**:620–623.

Broecker, W.S. 1982. Glacial to interglacial changes in ocean chemistry. *Prog. Oceanogr.* **11**:151–197.

Broecker, W.S. 1989a. The cause of the glacial to interglacial CO_2 change: a polar alkalinity hypothesis. *Glob. Biogeochem. Cyc.* **3**:215–239.

Broecker, W.S. 1989b. The salinity contrast between the Atlantic and Pacific oceans during the glacial time. *Paleoceanog.* **4**:207–212.

Broecker, W.S. 1990. Salinity history of the northern Atlantic during the last deglaciation. *Paleoceanog.* **5**:459–467.

Broecker, W.S, M. Andree, G. Bonani, W. Wolfli, H. Oeschger, M. Klas, A. Mix, and W. Curry. 1988. Preliminary estimates for the radiocarbon age of deep water in the glacial ocean. *Paleoceanog.* **3**:659–669.

Broecker, W.S., and G.H. Denton 1990. The role of atmosphere–ocean reorganizations in glacial cycles. *Geochim. Cosmochim. Acta* **53**:2465–2501.

Broecker, W.S., T.H. Peng, S. Trumbore, G. Bonani, and W. Wolfli 1990. The distribution of radiocarbon in the glacial ocean. *Glob. Biogeochem. Cyc.* **4**:103–117.

Broecker, W.S., and J. van Donk 1970. Insolation changes, ice volume and the ^{18}O record in deep-sea cores. *Rev. Geophys. Space Phys.* **8**:169–197.

Bryan, F. 1986. High-latitude salinity effects and interhemispheric thermohaline circulations. *Nature* **323**:301–304.

Cantrell, C.A., R.E. Shetter, A.H. McDaniel, J.G. Calvert, J.A. Davidson, D.C. Lowe, S.C. Tyler, R.J. Cicerone, and J.P. Greenberg 1990. Carbon kinetic isotope effect in the oxidation of methane by hydroxyl radical. *J. Geophys. Res.* **95**:22,455–22,462.

Chappellaz, J., M.M. Barnola, D. Raynaud, Y.S. Korotkevich, and C. Lorius 1990. Ice core record of atmospheric methane over the past 160,000 years. *Nature* **345**:127–131.

Cicerone, R.J., and R.S. Oremland 1988. Biogeochemical aspects of atmospheric methane. *Glob. Biogeochem. Cyc.* **2**:299–327.

CLIMAP Project Members. 1984. The last interglacial ocean. *Quat. Res.* **21**:123–224.

Crowley, J.T., and G.E. North. 1991. Palaeoclimatology. Oxford Monograph on Geology and Geophysics No. 16. Oxford: Oxford Univ. Press.

Crutzen, P.J. 1991. Methane's sinks and sources. *Nature* **350**:380–381.

Curry, W.B., and G.P. Lohmann. 1982. Carbon isotopic changes in benthic foraminifera from the western south Atlantic: Reconstruction of glacial abyssal circulation patterns. *Quat. Res.* **18**:218–235.

De Angelis, M., N.I. Barkov, and V.N. Petrov 1987. Aerosol concentrations over the last climatic cycle (160 kyr) from an Antarctic ice core. *Nature* **325**:318–321.

Deblonde, G., and W.R. Peltier. 1991a. A one-dimensional model of continental ice volume fluctuations through the Pleistocene: Implications for the origin of the Mid-Pleistocene climate transition. *J. Climate* **4**:318–344.

Deblonde, G., and W.R. Peltier. 1991b. Simulations of continental ice sheet growth over the last glacial-interglacial cycle: Experiments with a one-level seasonal energy balance model including realistic geography. *J. Geophys. Res.* **96**:9189–9215.

Dörr, H., L. Katruff, and I. Levin. 1992. Soil texture parameterization of the methane uptake in aerated soils. Paper presented at the NATO Advanced Research Workshop on Atmospheric Methane, Mount Hood, Oregon, October 1991. *Chemosphere*, in press.

Duplessy, J.C., M. Arnold, E. Bard, A. Juillet-Leclerc, N. Kallel, and L.D. Labeyrie. 1989. AMS ^{14}C study of transient events and of the ventilation rate of the Pacific intermediate water during the deglaciation. *Radiocarbon* **31**:493–502.

Duplessy, J.C., and L.D. Labeyrie. 1989. Global ocean circulation during the last ice age: enhanced intermediate water formation and reduced deep water circulation. In: Oceanography 1988, ed. A. Ayala Castanares, A.W. Wooster, and A. Yanez Arancibia. CITY: Conacyt–Unesco.

Duplessy, J.C., L. Labeyrie, A. Juillet-Leclerc, F. Maitre, J. Duprat, and M. Sarnthein. 1991. Surface salinity reconstruction of the North Atlantic Ocean during the last glacial maximum. *Oceanol. Acta* **14**:311–324.

Duplessy, J.C., and N.J. Shackleton. 1985. Response of global deep-water circulation to the Earth's climatic change 135,000–107,000 years ago. *Nature* **316**:500–507.

Duplessy, J.C., and N.J. Shackleton, R.G. Fairbanks, L.D. Labeyrie, D. Oppo, and N. Kallel. 1988. Deep water source variation during the last climatic cycle and their impact on the global deep water circulation. *Paleoceanog.* **3**:343–360.

Duplessy, J.C., N.J. Shackleton, R.K. Matthews, W. Prell, W.F. Ruddiman, M. Caralp, and C.H. Hendy. 1984. ^{13}C record of benthic foraminifera in the last interglacial ocean: Implications for the carbon cycle and the global deep water circulation. *Quat. Res.* **21**:225–243.

Fultz, D. 1961. Developments in controlled experiments on large-scale geophysical problems. *Adv. Geophys.* **7**:1–103.

Gallée, H., J.P. van Ypersele, Th. Fichefet, Ch. Tricot, and A. Berger. 1991. Simulation of the last glacial cycle by a coupled 2-D climate-ice sheet model. Part 1: The climate model. *J. Geophys. Res.* **96**:13139–13161.

Hays, J.D., J. Imbrie, and N.J. Shackleton. 1976. Variations in the Earth's orbit: Pacemaker of the ice ages. *Science* **194**:112–132.

Hyde and Peltier, W.R. 1985. Sensitivity experiments with a model of the ice age cycle: The response to harmonic forcing. *J. Atmos. Sci.* **42:**2170–2188.

Hyde and Peltier, W.R. 1987. Sensitivity experiments with a model of the ice age cycle: The response to Milankovitch forcing. *J. Atmos. Sci.* **44:**1351–1374.

Imbrie, J., J.D. Hays, D.G. Martinson, A. McIntyre, A.C. Mix, J.J. Morley, N.G. Pisias, W.L. Prell, and N.J. Shackleton. 1984. The orbital theory of Pleistocene climate support from a revised chronology of the marine $\delta^{18}0$ record. In: Milankovitch and Climate, ed. A. Berger et al., pp. 269–305. Dordrecht: Reidel.

Jouzel, J., G.L. Russell, R.J. Suozzo, R.D. Koster, J.W.C. White, and W.S. Broecker. 1987. Simulations of the HDO and $H_2{}^{18}O$ atmospheric cycles using the NASA/GISS general circulation model: The seasonal cycle for present-day conditions. *J. Geophys. Res.* **92:**14739–14760.

Kallel, N., L.D. Labeyrie, A. Juillet-Leclerc, and J.C. Duplessy. 1988. A deep hydrological front between intermediate and deep-water masses in the glacial Indian Ocean. *Nature* **333**:651–655.

Keller, M., T.J. Goreau, S.C. Wofsy, W.A. Kaplan, and M.B. McElroy. 1983. Production of nitrous oxide and consumption of methane by forest soils. *J. Geophys. Res.* **91(D11)**:11791–11802.

Labeyrie, L.D., J.C. Duplessy, and P.L. Blanc. 1987. Variations in mode of formation and temperature of oceanic deep waters over the past 125,000 years. *Nature* **327**:477–482.

Labeyrie, L.D., N. Kallel, M. Arnold, A. Juillet-Leclerc, F. Maitre, J.C. Duplessy, and N.J. Shackleton. 1990. Variabilite des eaux intermediaires et profondes dans l'ocean Pacifique Nord-Ouest pendant la derniere deglaciation. *Oceanologica Acta* **10**:329–339.

Maasch, K.A., and B. Saltzman. 1990. A low-order dynamical model of global climatic variability over the full Pleistocene. *J. Geophys. Res.* **95**:1955–1963.

Marotzke, J. 1989. Instabilities and steady states of the thermohaline circulation. In: Ocean Circulation Models, ed. D.L.T. Anderson and J. Willebrand, pp. 501–511. Dordrecht: Kluwer Academic Publ.

Marotzke, J., P. Welander, and J. Willebrand. 1988. Instability and multiple equilibria in a meridional-plane model of the thermohaline circulation. *Tellus* **40A**:162–172.

North, G.R., J.G. Mengel, and D.A. Short. 1983. Simple energy balance model resolving the seasons and the continents: Application to the astronomical theory of ice ages. *J. Geophys. Res.* **88**:6576–6586.

Peltier, W.R. 1982. Dynamics of the ice age Earth. *Adv. Geophys.* **24**:1–146.

Pollard, D. 1982. A simple ice sheet model yields realistic 100-kyr glacial cycles. *Nature* **296**:334–338.

Prentice, K.C., and Fung, I.Y. 1990. The sensitivity of terrestrial carbon storage to climate change. *Nature* **346**:48–51..

Raynaud, D., J. Chappelaz, M.M. Barnola, Y.S. Korotkevich, and C. Lorius. 1988. Climatic and methane cycle implications of glacial-interglacial CH_4 change in the Vostok ice core. *Nature* **333**:655–657.

Sarnthein, M., K. Winn, J.C. Duplessy, and M.R. Fontugue. 1988. Global variations of surface ovean productivity in low and mid latitudes: Influence on CO_2 reservoirs of the deep ocean and atmosphere during the last 21,000 yrs. *Paleoceanog.* **3**:361–399..

Sejrup, H.P., E. Jansen, H. Erlenkeuser, and H. Holtedahl. 1984. New faunal and isotopic evidence on the late Weichselian–Holocene oceanographic changes in the Norwegian Sea. *Quat. Res.* **21**:74–84.

Shackleton, N.J. 1967. Oxygen isotope analyses and Pleistocene temperatures re-addressed. *Nature* **215:**15–17.

Shackleton, N.J., A. Berger, and W.R. Peltier. 1990. An alternative astronomical calibration of the lower Pleistocene time scale based upon ODP Site 677. Transactions of the Royal Society of Edinburgh. *Earth Sci.* **81**:251–261.

Shackleton, N.J., J.C. Duplessy, M. Arnold, P. Maurice, M. Hall, and J. Cartlidge. 1988. Radiocarbon age of last glacial Pacific deep water. *Nature* **335**:708–711.

Shackleton, N.J., J. Imbrie, and M.A. Hall. 1983. Oxygen and carbon isotope record of East Pacific core V 19–30: Implications for the formation of deep water in the late Pleistocene North Atlantic. *Earth Planet Sci. Lett.* **65**:233–244.

Staub, B., C. Rosenzweig, and D. Rind. 1987. Global digital data sets of soil type, soil texture, surface slope, and other properties: documentation of archived data tape. *NASA Technical Memo.* 100685.

Stauffer, B., G. Fischer, A. Neftel, and H. Oeschger. 1988. Increase of atmospheric methane recorded in Antarctic ice core. *Science* **229**:1386–1388.

Stendler, P.A., R.D. Bowden, J.M. Melillo, and J.D. Aber. 1989. Influence of nitrogen fertilization on methane uptake in temperate forest soils. *Nature* **341**:314–316.

Stocker, T.F., and D.G. Wright. 1991. Rapid transition of the ocean's deep circulation induced by changes of surface water fluxes. *Nature* **351**:729–732.

Stommel, H.M. 1961. Thermohaline convection with two stable regimes of flow. *Tellus* **13**:224–230.

Streeter, S.S., and N.J. Shackleton. 1979. Paleocirculation of the deep North Atlantic: 150,000-year record of benthic foraminifera and oxygen-18. *Science* **203**:168–171.

Tushingham, A.M., and W.R. Peltier. 1991. ICE–3G: A new global model of late Pleistocene deglaciation based upon geophysical predictions of post glacial relative sea level change. *J. Geophys. Res.* **96**:4497–4523.

Tushingham, A.M., and W.R. Peltier. 1992. Validation of the ICE–3G model of Wurm–Wisconsin deglaciation using a global data base of relative sea level histories. *J. Geophys. Res.* **97**:3285–3304.

Vaghijani, G.L., and A.R. Ravishankara. 1991. New measurements of the rate coefficient for the reaction of OH with methane. *Nature* **350**:406–409.

Whalen, S.C., and W.S. Reeburgh. 1990. Consumption of atmospheric methane by tundra soils. *Nature* **346**:160–162.

17

Role of Orbital Forcing: A Two-Million-Year Perspective

J. IMBRIE[1], A. BERGER[2], and N.J. SHACKLETON[3]
[1]Department of Geological Sciences, Brown University, Providence, RI 02912, U.S.A.
[2]Université Catholique de Louvain, Institut d'Astronomie et de Géophysique G. Lemaître, Louvain-La-Neuve, Belgium
[3]Godwin Laboratory for Quaternary Research, Cambridge CB2 3RS, U.K.

ABSTRACT

Geological records of climate, extending back over the past half million years and beyond, provide an opportunity to identify and evaluate the causes of climate change that operate on time scales from 10,000 to 100,000 years. This opportunity is intriguing because over these time scales it is known that changes in orbital geometry modulate the seasonal radiation cycle at periods which match quite closely the dominant periods of major peaks in the spectrum of Pleistocene climate. The opportunity is also unique because, outside of the daily and annual cycles, it is the only place in the entire climatic spectrum where the exact temporal and spatial structure of an external forcing function can be calculated *a priori.* The main objectives in this chapter are to introduce the general reader to the subject and to identify some of the important unsolved problems.

INTRODUCTION

One objective of this workshop is to answer the following question: How can paleodata be used to evaluate the response of the climate system to *orbital forcing* of past and future climates? Implicit in this question is the idea that variations in the geometry of the Earth's orbit, with respect to the Sun, Moon, and other planets, exert influences on climate that operate at strengths and on time scales detectable in proxy records. Stated in this way, the question is much too broad to be addressed in a short paper, at least by these authors. Instead, we will focus on one set of orbitally driven influences that operate on time scales of 10 Ka to 400 Ka (Ka = 1000 years), and which have been extensively studied since the seminal work of Milankovitch (1930). These influences are variations in the pattern of incoming radiation that result from secular changes in

Global Changes in the Perspective of the Past
Edited by J.A. Eddy and H. Oeschger

the Earth's orbit around the Sun, namely the orbital cycles of eccentricity (~ 413 Ka, ~ 100 Ka), obliquity (~ 41 Ka) , and the precession of the equinoxes (~ 23 Ka and ~19 Ka) (Berger 1977).

How important have these influences been in forcing climate change? Geologic records of climate, extending back over the past half million years and beyond, provide an opportunity to find out. This opportunity is intriguing—perhaps seductive is a better word—because it is now known that changes in orbital geometry modulate the seasonal radiation cycle at periods which actually match quite closely the dominant periods of major peaks in the spectrum of Pleistocene climate (~ 19,000, ~ 23,000, ~ 41,000, and ~ 100,000 years) (Imbrie et al. 1989). The opportunity is also unique because it is the only place in the entire climatic spectrum, outside of the daily and annual cycles, where the exact temporal and spatial structure of an external forcing function can be calculated *a priori*.

The fact that this opportunity is both intriguing and unique does not mean it provides an easy way of evaluating forcing—either of climates in the past or climates in the future. Since the basic ideas discussed here were developed over 60 years ago, an extensive observational and modeling literature has grown up around the problem of identifying "the Milankovitch mechanism"—precisely because an important opportunity seemed to be at hand for constructing a quantitative model that would be capable of describing how the global climate system responds to a known external forcing (Hays et al. 1976; COHMAP 1988; Kutzbach and Gallimore 1988; Berger et al. 1990). Most observers would probably agree with our assessment that, although substantial progress has been made in understanding the physical mechanisms by which some parts of the climate system respond to orbital forcing, many—perhaps most—important questions remain to be answered. In general, progress has been greater in understanding how monsoons and other low-latitude features of climate are forced compared to ice sheets and CO_2 levels (McIntyre et al. 1989; Clemens et al. 1991).

Why has the problem proved so difficult? The answer, in a nutshell, is that we lack a satisfactory quantitative theory of climate that includes both the "fast physics" of the atmosphere and the "slow physics" of the oceans and cryosphere. Given such a theory, it would be a simple matter to change the orbital boundary conditions and make predictions that could be tested against geological data. Lacking such a theory, we must (a) make do with models that treat only parts of the total climate system explicitly and (b) use statistical correlations between data and potential forcing functions as a way of identifying which response mechanisms are important. Unfortunately, the normal *post hoc propter hoc* pitfalls of reasoning from statistical evidence are compounded by similarities in the shapes of different, potentially important radiative forcing functions.

Our main objectives in this chapter are to introduce the general reader to the subject and to explain why it is useful to examine the climate system's response in the 100-Ka eccentricity band separately from its response in the precession and obliquity bands. We limit our discussion of the geological record of climate change to a single climatic parameter, an isotopic index of global ice volume, marine $\delta^{18}O$ (Shackleton and Opdyke 1973), and we illustrate important points with diagrams rather than words. The first two sections below summarize how the Earth's orbit varies over time and describe how these changes are reflected in spatial and temporal patterns of radiation change. The next section reviews statistical evidence linking variations in the Earth's orbit with climate. A final section lists some of the unsolved problems.

HOW DOES THE EARTH'S ORBIT VARY OVER TIME?

Three orbital elements modulate the seasonal cycle of radiation (Berger 1977). These are eccentricity (e), obliquity (ε), and the precession of the equinoxes. The climatic effect of precession is expressed as an index, $\Delta e \sin \omega$, where e is eccentricity and ω is the longitude of perihelion measured from the moving vernal point. This index is approximately equal to the deviation from its 1950 value of the Earth–Sun distance in June, expressed as a fraction of the semi-major axis of the Earth's orbit. Over the past 600,000 years, the geometry of the Earth's orbit has varied significantly (Figure 17.1). Variations in eccentricity are concentrated in two narrow bands, one near 412 Ka and the other near 100 Ka. Variations in obliquity are concentrated near 41 Ka; variations in the precession index are concentrated near 23 Ka and 19 Ka.

HOW ARE CHANGES IN RADIATION RELATED TO ORBITAL VARIATIONS?

Spatial Patterns

The principal result of investigations by Milankovitch was an analysis of the way in which variations in eccentricity, obliquity, and precession drive changes in the incoming radiation (Milankovitch 1930). Figure 17.2 provides a convenient basis for examining this relationship. Two oscillatory modes of seasonal insolation are evident: one is driven by precession, the other by obliquity. The effect of eccentricity is to change slightly the total energy received by the whole Earth and to modulate the amplitude of the radiation pattern driven by precession ($e \sin \omega$). The annual insolation has only one mode, driven by obliquity. The spatial structure of these modes provides clues about the way in which the system might respond to Milankovitch forcing: (a) the largest changes occur in Arctic and Antarctic summer, with obliquity and precession making roughly equal contributions; (b) winter variation is an order of magnitude smaller in the polar regions; (c) obliquity drives a change in annual radiation that is significant in the polar regions, whereas precession does not; (d) the two orbital processes drive very different meridional patterns of change.

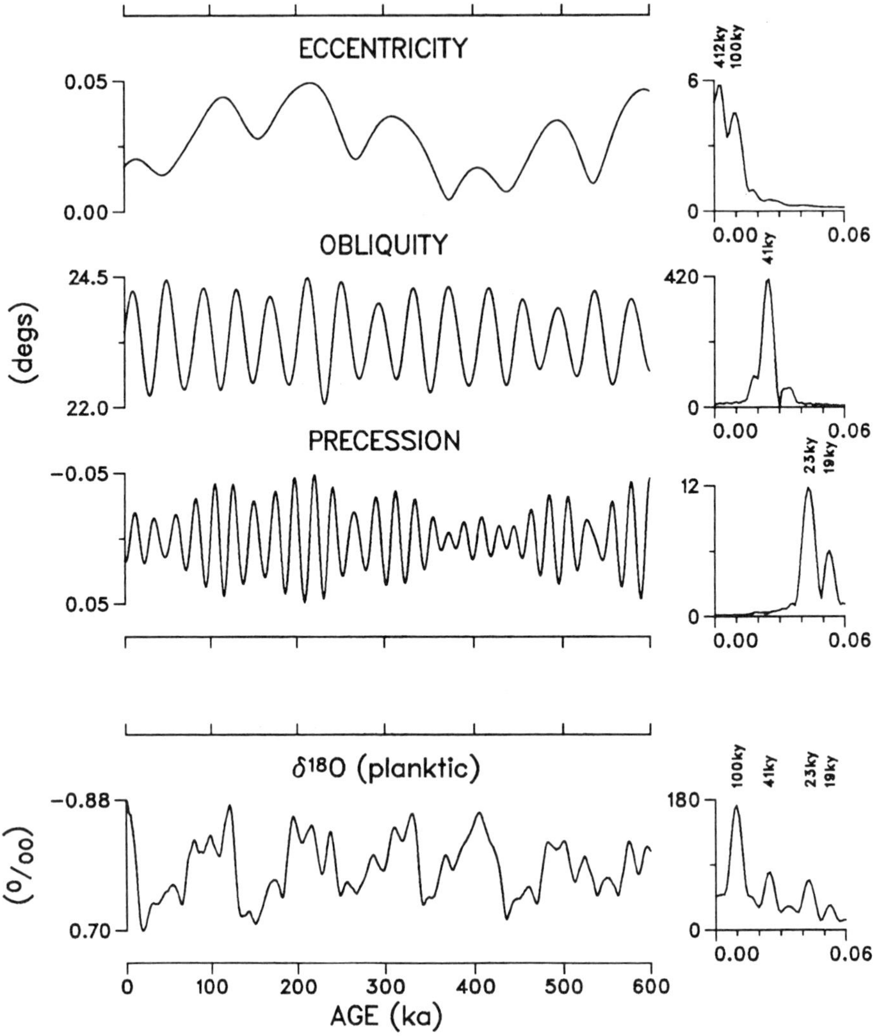

Figure 17.1 Orbital and climatic variations over the past 600,000 years (Ka = 1000 years ago). Variance spectra on the right, with their dominant periods indicated in thousands of years, are calculated from time series shown on the left. Top panel: variations in eccentricity, obliquity, and the normalized precession index ($\Delta e \sin \omega$), from calculations of Berger (1977; 1978). Bottom panel: variations in marine $\delta^{18}O$, from Imbrie et al. (1984).

Temporal Patterns

For any given latitude and season, the amount of incoming radiation at the top of the atmosphere can be calculated as a function of time, given time series of obliquity, ε, and the precession index, $e \sin \omega$ (Berger 1978). Although the shape of each insolation

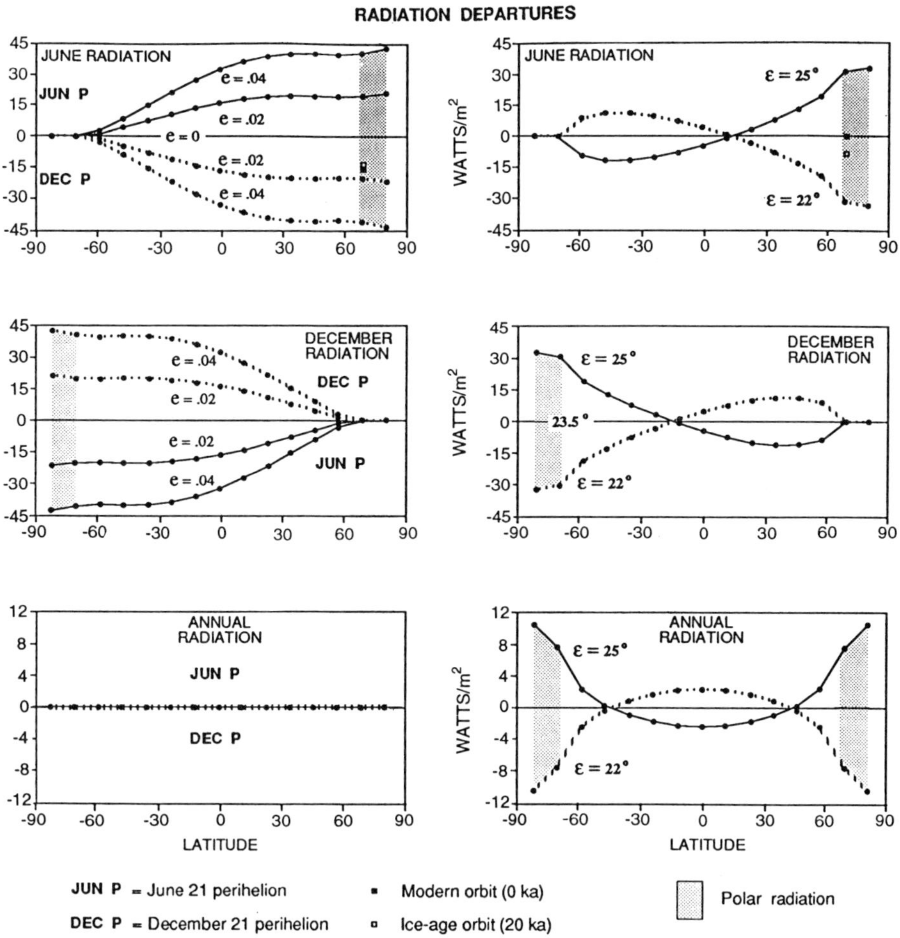

Figure 17.2 Incoming radiation departures as a function of orbital geometry. Values are expressed as departures from a reference orbit with eccentricity (e) = 0, and obliquity (ε) = 23.5°. Left: Radiation on June 21 is a function of eccentricity and perihelion position, with ε = 23.5°. Right: Radiation is a function of obliquity, with e = 0. Note that radiation values associated with the two different effects today and at the last glacial maximum are similar, as shown for June radiation at 70°N. The diagram is based on calculations made by John Kutzbach at the University of Wisconsin.

curve so calculated is distinct, that shape is (to a high degree of accuracy) simply a linear combination of an obliquity component, with a fixed phase, and a precession component whose phase is fixed by the chosen season (Imbrie and Imbrie 1980). This fact makes it difficult to distinguish statistically between one 23-Ka seasonal radiation cycle and another, unless the phase of the response is known from other evidence. Happily, the phase of the obliquity component of radiation is the same for all seasons (at a given latitude).

Figure 17.3 gives insolation curves for four different months at 65°N—a latitude band that many investigators, including Milankovitch himself, consider important in forcing changes in Northern Hemisphere ice sheets. Two features of this display are worth noting. First, summer variations have a much larger amplitude than those of winter. Second, the precession component of radiation shifts systematically in phase from one season to another, with each month advancing one-twelfth of a precession cycle (~ 2 Ka).

HOW ARE ORBITAL VARIATIONS STATISTICALLY RELATED TO CLIMATE?

The Past 600,000 Years

Many investigations have demonstrated that the spectrum of late Pleistocene climate is dominated by narrow-band cycles with periods near 19 Ka, 23 Ka, 41 Ka, and 100 Ka (Imbrie et al. 1984; McIntyre et al. 1989; Clemens et al. 1991). In time series of an $\delta^{18}O$ ice-volume index, the amplitude of the 100-Ka cycle is the largest, about equal to the sum of the other three cycles. The four main peaks in the $\delta^{18}O$ spectrum correspond closely with four peaks in orbital spectra (Figure 17.4). However, the 412-Ka peak in the eccentricity spectrum is *not* matched by a corresponding peak in the $\delta^{18}O$ spectrum (see Santer et al., this volume).

In order to search for evidence of an underlying physical relationship between orbital and climatic variations, it is necessary to go beyond the matching of spectral peaks. One important statistical strategy is to look for evidence that the correlations are coherent, i.e., to examine displays in the time domain and see if the amplitude (and length) of each individual oscillation in radiation is proportional to the amplitude (and length) of the corresponding oscillation in some climatic index. In Figure 17.5, this strategy is applied to one time series of radiation (June 65°N) and one climatic index ($\delta^{18}O$). We draw two conclusions from this display:

1. The climatic variations in the precession and obliquity bands are so strongly coherent with the corresponding radiative forcing functions that a causal relationship must exist. Thus, it should be possible to explain the 19-Ka, 23-Ka, and 41-Ka cycles as orbitally forced responses. The scientific challenge is to identify and model the the physical mechanisms of this forcing.

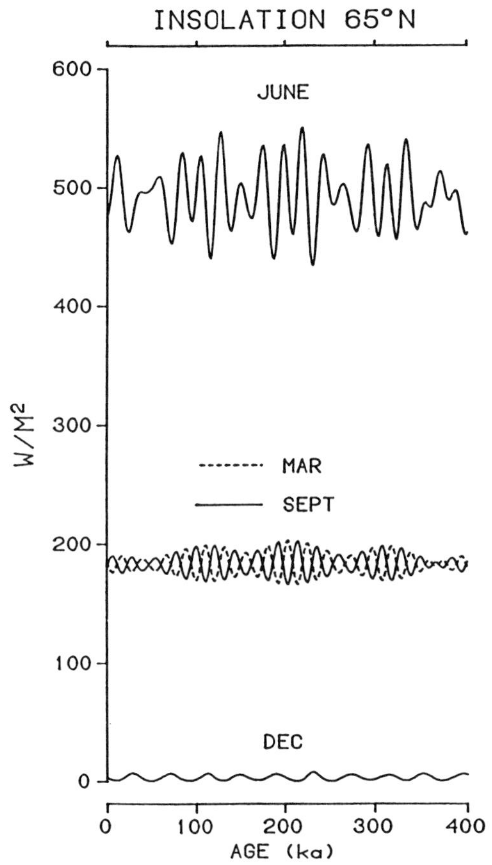

Figure 17.3 Mid-month values of incoming radiation at 65°N as a function of time, from Berger (1978).

2. Although 100-Ka oscillations in radiation do occur, their amplitude is so small, relative to those that occur in the precession and obliquity bands and relative to the large amplitude of the 100-Ka $\delta^{18}O$ cycle, that an explanation of this climatic cycle seems to require a mechanism that is very different from the mechanisms operating at 19 Ka, 23 Ka, and 41 Ka. Thus the strong statistical connection documented by the occurrence of 100-Ka peaks in spectra of eccentricity and climate masks a challenging problem in climate dynamics (Imbrie and Imbrie 1980).

Many theories have been advanced to account for the 100-Ka climate cycle (Imbrie et al. 1989). Some of these call on resonance mechanisms to amplify the relatively small radiative forcing in the 100-Ka band (Le Treut and Ghil 1983). Others treat the cycle as an internally driven oscillation that interacts in some way with the externally

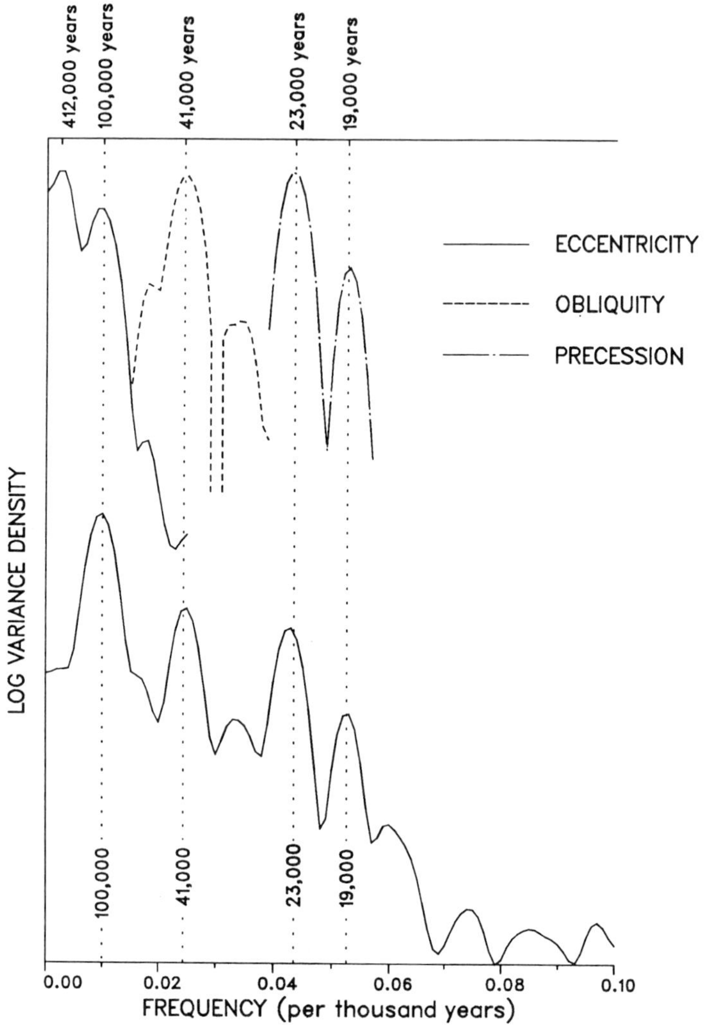

Figure 17.4 Variance spectra of orbital and $\delta^{18}O$ variations over the past 600,000 years (Figure 17.1). The periods of conspicuous peaks in the orbital spectra are indicated at the top. Dashed vertical lines identify conspicuous peaks in the $\delta^{18}O$ spectrum.

forced oscillations at 19 Ka, 23 Ka, and 41 Ka (Berger et al. 1990; Maasch and Saltzman 1990; DeBlonde and Peltier 1991).

The Past Two Million Years

In approaching the 100-Ka problem it is useful to look back over an interval long enough to see how this cycle evolves. To do this, we extend a stacked, 400-Ka benthic $\delta^{18}O$ record of glaciation intensity back over two million years (Figure 17.6). As is

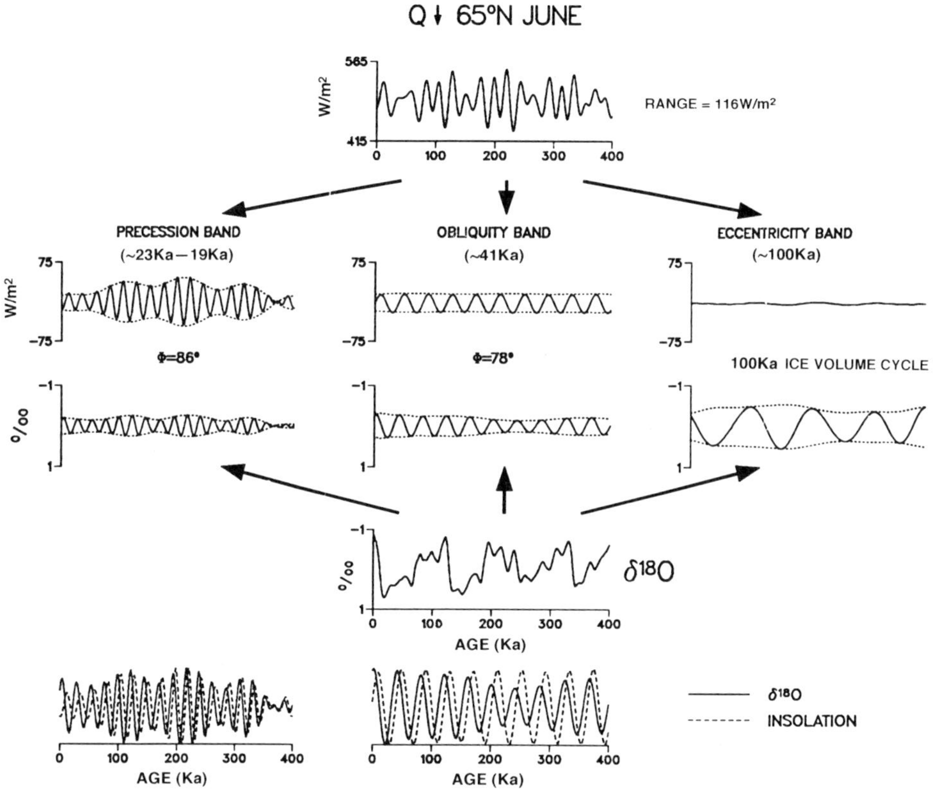

Figure 17.5 Dominant frequency components of time series of radiation and $\delta^{18}O$ over the past 400,000 years. Top four panels: using bandpass filters described in Imbrie et al. (1984), a radiation time series for June 65°N is partitioned into components occupying the ~100-Ka eccentricity band, the ~ 41-Ka obliquity band, and the entire precession band including the ~ 19-Ka and ~ 23-Ka cycles (Ka = 1000 years). The same filters are used to partition $\delta^{18}O$ into its 23-Ka, 41-Ka, and 100-Ka components (which in two bands have a significant phase lag ϕ with respect to radiation). Bottom panel: components in the obliquity and precession bands are normalized and replotted to dramatize the strong coherence between $\delta^{18}O$ and radiation.

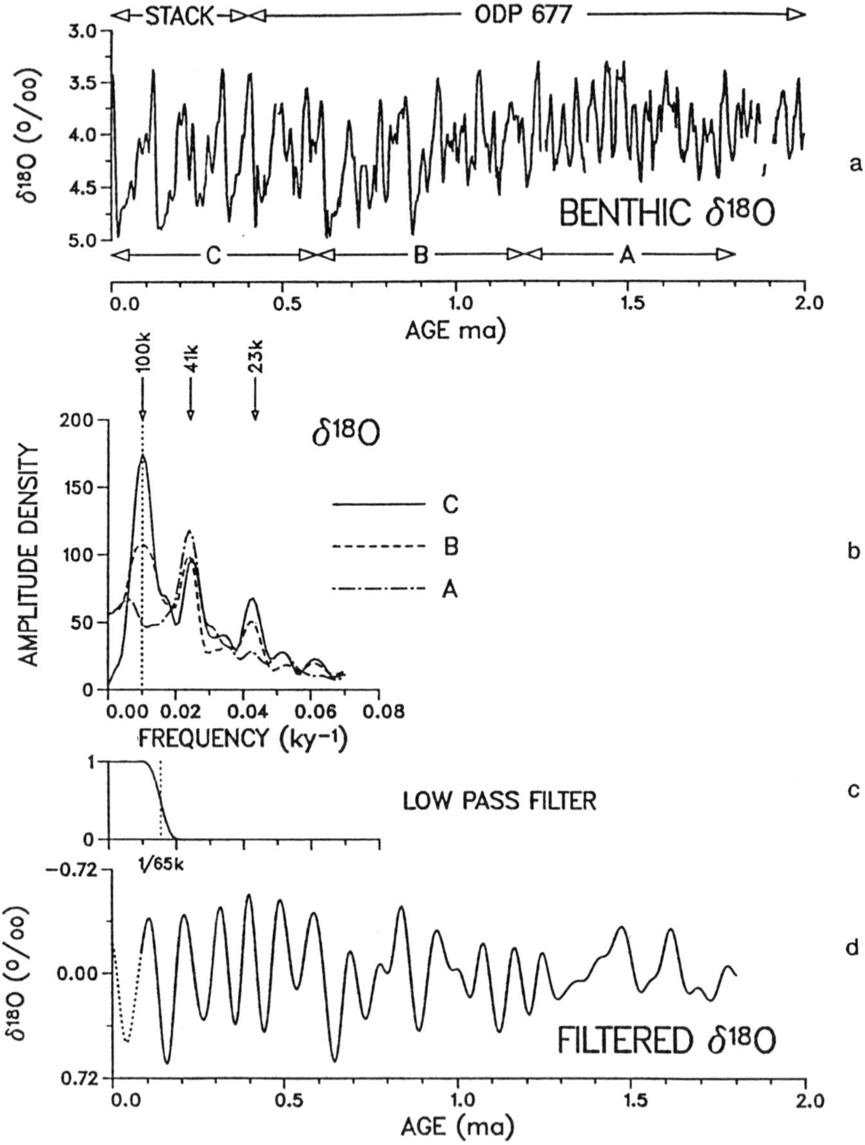

Figure 17.6 Glaciation cycles over the past 2 million years (Ma) based primarily on the $\delta^{18}O$ chronology of Shackleton et al. (1990) for ODP site 677. (a) A 0.4 Ma stacked benthic $\delta^{18}O$ record on the chronology of Imbrie et al. (1984) is extended with the data and chronology of ODP site 677. (b) Spectra of $\delta^{18}O$ for three intervals. (c) Frequency-response of the low-pass filter used to extract $\delta^{18}O$ oscillations with periods longer than 65 Ka in (d).

widely known (Pisias and Moore 1981; Ruddiman et al. 1989), the fairly regular, large-amplitude 100-Ka cycle of the past 0.5 million years evolved during the preceding 1.5 million years from a more irregular oscillation of much lower amplitude. The amplitude of the 23-Ka cycle also increased, whereas that of the 41-Ka cycle was remarkably constant. The onset of the 100-Ka regime, about one million years ago, was accompanied by a dramatic enrichment of the $\delta^{18}O$ values during the glacial phase of major oscillations, indicating that from then onward major ice ages had larger ice sheets and a colder boreal climate. The basic features of this pattern are not affected by present uncertainties about the exact chronology of the interval from 0.6 Ma to 2.0 Ma (Ma = one million years ago). The essential features of the analyses shown in Figure 17.6, which are based on an "alternate chronology" (Shackleton et al. 1990), do not differ from analyses based on the "standard" chronology (Raymo et al. 1990), shown in Figure 17.7.

What has perhaps not been sufficiently appreciated before is that the increase in the 100-Ka glaciation cycle around one million years ago is accompanied by a *decrease* in the 100-Ka eccentricity cycle, and an *increase* in the 412-Ka eccentricity cycle which, during the past 0.5 million years, has come to dominate the eccentricity spectrum (Figure 17.8). This fact argues strongly against those theories which explain the 100-Ka cycle as externally forced by changes in eccentricity, and supports those theories which explain it as an internal oscillation driven by positive and negative feedbacks within the system itself.

Measured over the same intervals, the amplitudes of the obliquity cycle and of the 23-Ka precession cycle remain essentially the same, whereas the amplitudes of the 19-Ka precession cycle decrease by about 50%.

UNSOLVED PROBLEMS

1. What physical mechanisms drive the orbitally forced responses to radiation that are associated with the precession and obliquity cycles (Kutzbach and Gallimore 1988; Broecker and Denton 1989; Imbrie et al. 1989)?
2. What internal feedbacks drive the 100-Ka cycle? How is the phase of this oscillation set (Clemens et al. 1991)?
3. How do the 19-Ka, 23-Ka, 41-Ka, and 100-Ka cycles interact (Pestiaux et al. 1988)? How do they interact with processes acting at higher frequencies in the climate system?
4. Do low-frequency trends driven by the combination of 19-Ka, 23-Ka, 41-Ka, and 100-Ka cycles cross system thresholds that trigger instabilities and result in episodes of abrupt climate change (Hughes 1987; Broecker and Denton 1989)?
5. What caused the onset of the 100-Ka regime about one million years ago (Ruddiman and Kutzbach 1989; Maasch and Saltzman 1990)?
6. Ignoring anthropogenic influences, is it possible that the system will behave differently in the future than it has in the past (Berger et al. 1991)?

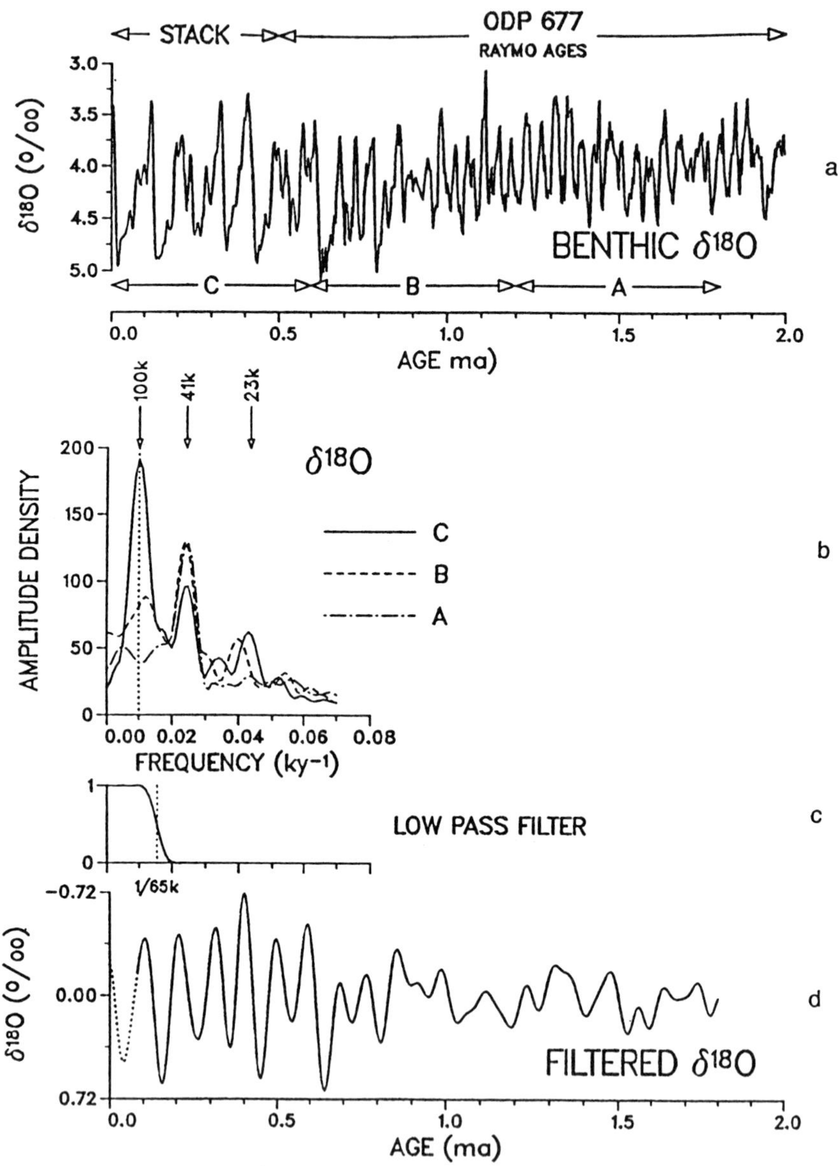

Figure 17.7 Glaciation cycles over the past 2 million years (Ma) based primarily on the $\delta^{18}O$ chronology of Raymo et al. (1990) for a record at ODP site 677. Otherwise as in Figure 17.6.

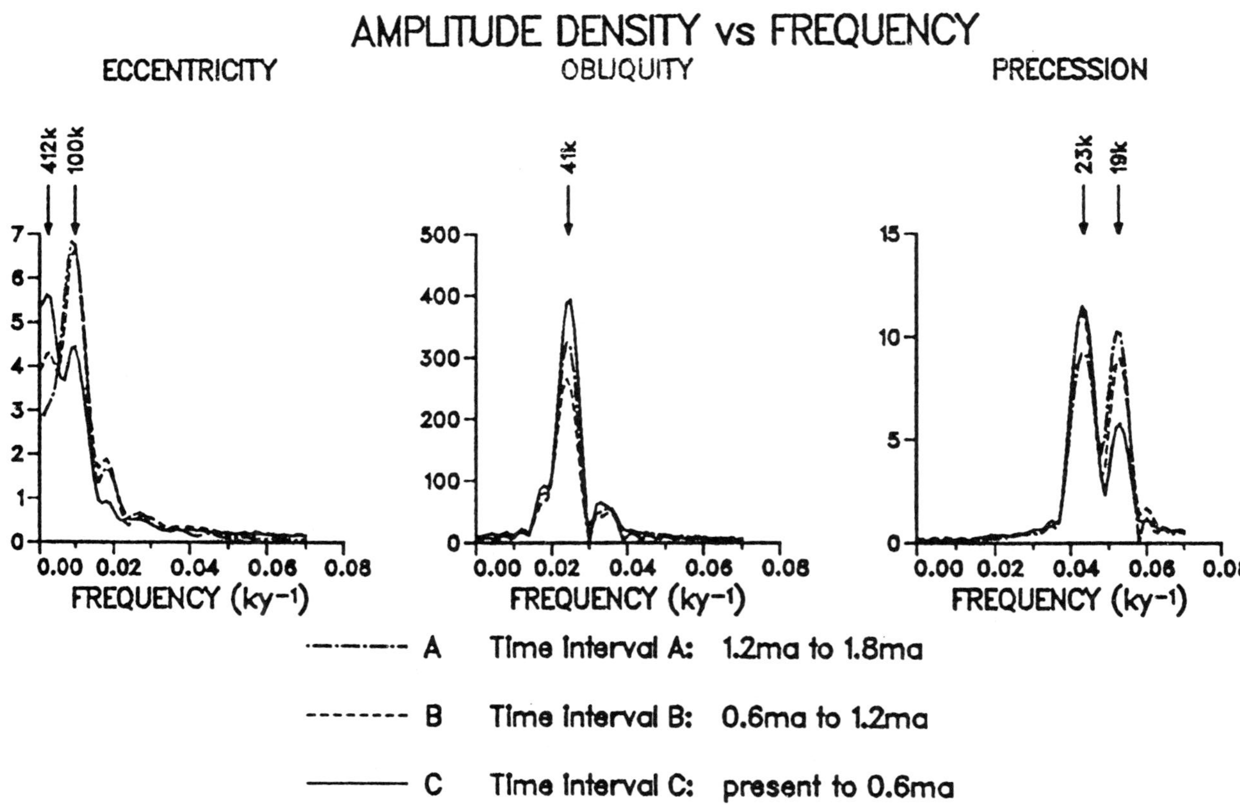

Figure 17.8 Orbital spectra for the three 600,000-year intervals indicated on Figure 17.6. Spectra estimated from time series of Berger and Loutre (1991).

REFERENCES

Berger, A. 1977. Long-term variations of the earth's orbital elements. *Celest. Mech.* **15**:53–74.

Berger, A. 1978. A Simple Algorithm to Compute Long Term Variations of Daily or Monthly Insolation. Louvain, Belgium: Univ. Catholique de Louvain.

Berger, A., H. Galleé, T. Fichefet, I. Marsiat, and C. Tricot. 1990. Testing the astronomical theory with a coupled climate-ice-sheet model. *Palaeogeog. Palaeoclim. Palaeoecol.* **89**:125–141.

Berger, A., H. Galleé, and J.L. Nélice. 1991. The Earth's future climate at the astronomical time scale. In: Future Climate Change and Radioactive Waste Disposal, ed. C.M. Goodess and J.P. Palutikof, pp. 148–165. Norwich, UK: Climatic Res. Unit, Univ. of East Anglia.

Berger, A., and M.F. Loutre. 1991. Insolation values for the climate of the last 10 million years. *Quat. Sci. Rev.* **10**:297–317.

Broecker, W.S., and G.H. Denton. 1989. The role of ocean-atmosphere reorganizations in glacial cycles. *Geochim. Cosmo. Acta* **53**:2465–2501.

Clemens, S., W. Prell, D. Murray, G. Shimmield, and G. Weedon. 1991. Forcing mechanisms of the Indian Ocean monsoon. *Nature* **353**:720–725.

COHMAP. 1988. Climatic changes of the last 18,000 years: Observations and model simulations. *Science* **241**:1043–1052.

DeBlonde, G., and W.R. Peltier. 1991. A one-dimensional model of continental ice volume fluctuations through the Pleistocene: Implications for the origin of the mid-Pleistocene climate transition. *J. Climate* **4**:318–344.

Hays, J.D., J. Imbrie, and N.J. Shackleton. 1976. Variations in the Earth's orbit: Pacemaker of the ice ages. *Science* **194**:1121–1132.

Hughes, T. 1987. Ice dynamics and deglaciation models when ice sheets collapsed. In: North America and Adjacent Oceans During the Last Deglaciation, ed. W.F. Ruddiman and H.E. Wright Jr., pp. 183–220. Boulder, CO: Geological Society of America.

Imbrie, J., J.D. Hays, D.G. Martinson, A. McIntyre, A.C. Mix, J.J. Morley, N.G. Pisias, W.L. Prell, and N.J. Shackleton. 1984. The orbital theory of Pleistocene climate: Support from a revised chronology of the marine $\delta^{18}O$ record. In: Milankovitch and Climate, Part 1, ed. A.L. Berger, J. Imbrie, J. Hays, G. Kukla, and B. Saltzman, pp. 269–305. Dordrecht: Reidel.

Imbrie, J., and J.Z. Imbrie. 1980. Modeling the climatic response to orbital variations. *Science* **207**:943–953.

Imbrie, J., A. McIntyre, and A. Mix. 1989. Oceanic response to orbital forcing in the late Quaternary: Observational and experimental strategies. In: Climate and Geo-sciences, ed. A. Berger, S. Schneider, and J.C. Duplessy, pp. 121–164. Boston: Kluwer.

Kutzbach, J.E., and R.G. Gallimore. 1988. Sensitivity of a coupled atmosphere/mixed-layer ocean model to changes in orbital forcing at 9000 years B.P. *J. Geophys. Res.* **93**:803–821.

Le Treut, H., and M. Ghil. 1983. Orbital forcing, climatic interactions, and glaciation cycles. *J. Geophys. Res.* **88**:5167–5190.

Maasch, K.A., and B. Saltzman. 1990. A low-order dynamical model of global climatic variability over the full Pleistocene. *J. Geophys. Res.* **95**:1955–1963.

McIntyre, A., W.F. Ruddiman, K. Karlin, and A.C. Mix. 1989. Surface water response of the equatorial Atlantic Ocean to orbital forcing. *Paleoceanog.* **4**:19–55.

Milankovitch, M. 1930. Mathematische Klimalehre und Astronomische Theorie der Klimaschwankungen. Berlin: Gebrüder Borntraeger.

Pestiaux, P., I. Van der Mersch, A. Berger, and J.C. Duplessy. 1988. Paleoclimatic variability at frequencies ranging from 1 cycle per 10000 years to 1 cycle per 1000 years: Evidence for nonlinear behaviour of the climate system. *Clim. Change* **12**:9–37.

Pisias, N.G., and T.C. Moore Jr. 1981. The evolution of Pleistocene climate: A time series approach. *Earth Planet. Sci. Lett.* **52**:450–458.

Raymo, M.E., W.F. Ruddiman, N.J. Shackleton, and D.W. Oppo. 1990. Evolution of Atlantic-Pacific $\delta^{13}C$ gradients over the last 2.5 m.y. *Earth Planet. Sci. Lett.* **97**:353–368.

Ruddiman, W.F., and J.E. Kutzbach. 1989. Forcing of late Cenozoic Northern Hemisphere climate by plateau uplift in southern Asia and the American West. *J. Geophys. Res.* **94**:18409–18427.

Ruddiman, W.F., M.E. Raymo, D.G. Martinson, B.M. Clement, and J. Backman. 1989. Pleistocene evolution: Northern Hemisphere ice sheets and North Atlantic Ocean. *Paleoceanog.* **4**:353–412.

Shackleton, N.J., A. Berger, and W.R. Peltier. 1990. An alternative astronomical calibration of the Lower Pleistocene timescale based on ODP Site 677. *Trans. R. Soc. Edin. Earth Sci.* **81**:251–261.

Shackleton, N.J., and N.D. Opdyke. 1973. Oxygen isotope and paleomagnetic stratigraphy of equatorial Pacific core V28–238: Oxygen isotope temperatures and ice volumes on a 10^5 year and 10^6 year scale. *Quat. Res.* **3**:39–55.

18

The Role of Solar Output Variations

J.A. EDDY[1] and H. OESCHGER[2]
[1]Consortium for International Earth Science Information Network,
2250 Pierce Road, University Center, MI 48710, U.S.A.
[2]Physikalisches Institut, Universität Bern, Bern, Switzerland

INTRODUCTION

The flow of radiation from the Sun supplies almost all the energy of the climate system. It also allows the existence of an atmosphere and liquid oceans, and the presence of all forms of life on the planet. Yet, as with every star, the solar outputs vary. One may assume, as scientists have for centuries, that solar variations should somehow be felt in the course of day-to-day weather or, in the longer term, of terrestrial climate. In spite of this, no solid connection has ever been established well enough to provide a reliable tool for practical weather or climate prediction (Eddy 1983a).

It could be that the Sun's intrinsic variations are insignificant when compared to other, more internal, climate drivers, and that they never rise above the level of noise. Perhaps the climate system responds in such a complex fashion–dependent, perhaps, upon preconditions–that we can never hope to identify, unequivocally, the marks, however strong, of a subtly varying Sun. It is also possible, as some now argue, that every one of the decade- to century-scale climatic changes over the last several thousand years, including the global warming of the last 100 years, is attributable to simple changes in solar irradiation (Friis-Christensen and Lassen 1991; Reid 1991; Wigley 1989; Eddy 1977). What is certain is that the question cannot be answered without (a) more quantitative information than is now available on the past and present range of variability of the total energy output of the Sun, (b) a much clearer history of the climate of the Earth, and (c) a fuller understanding of how the more variable, shortwave solar inputs spend their energy in the upper atmosphere.

Progress on the first two points will be paced, in large part, by the recovery and interpretation of paleodata, which supply proxy information on both climate and solar behaviors.

Global Changes in the Perspective of the Past
Edited by J.A. Eddy and H. Oeschger

THE LIMITS OF SOLAR VARIABILITY

Through what limits do solar outputs vary?

The only direct quantitative measurements available today pertain to a very brief segment of only the most recent moment of the life of a star. The still-young Sun is about 10 billion years old. By comparison, we now have almost 12 years of continuous measurements, from space, of the so-called "solar constant": the flux of energy at all wavelengths at the top of the atmosphere, corrected for varying Sun–Earth distance. In this period, the total irradiance was found to vary through a full range of about 0.06%, in part related to observed, solar surface changes (Willson and Hudson 1991). There are also scattered quantitative measurements, covering parts of the past 30 years, of the total flux of solar shortwave radiation in a number of spectral bands, exhibiting greater and more simply ascribed variations (see, e.g., White 1977).

What we do not know is how representative the periods of these brief glimpses are of the star's general behavior. Might we have been so lucky as to have captured the full range of the Sun's variability in our first, short sample of precise quantitative measurements? It seems unlikely that this could be, or that this kind of solar "uniformitarianism" should be applied to even the last 390 years of the Sun's history, the full span during which telescopic observations of its surface have been made (Eddy 1988).

However short, the recent radiometric data from space, or the few of them that are sufficiently calibrated and continuous, provide extremely useful insights; with certain caveats, they supply a template that can be used to relate observed changes in energy flux to more easily observed parameters of solar variability, such as the sunspot number. Caution is needed, however, in applying the template as a simple algorithm to convert observed sunspot numbers to changes in solar irradiance, where the range of variation corresponds to that measured by the SMM and Nimbus spacecraft between 1979 and 1991. Ignored are the facts that the total solar flux is modulated by more than sunspots and that solar activity varies on time scales longer than the 11-year cycle. It is entirely possible, given the relaxation time of the Sun's convective zone, that these longer periods may be related to more fundamental and greater-amplitude changes in the Sun's output of radiative energy.

The limits through which the Sun has varied in the distant past are now only crudely known: we can invoke evidence of the continuous presence of life on the planet over the last several billion years to establish that the total solar energy delivered to the Earth during this time has been sufficiently near the present value to prevent the oceans from either boiling or freezing solid. In terms of possible flux reduction, this implies that the radiation from the Sun has not fallen as much as 10% from its present value during this time. Even this sketchy datum must be qualified, however, by possible compensating changes in the chemical composition of the Earth's atmosphere. The existence of such not-unreasonable changes are indeed required to resolve the paradox of the "faint early Sun": if the atmosphere of two billion years ago were like that of today, the oceans would have frozen, for there is powerful evidence from the study of other stars to suggest that the Sun of that day, roughly 20% younger, was perhaps 30% dimmer than at the present (Newkirk 1980).

A useful tool that provides an indirect estimate of the effects of solar irradiance variations is the evidence for Milankovitch forcing of the major ice ages of the Quaternary period. Calculations based on known orbital and Earth-axis-inclination changes can here provide quantitative information on changes in past insolation as a function of position on the globe. When compared to what is known of past climate changes, these variations can be only used to sharpen our theoretical knowledge of the sensitivity of climate to solar forcing. This again provides a rather blunt tool, for the amplitude of orbitally induced changes in insolation, when considered regionally, is up to two orders of magnitude greater than any changes thus far measured in the total solar output. Moreover, as with the case of the "faint early Sun," the simple connection between insolation and surface temperature is complicated by the likely role of feedbacks in the climate system. We simply do not know whether, in the course of centuries or millennia, the flux released at the surface of the Sun varies through limits as wide as those that are imposed over much longer periods by orbital and Earth-axis changes.

In this chapter, we consider only changes in the solar output, i.e., changes intrinsic to the Sun.

SOLAR ACTIVITY AND SOLAR OUTPUT VARIATIONS

In terms of direct observations, almost all that we know about solar behavior in the period before the present century consists of the number and placement of dark spots on the visible solar surface of the star. Since about 1850, a continuous record has been available for the *relative sunspot number index*, R_Z, including estimated corrections for observing conditions and equipment (see Figure 18.1). This record has been extended, with decreasing reliability, backward in time to 1609: the year that the telescope came into use (Eddy 1980; McKinnon 1987). In addition, there are several hundred accounts of individual sunspots seen with the unaided eye, almost entirely from the Orient, which reach back to about the beginning of the Christian era (see, e.g., Stephenson and Clark 1978); however, these provide only sketchy information on possible solar behavior as the record is irregular, admittedly subjective, and grossly undersampled. The number of spots reported over 19 centuries is far less than what is observed within a single year of minimum activity today (Eddy et al. 1989).

Sunspots are now known to reveal concentrations in the surface magnetic field of the star. They are cooler, and hence darker, than the surrounding surface in that associated magnetic lines of force locally inhibit the escape of radiation flowing outward from the deep interior of the Sun. Individual spots exist for days to weeks. As such, their day-to-day effect is to deplete the flux of radiation, seen from the Earth, in proportion to the amount of the visible disk that sunspots cover. When the Sun is most spotted, e.g., during peak times in the well-known 11-year cycle, sunspots can

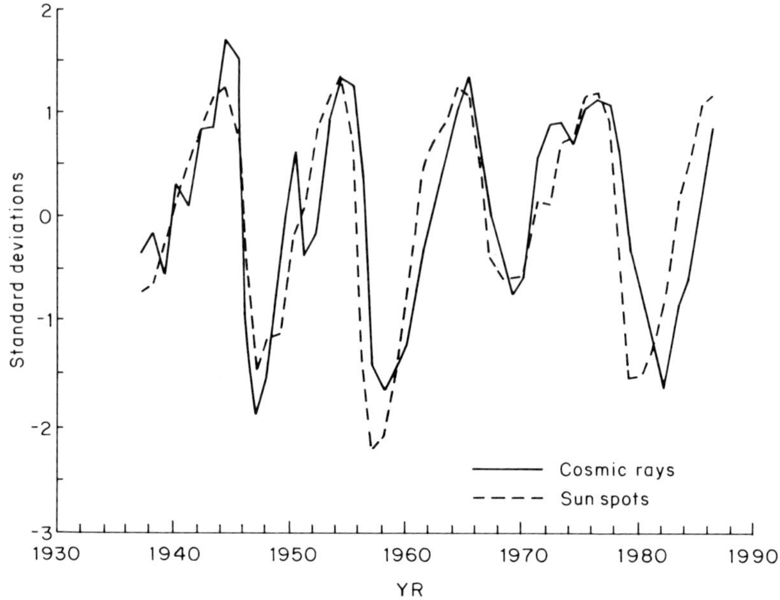

Figure 18.1 Flux of cosmic rays (neutron flux) at the Earth and observed annual mean sunspot number, R_z, for the period of 1938–1986. The sunspot number (dotted line) is plotted with increasing values downward to show the inverse relationship.

cover a few tenths of one percent of the visible hemisphere of the star. This postulated effect of reduction in total flux by sunspots was first confirmed when high-precision solar radiometers were first flown in space in the 1980s, on the SMM and Nimbus spacecrafts, which found sunspot-related depletions of up to 0.3%, lasting several day (Willson and Hudson 1991; Foukal 1990).

We must be able to acount for the suppressed flux. Less-distinguishable but also related to surface magnetic changes are the associated bright regions on the solar surface, where the absence of concentrated fields allows an enhanced release of energy, and hence a positive contribution to the total flux (Foukal 1990). Radiation that is held back by sunspots is not simply balanced by associated bright-region emissions. If this were the case, sunspots would have no effect, either short- or long-term, on total solar irradiance. The continuous radiometric observations on the Sun, made over the last 12 years, have revealed that the sunspot-blocking effect is, in the long run, more than compensated for by the opposing effect of accompanying but less easily detected brighter regions of the solar surface (Willson and Hudson 1991). During the period of observation, the greater, long-term effect of increased solar activity on total irradiance was to increase the total flux: the more direct blocking of radiation by sunspots modulated, but did not overcome, this overall enhancement. This finding lends credence to the supposition (Eddy 1976, 1977, 1983b) that during times of protracted suppression of solar activity, as during the Maunder Minimum of

1645–1715, the Sun's total radiation was slightly decreased, and that between the turn of the present century and the present time its brightness should have steadily increased (Reid 1991; Friis-Christensen and Lassen 1991).

This recent finding supplies, for the first time, a quantitative connection between solar activity, as measured in sunspot number, and the Sun's total output of radiative flux. Specifically, evidence indicates that the measured total irradiance systematically increased by about 0.06% between minima and maxima of the (most recent) 11-year solar cycle (Willson and Hudson 1991). In terms of climate sensitivity, this limited datum would discount changes in total solar radiative forcing as a significant climate driver, in that an irradiance change of this amount—if it represents the full range of solar output modulation—could alter the surface temperature by at most 0.1°C: smaller, by almost an order of magnitude, than required to explain the Little Ice Age/Maunder Minimum correlation or the global warming of about 0.6°C that has characterized the present century.

It is not at all certain, however, whether the results of recent measurements apply to the longer run of solar behavior. The power spectrum of sunspot numbers and other related indices of solar behavior possibly reveal a periodic modulation of 80 to 100 years in the overall level of solar activity (Eddy 1988)—the so-called "Gleissberg cycle"—which may be a distinct and more fundamental index of processes related to total solar flux. There could well be longer-term modulations in the solar flux that are unrelated to the 11-year cycle and of greater amplitude that those associated with this shorter-term modulation.

The Sun's subsurface convective zone is involved in both the modulation of solar activity and, potentially, in the longer-term modulation of the energy that is ultimately released in radiative form. To some extent the two processes are linked through the action of a solar dynamo, which couples vertical magnetic fields with the differential rotation of the solar surface to produce the 11-year cycle. Still, significant variations in solar output, driven by invisible changes in the convective zone, may not appear as any simple modulation of sunspot activity. Recently, Lean et al. (1991) reexamined what is known about longer-term changes in solar behavior and established a possible range of 0.3% in total solar irradiance over the last several hundred years, positively correlated with the level of solar activity. Adopting the canonical climate sensitivity factor of 1–2°C per 1% change in total solar flux, their conjectured change in total flux of 0.3% is marginally sufficient to explain both the Little Ice Age cooling of the late 17th century and the global warming of the last 100 years on the basis of solar forcing alone.

EXTENDING THE RECORD OF SOLAR ACTIVITY AND SOLAR OUTPUT VARIATIONS

The available record of sunspot activity can be extended through the use of a number of indirect or proxy indicators, most of which relate not to the stream of photons that emanate from the Sun but to its generally related output of charged particles and to disturbances in the extended magnetic field of the Sun that fills interplanetary space in the vicinity of the Earth. Examples of this are occurrences of the aurora, geomagnetic storms, the incidence of solar cosmic rays, and the quite separate modulation of galactic cosmic ray flux by the perturbations in the extended solar magnetic field. Many but not all of these disturbances are directly related to solar activity, as conventionally described in terms of the sunspot record; they can thus provide another proxy of that record.

In terms of possible solar influences on climate, variations in the nonradiative outputs of the Sun, no matter how great in terms of relative change, are almost certainly far less important than those in photon flux. The total energy supplied to the Earth by the Sun's nonradiative outputs is at least six orders of magnitude less than the radiative flux (White 1977). Moreover, they uniformly spend their energy in the uppermost regions of the atmosphere that are, at best, weakly coupled to the domain of weather and climate. What is more, the kinetic and thermal energy of the upper atmosphere, compared to the troposphere, is again many orders of magnitude weaker. While it is conceivable that energetically weak inputs into a weak system could trigger more energetic processes in the lower atmosphere, here we only consider solar particle and field variations, in terms of what they tell us, indirectly, about changes in activity and in the radiative outputs of the star.

Occurrences of the aurora borealis or australis are induced in the upper atmosphere by the impact of energetic solar particles. Those aurorae that occur in lower geomagnetic latitudes are well correlated with solar flares, whose energy is released from solar active region magnetic fields; since flare occurrence is correlated with sunspot number, it is not surprising that aurorae are as well. Auroral displays are more widely and easily observed than sunspots, and a number of chronicles recording their historical occurrence have been compiled. The 11-year cycle is easily identified; longer annals give evidence, as well, of the long-term changes in solar activity that appear in the sunspot number (Siscoe 1980).

Available records of auroral occurrence in Europe and North America extend at least 1000 years into the past, with records from the Orient reaching back another 1000 years. Such records have been used to develop a possible history of solar activity through this period, extending the direct, telescopic record (Schove 1955, 1983). Problems with auroral records are that (a) they provide no more than an indirect or proxy indicator of conditions on the Sun; (b) aurorae, particularly at high latitudes, commonly occur unrelated to solar activity (the result of high-speed particle streams from coronal holes); and (c) in subjective and sometimes mystical annals dating from before the Renaissance, it is often difficult to separate them from other sky phenomena.

The longest records of solar activity, and at present our best hope of recovering an objective record of solar behavior before the time of the telescope, are naturally archived records of cosmogenic isotope production, most particularly ^{14}C and ^{10}Be, described in the following section. It should be stressed at the outset, however, that records of isotope production are themselves convolved with other effects, including changes in the Earth's dipole moment, in climate, and, in the case of ^{14}C, in the biogeochemical cycling of terrestrial carbon (Stuiver et al. 1991; Siegenthaler and Beer 1988). Moreover, what they tell of the Sun is yet another step removed from changes in solar radiative outputs.

Terrestrial measurements of ^{14}C and ^{10}Be can be related to associated changes in solar radiative output through the following analytical steps:

1. measurements of relative isotopic abundance in terrestrial archives,
2. inferred rates of production,
3. model-derived reconstructions of solar activity,
4. empirically reconstructed changes in solar radiative output.

The connection between the first two steps depends on coincident knowledge of the confounding forces, cited above, that with the Sun modulate the production of ^{14}C and ^{10}Be. To this extent it relies, in part, on other paleorecords. The connection between steps 2 and 3 rests on an understanding of the mechanism through which effects of solar activity in the solar wind enhance the deflection of galactic cosmic rays away from the Earth, tested empirically by overlapping records of solar activity and cosmic ray flux in the modern era. The connection between the last two steps has only recently been illuminated by the spaceborne measurements of the total radiative flux from the Sun made in the last decade.

Although we have observations relating to this last step, albeit for a period of only 12 years, these new data, for the first time, put within our grasp a kind of solar Rosetta Stone, on which empirical data for each of these four steps are written for the same period of time. This kind of overlapping record has long been desired to clarify the assumptions of theory and the uncertainties in models. Were we certain that the period of overlap was representative of solar behavior in general, it could allow for a reading of a much longer history of the Sun.

THE HISTORY OF SOLAR MODULATION OF COSMIC RADIATION

High-energy cosmic radiation incident in the upper atmosphere of the Earth produce radioisotopes of various kinds by nuclear reactions with the atomic nuclei of gaseous N, O, and other elements that are present in the air. The production of these radioisotopes is modulated by variations in the heliomagnetic and geomagnetic shielding of galactic cosmic rays. Depending on geochemical behavior, these radioisotopes follow different pathways in the Earth system and are recorded in different natural archives. The recovery of such records and their interpretation, in terms of changes in the solar magnetic field in the vicinity of the Earth, offer an opportunity to recover a history of past solar variations, through the chain described in the preceding section. The most abundant of the cosmogenic isotopes, ^{14}C, is sequestered through photosynthesis in the cellulose of annual growth layers in trees; ^{10}Be, the next most abundant, falls in rain and snow to the surface of the Earth, where its rate of deposition can be recovered through the analysis of dated ice cores. Other cosmogenic isotopes have also been used to recover the history of the Sun. For the following, however, we restrict ourselves to studies of ^{10}Be and ^{14}C, which have given us, until the present, most of the insight on past solar variations.

Hot ^{14}C atoms, produced chiefly in the lower stratosphere and higher troposphere, are oxidized to $^{14}CO_2$ and mix with ambient atmospheric CO_2. Thus, ^{14}C production variations, as deduced from geochemical analyses of tree-ring wood, appear dampened and delayed as a result of atmospheric mixing and the exchange with the ocean and terrestrial biosphere. Moreover, not all ^{14}C variations in atmospheric CO_2 are due to changing production rates. They can also reflect variations in the carbon cycle, through processes of carbon exchange between atmosphere and biosphere and between atmosphere and ocean. In addition, production rates are also affected by secular changes in the strength of the geomagnetic moment, the history of which must also be known or assumed.

Newly produced hot ^{10}Be atoms become attached to the surface of aerosols, to be rained out and deposited on the ground after an average atmospheric residence time of the order of one or two years. Again variations found in the ^{10}Be concentration that are observed, e.g., in polar ice samples, are not entirely due to changes in production; as they are deposited in rain or snow, they also reflect regional and local meteorological conditions, such as changes in atmospheric transport or dilution with ambient water vapor.

Tree-ring ^{14}C has provided significantly more information on solar variations, to date, than has ice-borne ^{10}Be, chiefly due to the far greater availability of tree-ring samples and the longer and more extensive study of radiocarbon. The longer atmospheric residence time of ^{14}C presents a relative disadvantage, however, as is the relatively short half-life of the nuclide. The much shorter residence time of ^{10}Be offers greater potential as a monitor of year-to-year solar variations, including the 11-year cycle, both of which are effectively filtered out by atmospheric residence time in the tree-ring radiocarbon record. Moreover, the ^{10}Be half-life of about 1.5×10^6 years, compared to a few thousand years for radiocarbon, holds out the promise of recovering a far longer history of the ancient Sun. These advantages are to a degree offset, however, by the need to remove local precipitation effects in analyzing the ^{10}Be record.

The neutron flux in the atmosphere induced by cosmic radiation has been followed for many years by neutron monitoring stations distributed at high-altitude sites over the Earth. The data show the Schwabe cycle (often called the 11-year cycle) modulation of cosmic radiation (fewer neutrons at times of solar maxima), as well as the opposing effect of neutrons (more abundant at times of solar maxima) that are the product of solar flares. The variation in the neutron flux over a Schwabe cycle due to the former effect is of the order of 20%. Based on these direct observations (Figure 18.1) it should be expected that records of cosmogenic nuclides should also show a solar modulation signal.

In the following we give an overview of some of the present ^{10}Be and ^{14}C data and comparisons of records of the two isotopes.

Yearly ^{10}Be Data and the Schwabe Cycle

From an ice core taken at Dye 3 Station Greenland, measurements of yearly samples covering from A.D. 1780 through 1985 (Figure 18.2) exhibit clearly periodic variations that have the same character as the curve of annual mean sunspot number, including secular or longer-term trends (Beer et al. 1988). In order to remove the meteorological noise, a low-pass filter was applied. Cross correlation with the corresponding sunspot numbers confirms the presence of the Schwabe cycle, which is also visible in this combined plot of ^{10}Be concentrations and sunspot number, where sunspot number has here been plotted with higher numbers downward. Periods of low solar activity correspond to low shielding of galactic cosmic radiation and, therefore, higher isotope production. Records such as these give hope that ice-borne ^{10}Be data can eventually be used to rectify direct historical records of sunspot number for the less reliably known 17th, 18th, and early 19th centuries, as well as to extend that record many millennia into the past.

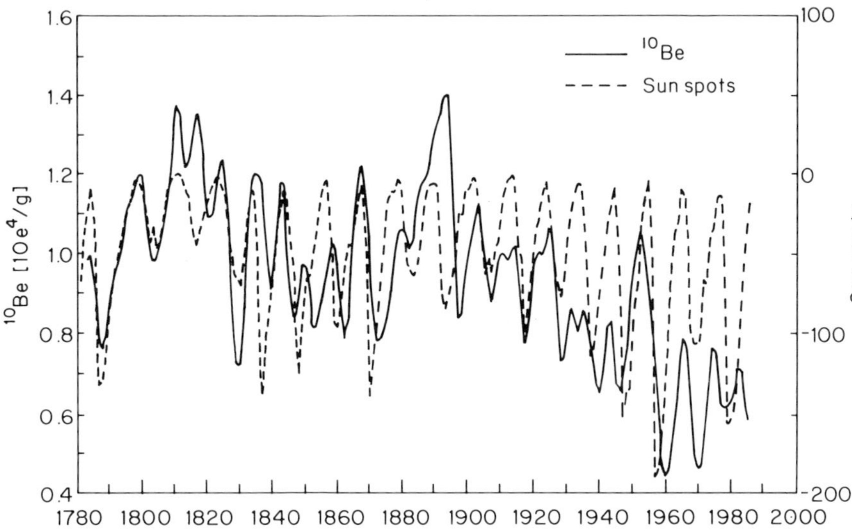

Figure 18.2 Yearly smoothed measurements of ^{10}Be from Dye 3 Station in South Greenland for the period of A.D. 1780 through 1985 (solid line) compared with historically obtained values of the annual mean sunspot number. As in Figure 18.1, the sunspot number is plotted with increasing values downward. From Beer et al. 1988.

Maunder Minimum-type Variations: Comparison of the Greenland and Antarctic Records

Figure 18.3 shows ^{10}Be records covering the Wolf (A.D. 1280–1350), Spoerer (A.D. 1420–1540), and Maunder (A.D. 1645–1715) solar minima obtained from ice cores from Milcent, Greenland, and the South Pole, Antarctica, together with a combined average curve. Three periods of maximum isotope production are clearly visible; they correspond to the mentioned periods of minimum solar activity, as these have been identified in historical records of sunspot occurrence and in analyses of tree-ring ^{14}C (Beer et al. 1991).

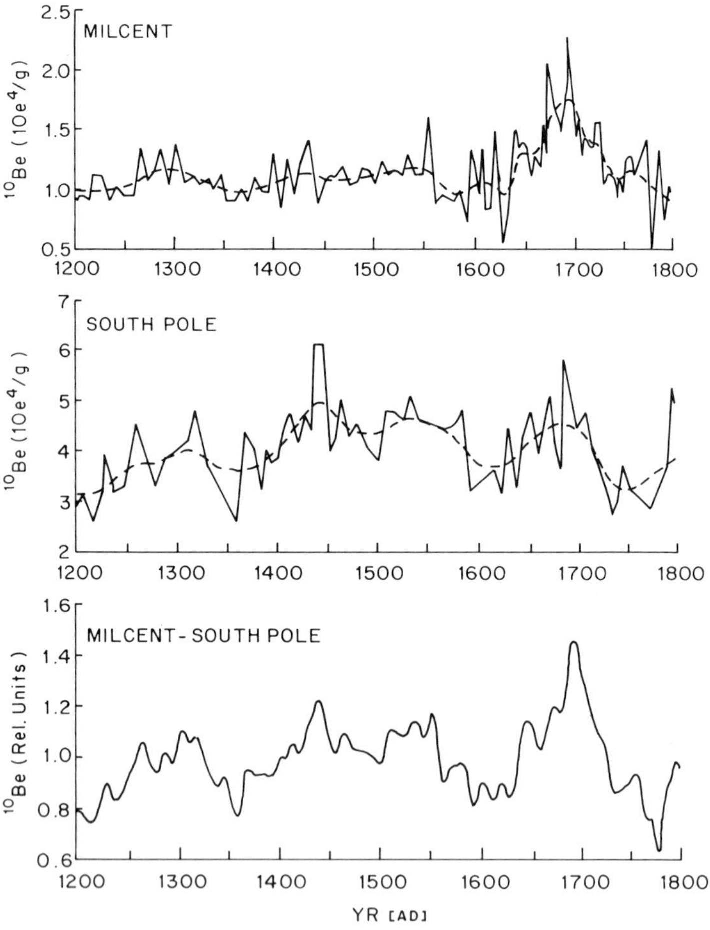

Figure 18.3 Measurements of ice-borne ^{10}Be from Milcent, Greenland and South Pole Station, Antarctica and as a combined average, in units of 10^4 grams per cm^3 of ice, as in Figure 18.2. From Beer et al. 1991.

Comparison of ^{10}Be and ^{14}C Records

Figure 18.4 compares tree-ring ^{14}C and ice-borne ^{10}Be records for the periods A.D. 1200–1800. A high correlation is observed, indicating that production variations are the dominant cause of the observed radioisotope variations, since it is highly improbable that terrestrial effects (climate variations) could modulate the two records in a similar way. As expected from model calculations, ^{10}Be reflects almost directly the production rate variations, whereas in the ^{14}C record they appear delayed and strongly dampened. ^{10}Be variations in the range of – 20 to + 20% correspond to ^{14}C variations of about – 10 to + 10‰, in agreement with the expected dampening of the radiocarbon records (Siegenthaler et al. 1980). The ^{10}Be variations on average lead the ^{14}C variations by about two decades, in agreement with the carbon cycle model estimates for the dampening and delay (in the carbon system) of ^{14}C production variations with an apparent period of about 200 years.

The Longer-term ^{10}Be and ^{14}C Variations

Samples from the ice core from Camp Century, Greenland have now enabled us to study the longer-term ^{10}Be variations. During the glacial period, the ^{10}Be concentrations

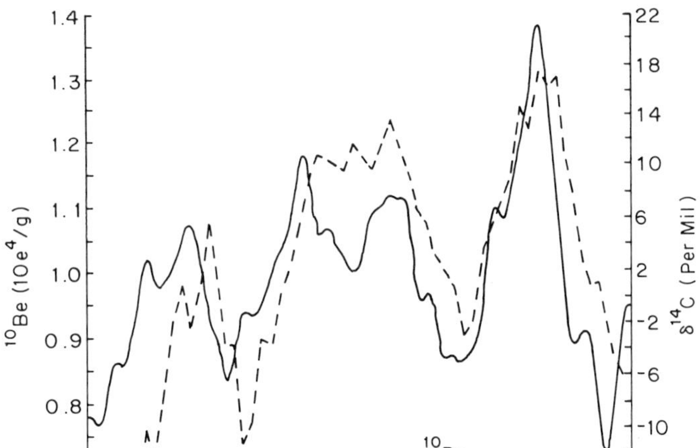

Figure 18.4 Comparison of ice-borne ^{10}Be from Dye 3 Station (10^4 grams per cm^3) with tree-ring ^{14}C (parts per mil relative to ^{12}C) for the period of A.D. 1200–1800. From Beer et al. 1991.

were, in general, considerably higher than during the ensuing Holocene. This can be explained by the assumption of a roughly constant ^{10}Be flux and decreased snowfall during the cold glacial period, when much of the global water supply was withheld from the water cycle in the form of polar ice sheets. During the Holocene the relatively small variations in ^{10}Be (of presumed solar origin) and δ18O (reflecting sea temperature) in the Camp Century records are not correlated with each other, except for the Maunder Minimum period. Were secular solar variations a significant driver of high-latitude climate, one would expect a more consistent correlation between the two.

Figure 18.5 shows the ^{10}Be concentrations, measured in the Camp Century ice core, from 7000 B.C. onwards, together with the measured atmospheric $\delta^{14}C$ data from

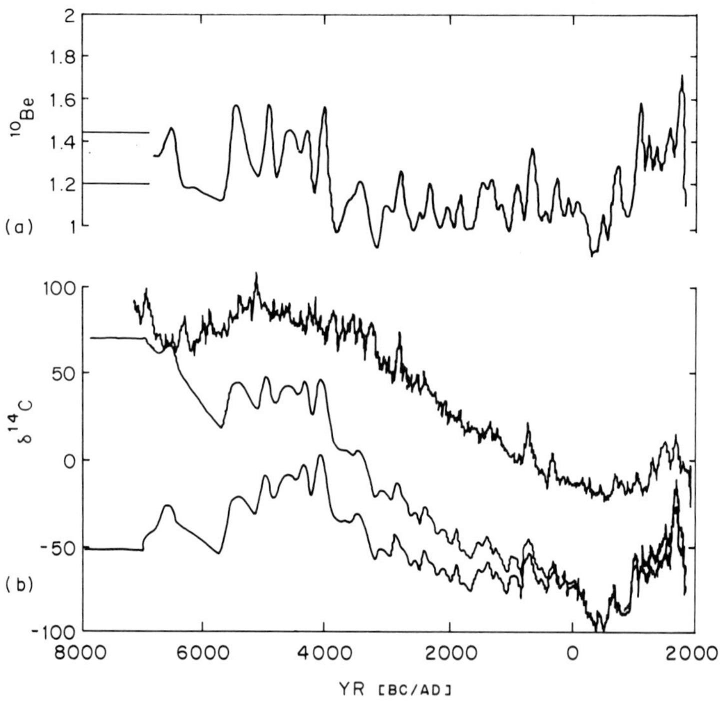

Figure 18.5 Comparison of ^{10}Be measurements from the Camp Century core (curve a, in 10^4 ions per gram of ice) with tree-ring radiocarbon (b1 [top], in parts per mil relative to ^{12}C) for the period from 7000 B.C. to the present. Note that for the period before ca. A.D. 500, the dominant low-frequency character of the ^{14}C record is not present in the ice-borne ^{10}Be record. Curves b2 and b3 are calculated values of $\delta^{14}C$ for two initial values of total carbon, using the ^{10}Be data (curve a) as input to a box diffusion model. The rough agreement of curves b2 and b3 with the general behavior of the measured ^{14}C values, demonstrates that the differences in the low-frequency characteristics of the ^{10}Be and ^{14}C curves can be accounted for in the first order by assuming a change in the distribution of carbon in terrestrial reservoirs during the period. From Beer et al. 1988.

tree-rings. Both data sets show variations of the Maunder Minimum type throughout the entire period. In the case of ^{14}C, they are superimposed on a general long-term trend, which is now explained as due to the combined effects of variations in carbon cycle dynamics and in geomagnetic shielding of cosmic radiation.

In Figure 18.6, the same data are shown for the period of 3000 B.C. to the present, after removal of the long-term trends by a high-pass filter. To facilitate a comparison, the ^{10}Be variations are transformed into model-calculated ^{14}C variations. A high degree of similarity is observed, confirming once more that the century-scale variations in both records are likely of external origin. As an example. the prominent maxima of the tree-ring ^{14}C record (corresponding to a 150-year depression in solar activity) at about 700 B.C. also appears in the ^{10}Be-based curve.

Based on this information, it now seems certain that the changes in measured concentrations of ^{10}Be and ^{14}C reflect production variations which, during the period of direct solar observation, are also highly correlated with the sunspot number record.

^{14}C and ^{10}Be studies demonstrate that throughout the Holocene, the Sun experienced a series of periods of "quiet Sun" behavior, such as that observed directly through the telescope during the Maunder Minimum (A.D. 1645 through 1715). These

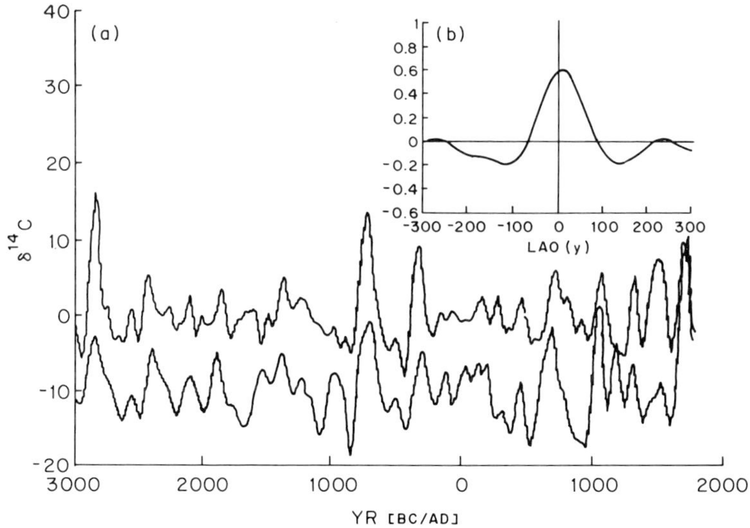

Figure 18.6 The most recent 5000 years of the tree-ring radiocarbon record (upper curve, enlarged from the data shown in Figure 18.5) compared with calculated ^{14}C (lower curve), based on the ^{10}Be data for the same period from the Camp Century ice core. Long-term trends in the ^{14}C data have been removed by a high-pass filter. Inset (b) is the cross-correlation coefficient between the two curves, as a function of lag time in years. From Beer et al. 1988.

century-scale excursions are well correlated and easily identified in both records. The longer-term trend found in the tree-ring ^{14}C record, originally presumed to be the reflection of a roughly 10,000-year cycle of geomagnetic origin, does not appear in the ^{10}Be records obtained to date. Since geomagnetic changes should affect the production of both of these isotopes, it is presumed that the longer-term ^{14}C excursions are due largely to changes in the terrestrial carbon cycle.

^{14}C and ^{10}Be in the Last Glacial Period

More difficult is the interpretation of the records of the glacial-interglacial transition and of the glacial periods. ^{14}C reconstructions, not shown here, indicate that during the last glacial, the atmospheric $^{14}C/^{12}C$ ratio was higher by about 40%. This can be explained by either a change in the state of the carbon cycle or a decreased geomagnetic shielding.

One might also speculate that during parts of the last glaciation, the Sun was in the state of the "quiet Sun." A higher $^{14}C/^{12}C$ ratio found in the record at about 20,000 yr B.P. is indeed of the same magnitude as the higher radioisotope production during a Holocene "quiet Sun" period (Suess 1968).

CONCLUSION

Although we stand at the brink of securing information on the history of solar activity over several hundred thousand years, its translation into a record of solar output variations will depend on a number of questions that are as yet unanswered:

1. Through what limits have the radiative outputs of the Sun varied over the past 100 years? Over the last 10,000 years?
2. In what way or ways are these variations related to solar activity (the more easily recovered parameter)?
3. Is the 22-year (magnetic) cycle of the Sun reflected in solar radiative outputs? The so-called "Gleissberg cycle" of 80–100 years?
4. Are there significant variations in solar radiation that are independent of detectable solar activity changes?
5. To what extent have changes in the Earth's dipole moment affected production of comogenic isotopes? To what extent has climate itself been a factor? To what extent are changes in the carbon cycle reflected in the ^{14}C record?
6. What significant climatic changes have taken place during the last 1000 years? The last 100,000 years? To what extent was the Little Ice Age a global happening? The so-called Medieval Warm Epoch? The Younger-Dryas and Allerod-Bolling events?

REFERENCES

Beer, J., A. Blinov, G. Bonari, R.C. Finkel, H.J. Hofmann, B. Lehmann, H. Oeschger, A. Sigg, J. Schwander, T. Staffelbach, B. Stauffer, M. Suter, and W. Wolfli. 1990. Use of ^{10}Be in polar ice to trace the 11-year cycle of solar activity. *Nature* **347**:164–166.

Beer, J., G.M. Raisbeck, and F. Yiou. 1991. Time variations of ^{10}Be and solar activity. In: The Sun in Time, ed. C.P. Sonett, M.S. Giampapa, and M.S. Matthews, pp. 343–359. Tucson, AZ: Univ. of Arizona Press.

Beer, J., U. Siegenthaler, G. Bonani, R.C. Finkel, H. Oeschger, M. Suter, and W. Wolfli. 1988. Information on past solar activity and geomagnetism from ^{10}Be in the Camp Century ice core. *Nature* **331**:675–679.

Eddy, J.A. 1976. The Maunder Minimum. *Science* **192**:1189–1202.

Eddy, J.A. 1977. Climate and the changing Sun. *Climatic Change* **1**:173–190.

Eddy, J.A. 1980. The historical record of solar activity. In: The Ancient Sun, ed. R.O.Pepin, J.A. Eddy, and R.B. Merrill, pp.119–134. New York: Pergamon Press.

Eddy, J.A. 1983a. An historical review of solar variability, weather, and climate. In: Weather and Climate Responses to Solar Variations, ed. B.M. McCormac, pp. 1–15. Boulder: Colorado Ass. Univ. Press.

Eddy, J.A. 1983b. The Maunder Minimum: A reappraisal. *Solar Physics* **89**:195–207.

Eddy, J.A. 1988. Variability of the present and ancient Sun: A test of solar uniformitarianism. In: Secular Solar and Geomagnetic Variations, ed. F.R. Stephenson and A. Wolfendale, pp. l–23. Dordrecht: D. Reidel.

Eddy, J.A., F.R. Stephenson, and K.K.C. Yau. 1989. On pre-telescopic sunspot records. *Quar. J. RAS* **30**:65–73.

Foukal, P. 1990. Solar luminosity variations over timescales of days to the past few solar cycles. In: The Earth's Climate ard Variability of the Sun over Recent Millennia, ed. J.-C. Pecker and S.K. Runcorn, pp. 193–201. London: The Royal Society.

Friis-Christensen, E., and K. Lassen. 1991. Length of the solar cycle: An indicator of solar activity closely associated with climate. *Science* **254**:698–700.

Lean, J., A. Skumanich, and O.R. White. 1992. Estimating the sun's radiative output during the Maunder Minimum. *Geophys. Res. Lett.* **19:**1591–1594.

McKinnon, J.A. 1987. Sunspot Numbers: 1610–1985. Boulder: World Data Center A for SolarTerrestrial Physics. 12pp.

Newkirk, G., Jr. 1980. Solar variability on time scales of 10^5 to $lO^{9.6}$ years. In: The Ancient Sun, ed. R.0. Pepin, J.A. Eddy, and R.B. Merrill, pp. 293–320. New York: Pergamon Press.

Raisbeck, G.M., and F. Yiou. 1985. ^{10}Be in polar ice and atmospheres. *Ann. Glaciology* 7:138–140.

Raisbeck, G.M., and F. Yiou. 1988. ^{10}Be as a proxy indicator of variations in solar achvity and geomagnetic field intensity during the last 10,000 years. In: Secular Solar and Geomagnetic Variations, ed. F.R. Stephenson and A. Wolfendale, pp. 287–296. Dordrecht: D. Reidel.

Reid, G.C. 1991. Solar total irradiance variations and the global sea surface temperature record. *J. Geophys. Res.* **96**:2835–2844.

Schove, D.J. 1955. The sunspot cycle, 649 B.C. to A.D. 2000. *J. Geophys. Res.* **60**:127–146.

Schove, D.J., ed. 1983. Sunspot Cycles. Stroudsburg, PN: Hutchinson Ross Publ. Co.

Siegenthaler, U., and J. Beer. 1988. Model comparison of ^{14}C and ^{10}Be isotope records. In: Secular Solar ard Geomagnetic Variations, ed. F.R. Stephenson and A. Wolfendale, pp. 315–328. Dordrecht: D. Reidel.

Siegenthaler, U., M. Heimann, and H. Oeschger 1980. ^{14}C variations caused by changes in the global carbon cycle. *Radiocarbon* **22**:177–191.

Siscoe, G.L. 1980. Evidence in the auroral record for secular solar variability. *Rev. Geophys. Space Phys.* **18**:647–658.

Stephenson, F.R., and D.H. Clark. 1978. Applications of Early Astronomical Records. London: Adam Hilger.

Stuiver, M., T.F. Braziunas, G. Becker, and B. Kromer. 1991. Climatic, solar, oceanic, and geomagnetic influences on Late Glacial and Holocene atmospheric $^{14}C/^{12}C$ change. *Quat. Res.* **35**:1–24 .

Stuiver, M., and P.D. Quay. 1980. Changes in atmospheric carbon-14 attributed to a variable sun. *Science* **207**:11–19.

Suess, H.E. 1968. Climatic changes, solar activity, and the cosmic-ray production rate of natural radiocarbon. *Meteor. Mono.* **8**: 146–150.

Wigley, T.M.L., and P.M. Kelly. 1990. Holocene climatic change, ^{14}C wiggles and variations in solar irradiance. In: The Earth's Climate and Variability of the Sun over Recent Millennia, ed. J.-C. Pecker and S.K. Runcorn, pp. 149–160. London: The Royal Society.

White, O.R., ed. 1977. The Solar Output and Its Variation. Boulder: Colorado Assoc. Univ. Press.

Willson, R.C., and H.S. Hudson. 1991. The Sun's luminosity over a complete solar cycle. *Nature* **351**:42–44.

19

Climatic Evolution During the Last Millennium: What Can We Learn from It?

H. FLOHN
Meteorologisches Institut, Universität Bonn, Auf dem Hügel 20,
5300 Bonn 1, F.R. Germany

ABSTRACT

After a selected review of available sources, a short climatic history of the last millennium is given. The most significant change was the transition between the Medieval Optimum (about A.D. 900–1200) and the Little Ice Age (1570–1850). It was marked by two periods of alternating cold/wet cycles, each with a strong meridional exchange. Their maxima occurred around 1320 and 1440/1460.

A new ice core from the Peruvian Andes suggests the existence of an equatorial rainy period between 1490 and 1700 and increased volcanic dust between 1550 and 1880. Weak changes of equatorial oceanic surface temperature can be responsible for large intensifications of the hydrological cycle and (as a nonlinear feedback) of extratropical circulations.

INSTRUMENTAL PERIOD

Instrumental observation series began shortly after the invention of the main meteorological instruments in the 17th and 18th centuries. The first thermometric records were introduced in Italy, France, Germany, Great Britain, and the Netherlands between 1650 and 1700. However, in the first decades, the position of the instruments, their protection against solar radiation, their ventilation, the introduction of fixed hours of observation, etc., were rather different, even if various scientific organizations took responsibility for their accomplishments. At an international level, the Societas Meteorologica Palatina of Mannheim took initiative, prepared instructions and instruments, and published yearbooks as early as 1781; political changes after the French Revolution, however, led to its closing.

Global Changes in the Perspective of the Past
Edited by J.A. Eddy and H. Oeschger © 1993 John Wiley & Sons Ltd.

Hemispheric reconstructions of early temperature records—excluding Chinese sources—were given by Groveman and Landsberg (1979) (since 1579, based mainly on European records) and Jacoby and d'Arrigo (1989) (reconstructed from tree-rings since 1671). Unfortunately, great differences are visible even in the early 19th century.

Only governmental organizations—the first founded by J.W. von Goethe in 1822, in Weimar, which existed until his retirement in 1831—were able to organize and maintain permanent networks with homogeneous station records. This took place in European states and in the U.S.A. between 1850 and 1880, and primarily at the beginning of the 20th century in most European colonies and other independent countries. Maritime observations were collected on a national level since about 1865. Several monographs and other papers have reviewed fluctuations during the instrumental period (Bottomley et al. 1990; Ellsaesser et al. 1986; von Rudloff 1967) and during the last millennium (Lamb 1977, 1982; Ladurie 1971; Mörner and Karlén 1984; Pfister 1984; Wigley et al. 1981).

PROXY DATA

In many areas, isolated instrumental records from the late 17th to the early 19th centuries can be extrapolated by proxy data, which partly allow an approximate numerical calibration based on extensive comparisons of their recent records with contemporaneous measurements. For example, this is possible with (a) a series of wine harvesting, representing the temperature of the vegetation period (Figure 19.1)

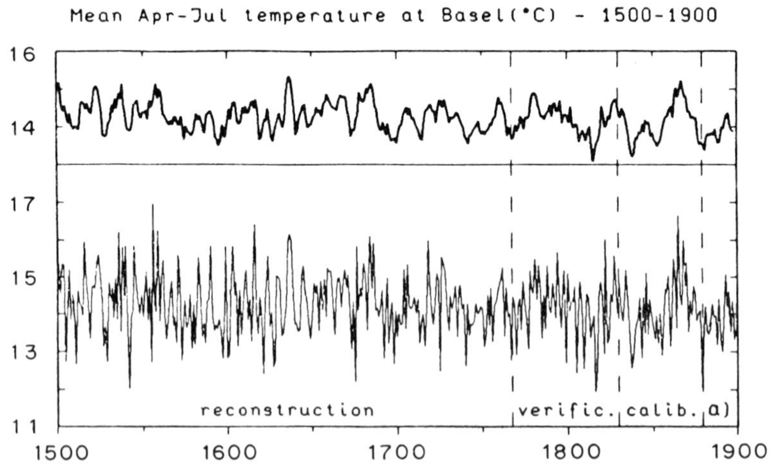

Figure 19.1 Mean temperature vegetation period at Basel, reconstructed after area-averaged wine harvest data at Dijon; 50-yr correlations between 1768 and 1879, 0.77–0.88, very stable (Burkhardt and Hense 1984).

(Burkhardt and Hense 1984), (b) the closing of harbors or canals by ice (Netherlands, Baltic), representing winter temperature, and (c) tree-ring data (Hughes et al. 1982; Jacoby and d'Arrigo 1989) at their boundaries with respect to height or aridity. Regular historical sources (noninstrumental weather diaries, seasonal reports to a central government) can also be quite useful, even if they are only evaluated in the form of unquantifiable indices. Such data are available—without gaps—for summer rainfall in China (Figure 19.2), wine harvests in eastern France (since 1475), and various sources for all seasons in Switzerland since 1525 (Pfister 1984). A good example of reconstructed temperatures over a period of 800 years has been given from Mediterranean trees by Serre-Bachet and Guiot (in Berger and Labeyrie 1987). The Chinese data (Wang et al. 1991) are expected to appear for the whole millennium; weather diaries from Japan also exist from the 12th century onwards. Future compilations based on more complete proxy data for the whole Northern Hemisphere are expected, at least back to about 1200 (cf. Figure 19.3).

Written historical sources, such as chronicles, exist for Europe since about 1100 (Alexandre 1987); the history of glaciers (their advances and retreats) has contributed much to our knowledge of climatic history, at least at a decadal or century scale (Figure 19.4). The climatic history of Europe, of parts of Eastern Asia, and the western U.S.A. can be written as a first approximation back to 1100 or 1200. The history of rivers and lakes without outlets (Nile River, Dead Sea, Lake Chad [see Figs. 19.4 and 19.5 as well as Klein and Flohn 1987; Maley 1981] and probably Caspian Sea) gives useful supplements, representing the hydrological balance of large areas. Another source of summer temperatures seems to be isotope ($\delta^{13}C$) data from late-wood tree-rings (see Figure 19.6; data based here on one single tree).

Prior to A.D. 1000, the number of reliable proxy sources declines rapidly. They are primarily available for isolated areas and are of varying quality and representative nature. In addition, during the last millennium, many areas were virtually empty: most of the tropical countries, including India, the polar regions (except Iceland and northern Canada), and most of the Southern Hemisphere (except New Zealand and South Africa). This fact prohibits serious investigation of (even rather crude) spatial relationships. Therefore, this review can present only an educated guess. Longitudinal differences are typical only for recent climate evolution.

Moreover, a serious difficulty has emerged: in addition to spatial differences, we must also take into account the seasonal limitation of many proxy data. For example, the high stands of the Dead Sea, dependent only on winter rainfall (Figure 19.3), do not coincide with the advances of Alpine glaciers caused by cool, cloudy summers with frequent snowfall in high altitudes (Flohn 1991).

CLIMATIC HISTORY OF THE LAST MILLENNIUM

Most historical and proxy data have been evaluated from European countries. Here, a fairly reliable outline of the climatic evolution can be given, with increasing accuracy

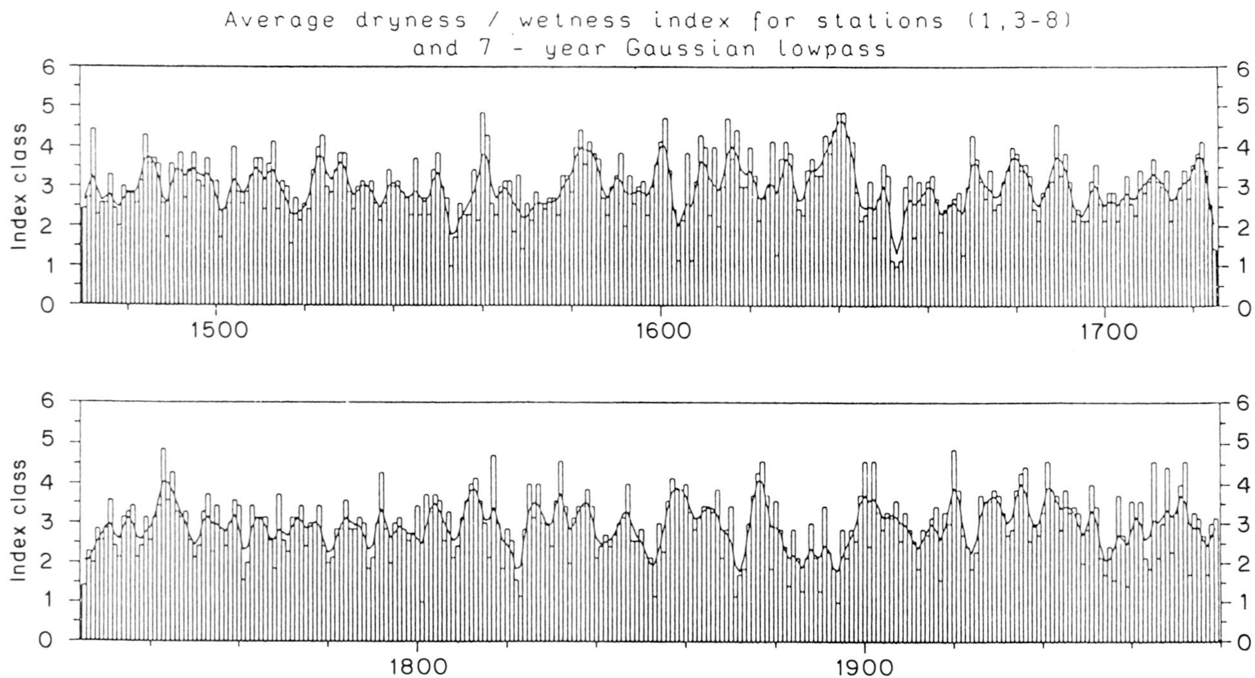

Figure 19.2 Rainfall index (after 5 index classes of dryness–wetness) in northern China, averaged for seven sites near Beijing; individual years and 7-yr Gauss filter (Lu and Hense, pers. comm.).

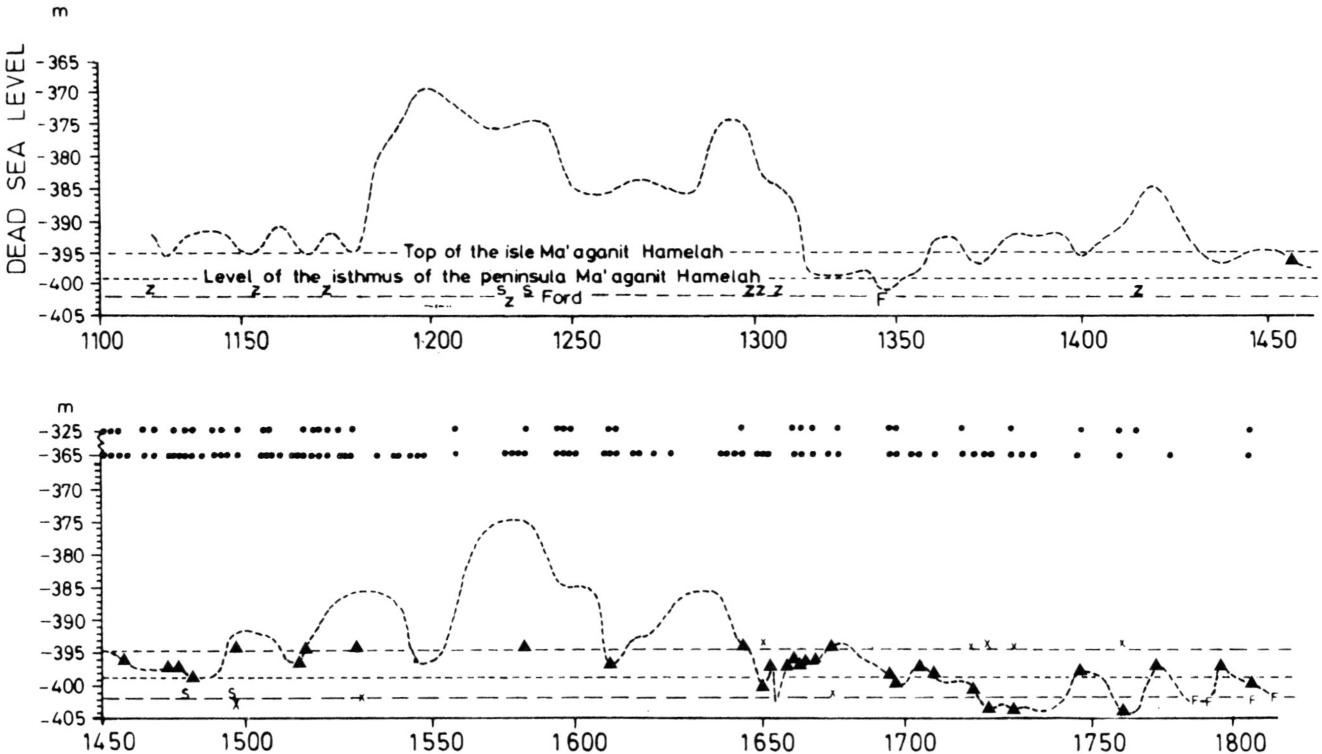

Figure 19.3 Variations of the Dead Sea level between 1115 and 1800 after historical (dots) and proxy sources (tree rings); reconstructed from C. Klein, Jerusalem (data after 1800, see Klein and Flohn 1987).

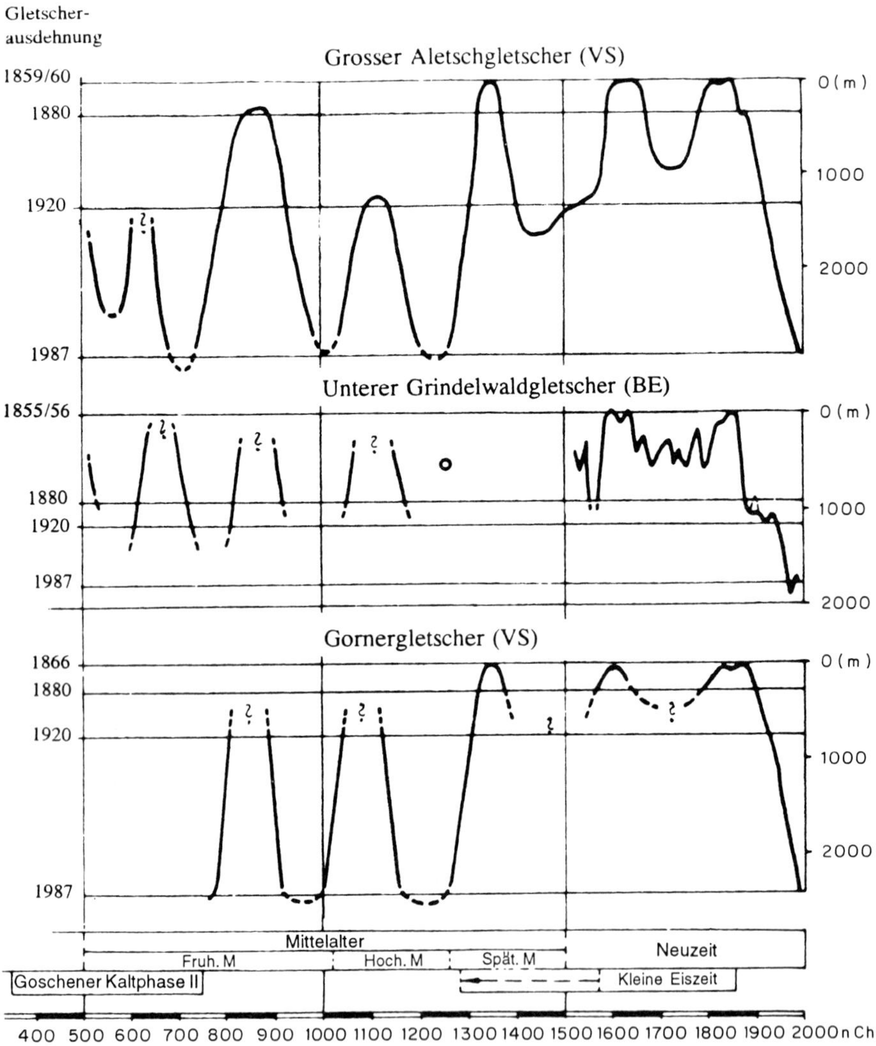

Figure 19.4 Variations of three Swiss glacier tongues since A.D. 1000, between the levels of 1850 (above) and 1987 (below) (Furrer 1991). Open circle equals large expansion.

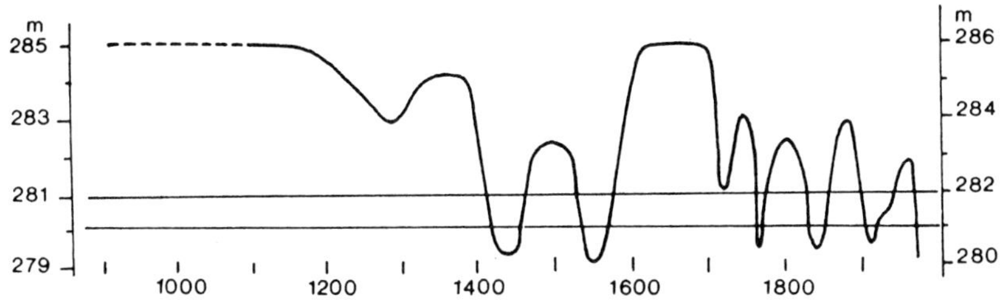

Figure 19.5 Variations of the Lake Chad level during the last millenium (after 1980, shrinking to 278 m !) (after Maley 1981).

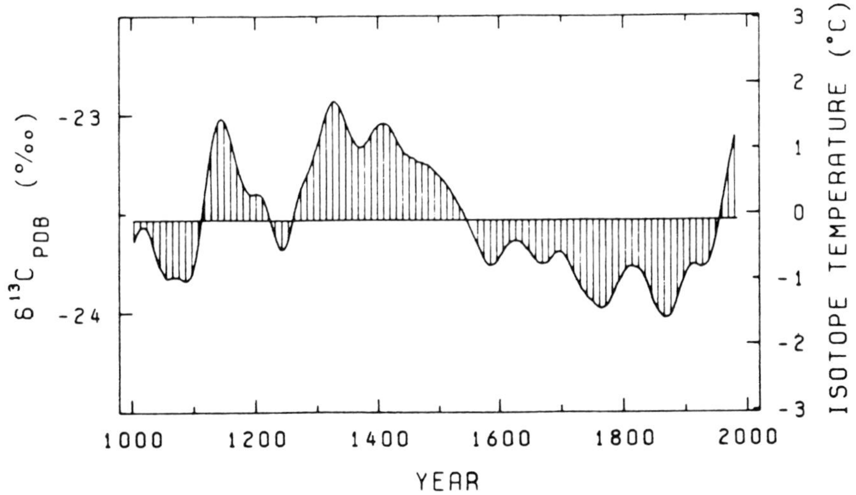

Figure 19.6 Isotope ($\delta^{13}C$) temperatures, smoothed by a low-pass (100a) Gauss filter, derived from late wood cellulose of firs from the Black Forest (S. Germany) between A.D. 1004 and 1980, representing August temperatures (Lipp et al. 1991). After 1850, nonlinear corrections are due to human impact.

after about A.D. 1500. The first two centuries of this millennium appear to have been relatively warm, partly dry, with only a few severe winters. Around 1160, climatic unrest appears to have increased; severe gales were observed in the North Sea (Lamb 1991) where the coasts, at that time, were not yet protected by dikes; after a long period of retreat since the ninth century, alpine glaciers advanced again. Between 1160 and 1350, alpine glaciers advanced, perhaps with a peak in a short cold period after 1310. In the 14th and 15th centuries, there were several intense fluctuations on a decadal scale, including (a) catastrophic cool/wet summers around 1315 and 1340, (b) a unique fast sea-ice cover in the Baltic (1322), which allowed non-ship traffic from Riga and Rostock to Sweden, and (c) a record summer flood in June, 1342, throughout the greatest part of Germany. The great plague of 1348 along with frequent locust swarms

(Camuffo and Enzi 1991), which reached Central Europe and Italy from the southeast, reduced the population by 25%–40%. While a direct climatic cause seems unlikely, the passive migration of locusts with the wind indicates a predominance of easterly winds during summer, which are rare today.

Especially between 1420 and 1460, a number of excessive climatic anomalies occurred (Lamb 1987). They were accompanied by large-scale anomalies, just like the previous century.

After a more or less normal period between 1480 and 1550 (Flohn 1979), the first maximum of the "Little Ice Age"—excluding here its forerunners around 1320 and 1430—started around 1563 (Flohn 1949) with a series of advancing glaciers, cool/wet summers, and cold winters. It reached its first peak in the Alps between 1620 and 1640. After a short recession (for complete details, see Pfister 1984), the years 1680–1699 brought an unprecedented series of very cold winters, accompanied partly by normal or even warm summers. The 18th century saw several new glacier advances, which were caused by cool and wet summers with frequent snow at high levels, especially around 1755 and 1770. In Europe and North America, unusually hot summers were followed by a unique series of five cold summers, 1813–1817. The last two were probably triggered by the greatest historical volcanic eruption, i.e., Tambora (Indonesia) in 1815. This evolution led to the last peak of the Little Ice Age, between 1820 and 1850, with Alpine glaciers reaching the same level as around 1640.

After 1855, a period of general warming commenced: it was indeed world-wide, as evidenced by the retreat of glaciers (and permafrost) in nearly all high mountains and, at least since 1900, in tropical mountains as well. In Europe, it was briefly interrupted by two cooler episodes, around 1890 and 1920. The latter coincided with a unique warming trend in the Arctic, around 1921. This general warming peaked around 1940 and was followed by a somewhat cooler period in the Northern Hemisphere, which ended around 1970. Since then, most regions of the globe have been affected by a continuous warming, which is now accompanied by an increasing atmospheric water vapor content and an intensification of the hydrological cycle and the circulation of the entire troposphere (Flohn et al. 1992).

Long proxy series from North America show some positive correlations but, before 1600, not in all details; annual maps of temperature anomalies for Europe and western North America have been published by Schweingruber et al. (1991). Climatic evolution in the subtropical and tropical belt, as well as in the Far East, differs from that in Europe in many details: European climatic evolution cannot be generalized to include the Southern Hemisphere, with the exception of New Zealand (Figure 19.7). This is probably due to the most widespread and effective climatic anomaly: the El Niño Southern Oscillation, which controls the climate of the greater part of the tropical and subtropical belts for periods of one to two years, at 3- to 7-yr intervals. Its influence on Europe, however, is in most cases remarkably weak. Its frequency and severity is difficult to quantify before 1850; the most severe case, 1982/1983, has hardly been surpassed, and since 1525, only seven cases have been described as being of comparable intensity (Quinn et al. 1987). Available data suggest that the event 1877/1878

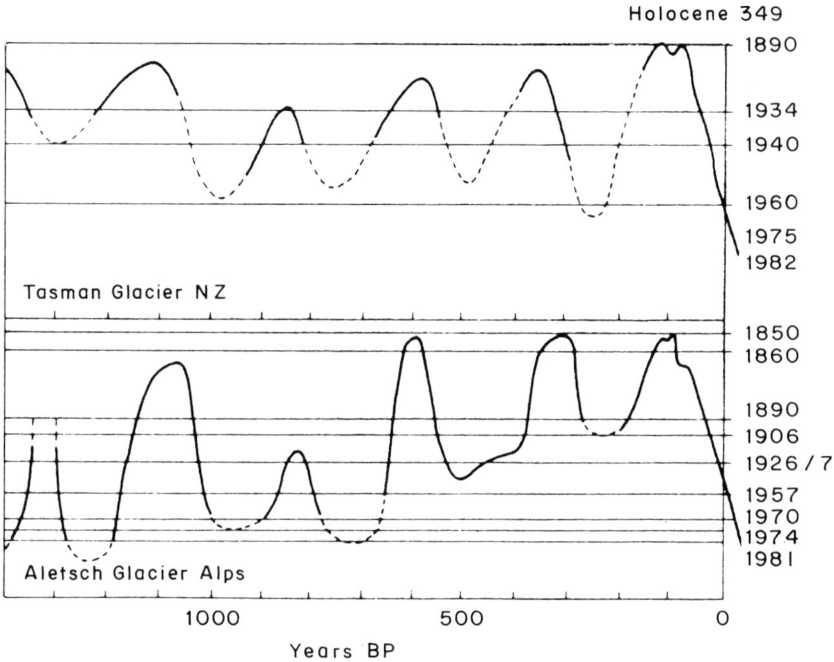

Figure 19.7 The fluctuations of the Tasman glacier, New Zealand, over the last 1500 years compared with those of the Grosser Aletsch (from Gellatly et al. 1985).

was of about equal intensity, with 1891 and 1925/1926 being distinctly weaker; very little is known about earlier events (1578, 1728, 1791, and 1828). At a first glance, the ice core record from Quelccaya (Peru, 14°S) seems to indicate some coincidence, but the century-long fluctuations (Figure 19.8) are certainly not coherent with the Pacific events; prevailing westerly winds suggest a moisture source over the Atlantic.

Comparative interpretation of existing data shows, at the century scale, a fairly comparable evolution of climate during the last millennium (cf. Grove 1988). The Medieval Optimum (probably A.D. 900–1200) was followed by a period of alternating cold/wet periods with strong meridional exchange—indicated by a great frequency of anomalous years with opposite characteristics—between about 1250 and 1550. From 1570 until 1850, the Little Ice Age dominated, with several peaks (1570–1620, 1680–1700, around 1755, and 1810–1850), interrupted by several more genial periods. After 1850, the transition to a warmer period began; it culminated around 1940 and was followed by a rather dramatic evolution after about 1960, which was probably triggered by growing anthropogenic effects.

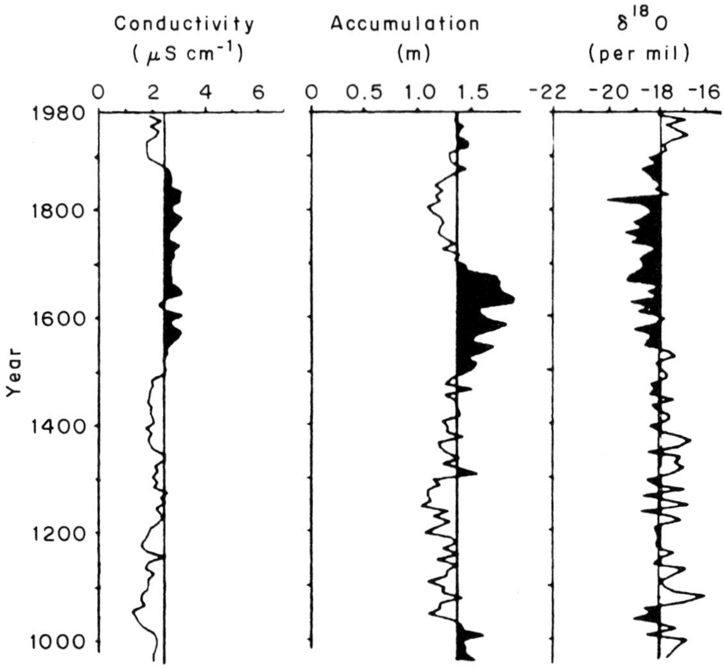

Figure 19.8 Decadal variations in conductivity, accumulation, and $\delta^{18}O$ for the past 1000 years at Quelccaya (after Thompson et al. 1986).

SURVEY OF PALAEOCLIMATIC THEORIES

Since this report is limited to the last thousand years, all longer time scales—including variations of the Earth's orbit—can be safely excluded. Here, the essential time scale is that of decades and centuries. Therefore, a gradual cooling of 5°C over 10,000 years amounts to a marginal change of 0.05°C per century (4–5 human generations), far below any perceptible level.

One must not, however, fail to consider the occurrence of *abrupt* climatic changes (here mainly on a time scale of 1–10 years), as has been observed (Flohn 1986), e.g., in the long and famous records of the Nile floods. Such events were also recorded in the instrumental period: short-lived cooling of most oceans around 1902, Arctic warming around 1921. The most recent event of this kind was the great Atlantic salinity anomaly in 1968–1982 (Dickson et al. 1988), which perhaps had some predecessors since 1962. All of these events demonstrate that "chaotic" properties of nonlinear dynamics control the atmosphere, the ocean, and its interface (see Appendix 19.1).

On the other hand, it is impossible to view an expansion of Arctic drift ice or a deviation of ocean currents as "causes" of climatic change. The climatic system combines atmosphere, hydrosphere (including oceans), cryosphere (and, to some degree, terrestrial and marine biosphere), and their multiple interactions into *one* system. Each of these interactions is an internal process *within* the system. External effects include volcanic eruptions (Lamb 1977) and, in particular, anthropogenic changes (Lamb 1982; Flohn and Fantechi 1984) such as the conversion of forests into arable land, the drainage of bogs and swamps, the creation of huge water reservoirs, or the burning of fossil fuels.

POSSIBLE PHYSICAL PROCESSES

The sequence given above (see section on CLIMATIC HISTORY OF THE LAST MILLENNIUM) was observed (with minor alterations) in the area surrounding the Northern Atlantic, including eastern North America and Europe. Yet one must not overlook the fact that this area embraces only about 10% of the entire globe. The interpretation of the sequence depends, to a large extent, on *internal* processes in the area of the Northern Atlantic and the adjacent Arctic; no evidence exists (until now!) for effects arriving from the Tropics, even if their area is much greater than that of the Arctic (see Appendix 19.1).

Three *abrupt events* characterize the 20th century climatic evolution of the Atlantic/Arctic area outlined above. The first is a general cooling of the ocean surface, which peaked around 1902 (Bottomley et al. 1990) with an amplitude on the order of 0.3° to 0.6°C—in some areas even more. Large portions of the Pacific were similarly affected, though it has been suggested (cf. Ellsaesser et al. 1986) that this was probably caused by the fast transition from sailing to steam ship traffic. Regional differences and comparison of several parameters, however, do not support this hypothesis; the simultaneous drop of the Nile runoff by 25% (around 1899), representing large-scale changes of rainfall in Ethiopia, seems to indicate that this event may be real.

The second event was the rapid warming of the Arctic between 1918–1922. This warming was especially impressive along the western coasts of Greenland, where winter temperatures rose about 6°–8°C. To a lesser degree, this warming spread all over the Arctic during the following years and also initiated the marked warming of the Northern Hemisphere, which lasted until 1938 (Scherhag 1936, 1939).

The third event occurred in 1961/1962. Here, a sudden change in salinity and in the potential temperature of the Labrador Sea below the mixing layer—probably initiating the "Great Salinity Crisis" of the North Atlantic (Dickson et al. 1988) between 1967 and 1982—occurred simultaneously with an abrupt drop in temperatures of the lower troposphere (500/1000 mb layer) in the entire Arctic (Flohn 1986), lasting until 1975. Even more conspicuous was the unique occurrence of a 6- to 8-month anomaly in the equatorial Indian Ocean and East Africa. This event was similar to the El Niño Southern Oscillation of the Pacific. During an entire rainy season, East Africa received 200%–600% of the normal rainfall over an area of at least

$6 \times 10^6 km^2$, extending even to the Seychelles and Zanzibar, simultaneous with unusually dry conditions in western Indonesia and with anomalous winds over the ocean (for details, see Flohn 1987). No parallel event is known since the first regular observations at the coast of East Africa (1891). The simultaneity of such very rare but large-scale extreme events deserves a more detailed study, even if no geophysical teleconnection mechanism can be suggested as yet. During January/February 1963, a quite unusual reversal of the atmospheric circulation above the Atlantic (Icelandic High, Azores Low) occurred, together with a long-lasting severe European winter.

In recent years, several hints as to unexpected processes, which may have important climatic consequences, have been found in the subarctic and arctic Atlantic. Under high wind conditions, open water is strongly mixed—with cooling, increasing salinity and density. A broken sea-ice cover (usually with less than 20% open leads) tends to freshen the surface water and to protect its heat content. These processes lead to oceanic frontogenesis in the marginal ice zone, and together with Ekman pumping in heavy gales, this leads to a deep vertical mixing and to the formation of the North Atlantic Deep Water, which follows a path along the western side of the Atlantic into the intermediate depths of the large-scale abyssal circulation. This southward flow occurs only stepwise, due to the marked sill of the Faroe-Greenland ridge at depths not exceeding 400–600 m. Unfortunately, our knowledge of these processes is hampered by the unavailability of long observational series from the deep ocean.

Over the last few years, Labitzke and van Loon (1990) have shown that a combination of the 11-yr solar activity cycle and the quasibiennial period first observed in the stratospheric wind pattern in the Tropics does indeed have a noticeable effect on climate, at least in some regions. More important, century-scale fluctuations of solar activity have been discovered by their simultaneous modulation of the production of radionuclides (^{10}Be, ^{14}C) from cosmic rays during the last 8–10,000 years (Pecker and Runcorn 1990; Stuiver et al. 1991). During the last millennium, three marked minima of sunspot numbers have been noted and named after Maunder (A.D. 1645–1715), Spörer (1420–1520), and Wolf (1290–1340).

Regarding a possible correlation between these ^{14}C maxima (sunspot minima) and climate, Wigley and Kelly (1990) noted nearly simultaneous cold periods in the European area with peaks around 1330, 1460, and 1680. Based on a global survey of Holocene glacier fluctuations (Röthlisberger 1986), they found 14 century-scale cool periods with advancing glaciers similar to the Little Ice Age. However, only in about half of them could a significant relation to ^{14}C maxima be found. They considered this result to be suggestive of a ^{14}C-climate link, but their argument is not convincing. Figure 19.9 shows independent measurements of $\Delta^{14}C$ from tree-rings and of ^{10}Be from a South Pole core.

Using the same glacier data, but with a better time resolution, Stuiver et al. (1991) found *no* unequivocal confirmation of a solar–climate link. Indeed, the variation of the solar–constant with sunspot activity shows a quite weak change of only 0.03%,

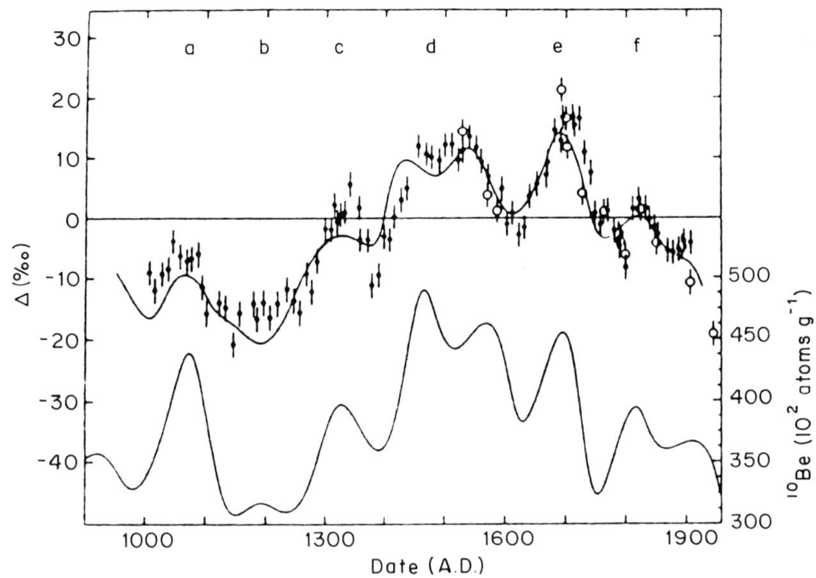

Figure 19.9 Letters a–f: solar epochs named after Stuiver (a), Medieval Warm (b), Wolf (c), Spörer (d), Maunder (e), and Dalton (f). Upper curve: Courier analysis of 1200 ^{14}C measurements; dots = high precision data of Stuiver and Quay (1980); vertical error bars = 1 σ. Lower curve: 10Be data from an Antarctic ice core (Raisbeck et al. 1990), after McHargue and Damon (1991).

i.e., much smaller than needed for temperature variations in the order of those between the Little Ice Age and this century. Any hypothesis of an internal amplification (within the climate system) of this small effect remains speculative.

Of particular interest is the possible role of large volcanic eruptions, especially of the sulfur gases injected into the stratosphere (Junge layer at about 20–25 kms altitude), where they are converted into small particles with a lifetime of 1–3 years. In contrast to this, mass ejections of silicate-rich ashes and lava are only of local or regional significance; this omits some climatically ineffective eruptions from the catalogues, while others with a significant effect remain unknown. These cases can be detected by the acidity of particles preserved in the ice cores of Greenland (Hammer et al. 1980) and the Antarctic (Moore et al. 1991), measured by electrical conductivity or pH content. Unfortunately, the dominant zonal transport of these particles causes a strong dependency on latitude, e.g., large (and effective) eruptions of tropical volcanoes only give a weak signal in these polar records. Cress and Schönwiese (1990) critically investigated the relations between surface temperature and several volcanic indices, which have varied during the last century around – 0.5°C and are thus significant. Figure 19.10 shows a space–time pattern of the volcanic signal, indicating significant temperature effects only in higher latitudes, not in the Tropics. Figure 19.11

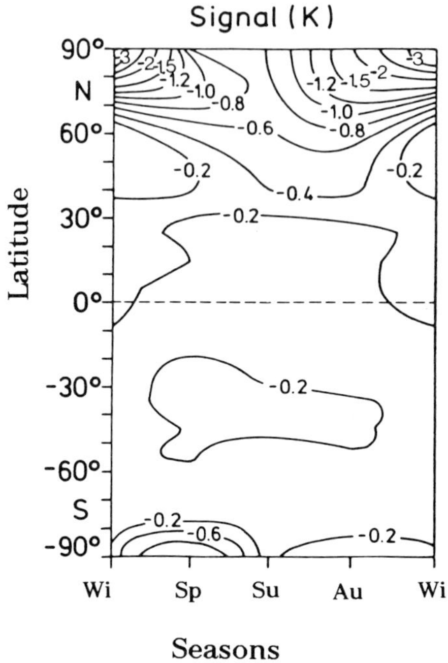

Figure 19.10 Statistical estimates of volcanic-induced maximum surface temperature signals (in K) for the period 1880–1984, based upon a 5-dimensional regression model (Cress and Schönwiese 1990).

gives a latitude-corrected acidity series from Greenland for the period since 1500. The peaks of the filtered thick lines show several peaks between 1570 and 1680, as well as around 1820 and 1890, which correlate rather well with the coldest periods of the Little Ice Age in Europe and North America. The climatic role of the greatest historical eruption (Tambora, in 1815) in the two following summers has been frequently quoted. If the conductivity record from Quelccaya (Figure 19.8) is representative for the Tropics, then the sudden transition from low to high values around 1550, the latter lasting until 1900, seems to indicate a drastic increase of volcanic activity around 1550.

CONCLUSIONS: WHAT CAN WE LEARN FROM THE RADICAL CLIMATIC BREAK IN THE FOURTEENTH CENTURY?

While climatic evolution during the last millennium is fairly well documented for mid-latitude continents at both flanks of the Atlantic, long series from the tropical–subtropical latitudes and from the Southern Hemisphere are rare. One of the best examples of coincidence is given by Figure 19.7, which compares the fluctuations of

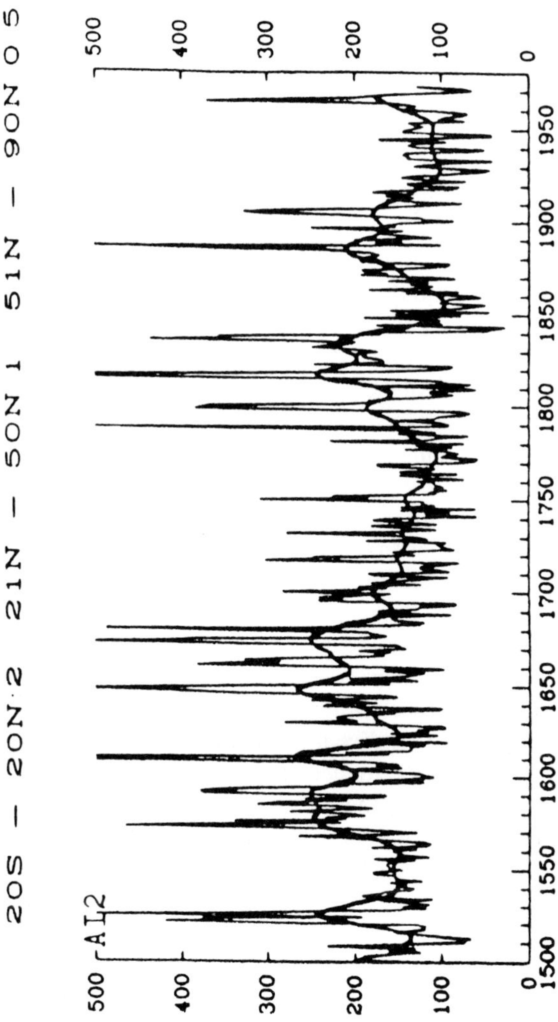

Figure 19.11 Annual averages and 30 point low-pass filtered values of the Crete acidity index (A12) for the period 1500–1972 (latitudinal corrected values, as above the curve) (Cress and Schönwiese 1990).

the Tasman Glacier from New Zealand (lat. 44°S) with the largest Alpine glacier (relative intensity of the peaks). Disregarding minor deviations at the decadal time scale, the synchronism of glacier advances and retreats, at a century-scale, confirms a fairly good global correlation. Unfortunately it is impossible to reconstruct the atmospheric and oceanic circulation before the beginning of our century with any degree of confidence.

The historical data from Greenland and Iceland seem to indicate a relatively genial climate during the first centuries, which allowed the Viking colonization of Iceland (which was then widely forested) and southwestern Greenland, as well as the simultaneous Eskimo expansion into northern Greenland (Thule culture, based on whale fishing) in the 10th century. Severe ice advances in the East Greenland Current, around 1330, interrupted the traffic between the colonies of Iceland and Greenland, while disastrous clashes between the advancing Eskimos and Vikings occurred. Together with famine and epidemics, this ultimately led to the tragic extinction of the early isolated Viking settlements. This apparently regional climatic shock marked the end of the Medieval Optimum and the first forerunners of the Little Ice Age. According to our present knowledge, all events between 1310 and 1350 (see the above section on CLIMATIC HISTORY OF THE LAST MILLENNIUM) could have been triggered by first a local, then regional advance of the Arctic drift-ice over an estimated distance of more than 1000 km. Yet what mechanism was responsible for that? Its consequence could have been a marked increase in baroclinicity above the Atlantic, with a high frequency of cold air outbreaks, accompanied by increasing storminess.

Is it possible that the Medieval Optimum, during which the Arctic sea-ice cover was probably even smaller than in the 1940s or now, could be a likely scenario for the expected global warming? Indeed, we know very little of the state of the Tropics at this time. It is unlikely, at this time, that the hydrologic cycle was intensified as much as now; the relatively dry climate of Europe, documented in the 11th century, does not support such a hypothesis. Furthermore, there are no implications nor physical causes at hand to support higher sea surface temperatures in lower latitudes at that time.

The occurrence of decade- and century-scale climatic fluctuations, as evidenced by the Little Ice Age with its forerunners and its maxima around 1680 and 1820, had a great impact on society. They occurred (at least in areas on the order of 10%–20% of the global surface) nearly simultaneously (accompanied by temperature drops in the order of 1–2°K during the vegetation period) and lasted (with minor decade-scale fluctuations) about 300 years. During these cold periods, an increasing frequency of extreme years has been documented (e.g., Pfister 1984); they are separated by more quiet, genial intervals, like the Medieval Optimum, lasting perhaps somewhat longer. An external influence by solar activity or volcanism is—if it exists at all—weak at the century scale and much smaller than the expected climate changes to be triggered by humans during the next century. The present warming can hardly be compared with the Medieval Optimum, since the present climatic evolution (Flohn et al. 1990, 1992) occurs under different geophysical conditions.

A possible natural mechanism for abrupt climatic changes, at an interannual to decadal scale, has been suggested within the subarctic Atlantic and around Greenland, where intense vertical exchange processes with distinct variations of salinity, density, and potential vorticity in the upper 200–600 m have been found. These processes interact with the drifting sea-ice and thus with the surface air temperatures of the Arctic. Due to the lack of long-term observations and a coherent network of soundings below the ocean surface, no details of this mechanism can be given.

For further discussion, several theses shall be formulated:

1. Evidence for nearly simultaneous century-scale global fluctuations of climate is available, with a temperature amplitude of the order 1°–2°K and regional time deviations of a few decades.
2. No single cause can be suggested for these fluctuations, in spite of some coherence with periods of increased volcanic activity and decreased solar activity. A complex origin should be discussed, including an effect of small changes of the surface temperature of tropical oceans via water vapor content.
3. Evidence is available for abrupt climatic changes on a time scale of 20–50 years for hemispheric changes (Dansgaard et al. 1989; Thompson et al. 1986, 1989), and of 2–6 years for regional changes (Flohn 1986, 1987). This is probably the most important problem to be investigated.

APPENDIX 19.1 Abrupt Changes and Extreme Events

Several authors found that the occurrence of extreme events, which must be (at any standard) described as catastrophic, was greater during the cold periods of the Little Ice Age than before or after. Several examples were described by Pfister (1984). Here, I have selected the sequence of five unusually cool/wet summers in Europe, which existed at least in the entire Northern Hemisphere and in the Tropics. They can usually be related to the greatest historically documented volcanic eruption (Tambora, 1815), which was—according to new ice core data from the Antarctic (Moore et al. 1991)—preceded by a hitherto unknown, but likewise severe, eruption in 1808. This might perhaps explain the beginning of extreme years before 1815 and the unusual period of El Niño Southern Oscillation events since 1812, as well as at the Quelccaya ice field in Peru, between 1816 and 1820 (given the chronology is correct). The year "without a summer" in North America and Europe was 1816; in the Alps, the summertime snow-line remained as low as 1800 m locally (2700 m being normal). Earlier examples include the period of 1313–1322, with a sequence of famines (peak 1315) and record-breaking winters (1322/1323), or the peak of the Little Ice Age between 1693 and 1699.

The Little Ice Age occurs, in all detailed records, as a sequence of warmer and colder periods; no objective definition of its beginning and end can be found. From European data, further cold periods between 1425 and 1438 (Lamb 1987), and between 1563 and 1580 (Flohn 1949) are frequently documented. The latter cold period (documented by the famous astronomer Tycho de Brahe, at his observatory near Copenhagen) was accompanied by predominant winds from the southeast, indicating a southward shift of cyclone tracks and blocking highs. Since the temperatures between 1510 and 1530 (Flohn 1979) are similar to recent values, many authors place the beginning of the Little Ice Age around 1550, with the end around 1850, the date of the Last Glacier Maximum in the Alps (Furrer 1991). Then the periods of really severe cold years around 1320 and 1430 can be defined as (pronounced) precursors of the Little Ice Age.

New thoughts are provoked by a high-resolution ice core from the tropical Andes (lat. 14°S, 5670 m, Quelccaya; cf. Thompson et al. 1985, 1986). Estimates of temperature fluctuations are available from $\delta^{18}O$ data and the seasonal precipitation from net snow accumulation rate separated by dry-season (winter) dust layers. Between 1000 and 1490, the annual accumulation rate remains around 120–130 cm water equivalent; it increases abruptly, around 1490, to about 155 cm until 1600 and even to 170 cm between 1600 and 1680; then it decreases between 1710 and 1850 again to values near 120 cm/yr. After 1590, a very high correlation between decadal $\delta^{18}O$ values and Northern Hemisphere temperature estimates (Groveman and Landsberg 1979) has been found, with pronounced minima around 1815 (as above). A universal linear relation between $\delta^{18}O$ and temperature (of the low clouds in which most snow is formed) seems to be doubtful. An application of a relation valid for Greenland suggests cold-season temperature changes between five-year averages near 10°K; this appears to be too high. In contrast, Figure 19.8 suggests that a realistic cooling of ca. 1.3°K occurred on a decadal scale around 1815.

This core confirms the global coincidence of such extreme events, e.g., after 1815. It also indicates that the Little Ice Age began in the Tropics with high precipitation about 50 years earlier and with sinking temperatures about 20 years earlier than in Europe, while its first peak with high winter-time particle fall (caused by high wind speed) began almost simultaneously (~1580). The period of high precipitation lasted only until the cold peak around 1690; later cold peaks around 1750–1770, often the maximum in North America and Greenland, coincided in the Andes with a longer dry period.

If the conductivity record at Quelccaya (Figure 19.8) represents volcanic activity at tropical latitudes, the coincidence of high values between 1570 and 1850 with the Northern Hemisphere Little Ice Age suggest a causal relationship. As a working hypothesis, a cooling of the tropical sea-surface temperatures by about 0.2°–0.3°C (see Figure 19.10) could be accompanied by a drop of tropical ocean evaporation in the order of 7%–10%. This would be inversely comparable with the recent rise of tropical sea-surface temperatures near 0.5°C, of evaporation near 16%, and of precipitable water (= vertically integrated water vapor content of the air) +20%–30% (Flohn et al. 1992). However, this hypothesis does not agree with accumulation (Figure 19.8).

Experience over the last 40 years has shown that a small increase in tropical ocean temperatures (on the order of 0.5°C) leads to a marked increase in precipitable water in the atmosphere and oceanic evaporation, both on the order of 20%. These increases have been triggered, as a natural feedback, by anthropogenic input of "dry" greenhouse gases (such as CO_2, CH_4, N_2O, O_3, and the chlorofluoromethanes) and lead, via an increasing tropospheric temperature gradient equator-pole, towards rising baroclinic processes in the atmosphere and an intensification of the atmospheric circulation and tropospheric winds (Flohn et al. 1990). The role of the Tropics in these processes has been frequently underrated; we should never forget that the tropical oceans cover (between lat. 20°N and 20°S) 78% of an area near 174×10^6 km^2, compared with only 31×10^6 km^2 of both polar caps at lat. 70°–90.° Even more essential is the exponential rise of water vapor content at sea-surface temperatures of 27°–29°C, due to the Clausius–Clapeyron equation; minimal changes of 0.5°C are responsible for this intensification of the hydrologic cycle.

Failure to find a convincing explanation for global temperature variations on the order of 1°–2°C, as during the last millennium before the Industrial Revolution, suggests a substantial contribution of the tropical hydrologic cycle (even at the interannual scale, as indicated by the irregular occurrence of El Niño Southern Oscillation events) to the decade- and century-scale variation of climate. It is hardly possible to answer the question pertaining to the ultimate cause. In such a meshwork of (mostly nonlinear) internal feedback mechanisms, the original trigger may be quite small and hardly detectable, especially (as in these cases) in the Southern Hemisphere. This ad-hoc working hypothesis may open a new perspective on the climate system, in addition to the unconvincing hypotheses of an (external) solar or volcanic effect.

The strong positive feedback above the equatorial oceans (via convective cloud heights, precipitation, mid-tropospheric warming, and decreasing static stability)—confirmed by model simulations (Hansen et al. 1984)—needs only a weak triggering within white noise. Once initiated, the warming process may be maintained for a long time, simultaneous with periods of frequent and strong El Niño events. Indeed, such periods of extraordinary rains—with daily area averages of 20 cm/day above areas up to several 10^6 km^2 (Flohn 1987), releasing latent energy up to 7 times larger than the extraterrestrial solar constant (340 W/m^2), maintained by fields with convergent, moisture-laden low-level winds—can trigger repeated near-global circulation anomalies as in recent years.

However, the climatic system has many degrees of freedom. The second half of the mid-latitude Little Ice Age coincides with a dry period at Quelccaya. The recent increase of moisture content and evaporation at tropical oceans leads to an intensification of the mid-latitude circulation with prevailing zonal winds and continental warming instead of a prevailing meridional exchange and cooling as during the Little Ice Age. It is not to be expected that a short-living anomaly above a limited area should lead, in each case, to similar consequences (teleconnections) all over the globe.

REFERENCES

Alexandre, P. 1987. Le Climat en Europe au Moyen Age. Vol. 24 of Recherches d'historie et de sciences sociales. Paris: Ed. de l'École des Hautes Études en Science Sociales.

Berger W.H., and J. Labeyrie, eds. 1987. Abrupt Climatic Changes: Evidence and Implications. Dordrecht: Reidel.

Bottomley, M., C.K. Folland, J. Hsiung, R.E. Newell, and D.E. Parker. 1990. Global Ocean Surface Temperature Atlas (GOSTA). London: HMSO.

Burkhardt, T., and A. Hense. 1984. On the Reconstruction of Temperature Records from Proxy Data in Mid-Europe. *Arch. Met. Geoph. Biocl. B* **35**:341–359.

Camuffo, D., and S. Enzi. 1991. Locust invasion and climatic factors from the middle ages to 1800. *Theor. Appl. Clim.* **43**:43–73.

Cress, A., and C.-D. Schönwiese. 1990. Vulkanische Einflüsse auf die bodennahe und stratosphärische Lufttemperatur der Erde. *Berichte d. Inst. für Meteorol. und Geophys. Univ. Frankfurt am Main*, No. 82.

Dansgaard, W., J.W.C. White, S.J. Johnson. 1989. The abrupt termination of the Younger Dryas climate event. *Nature* **339**:532–534.

Dickson, R.R., Y. Meincke, S.-A. Malmberg, and A.J. Lee. 1988. The great salinity anomaly in the northern North Atlantic, 1968–1982. *Progr. Oceanogr.* **20**:103–151.

Ellsaesser, H.W., M.C. McCracken, J.J. Welton, and S.L. Grotch. 1986. Global climatic trends as revealed by recorded data. *Rev. Geophys.* **24**:745–792.

Flohn, H. 1949. Klima- und Witterungsverlauf in Zürich im 16. Jahrhundert. *Vierteljahresschr. Naturforsch. Ges. Zürich* **94**:28–41.

Flohn, H. 1979. Zwei Bayrische Wetterkalender aus der Reformationszeit. *Wiss. Mitt. Meteor. Inst. Münch.* **35**:173–177.

Flohn, H. 1986. Singular Events and Catastrophes Now and in Climatic History. *Naturwissen.* **73**:136–149.

Flohn, H. 1987. East African rains of 1961/1962 and the abrupt change of the White Nile discharge. *Palaeoecol. Afr.* **18**:3–18.

Flohn, H. 1991. Towards a physical interpretation of the end of Holocene moist period in the Near East. *Erdkunde* **45(3)**:163–167.

Flohn, H., and R. Fantechi, eds. 1984. The Climate of Europe: Past, Present and Future. Dordrecht: Reidel.

Flohn, H., A. Kapala, H.-R. Knoche, and H. Mächel. 1990. Recent changes of the tropical water and energy budget and of midlatitude circulations. *Clim. Dynam.* **4**:237–252.

Flohn, H., A. Kapala, H.-R. Knoche, and H. Mächel. 1992. Water vapour as an amplifier of the greenhouse effect: new aspects. *Meteor. Zeitschr. N.F.* **1:**122–138.

Furrer, G. 1991. 25,000 Jahre Gletschergeschichte. *Vierteljahresschr. Naturforsch. Ges. Zürich* **136:**1–51.

Gellatly, A.F., F. Röthlisberger, and M.A. Geyh. 1985. Holocene glacier variations in New Zealand (South Island). *Z. f. Gletscher. Glazialgeol.* **21**:265–273.

Grove, J.M. 1988. The Little Ice Age. London: Methuen.

Groveman, B.S., and H.E. Landsberg. 1979. Reconstruction of the Northern Hemisphere Temperature: 1579–1880. Meteorology Program, Publ. No. 79–181. College Park, MD: Univ. Maryland.

Hammer, C.U., et al. 1980. Greenland ice-sheet evidence of postglacial volcanism and its climatic impact. *Nature* **288**:230–235.

Hansen, J.E., A. Lacis, D. Rind, G. Russel, P. Stone, I. Fung, R. Ruedy, and J. Lerner. 1984. Climate sensitivity: analysis of feedback mechanisms. In: Climate Processes and Climate Sensitivity, ed. J.E. Hansen and T. Takahashi. *Geophys. Monog.* **29**:130–163. Washington D.C.: American Geophysical Union.

Hughes, M.K., P.M. Kelly, J.R. Pilcher, and V.C. La Marche, eds. 1982. Climate from Tree Rings. Cambridge: Cambridge Univ. Press.

Jacoby, G.C., and R. d'Arrigo. 1989. Reconstructed Northern Hemisphere annual temperature since 1671 based on high-latitude tree-ring data from North America. *Clim. Change* **14**:39–59.

Klein, C., and H. Flohn. 1987. Contributions to the knowledge of the fluctuations of the Dead Sea level. *Theor. Appl. Clim.* **38**:151–156.

Labitzke, K., and H. van Loon. 1990. Sonnenflecken und Wetter. *Geowissen.* **8**:1–6.

Ladurie, E.R. 1971. Times of Feast, Times of Famine: A History of Climate since the Year 1000. New York: Doubleday.

Lamb, H.H. 1977. Climate: Present, Past and Future, vol. 2. London: Methuen.

Lamb, H.H. 1982. Climate, History and the Modern World. London: Methuen.

Lamb, H.H. 1987. What can historical records tell us about the breakdown of the medieval warm climate in Europe in the fourteenth centuries—an experiment. *Contr. Atmos. Phys.* **60(1)**:131–143.

Lamb, H.H. 1991. Historical Storms of the North Sea, British Isles and Northwest Europe. Cambridge: Cambridge Univ. Press.

Lipp, J., P. Trimborn, P. Fritz, H. Moser, B. Becker, B. Frenzel. 1991. Staple isotopes in tree ring cellulose and climatic change. *Tellus* **43B**:322–330.

Maley, J. 1981. Etudes palynologiques dans le bassin du Tchad et paléoclimatology de l'Afrique nord-tropicale de 30,000 ans a l'époque actuelle. Trav. Doc. ORSTOM 129. Paris: Organisation de Recherchés Scientifiques dans les Territoires d'Outre-Mer.

McHargue, L.R., and P.E. Damon. 1991. The global Beryllium 10 cycle. *Rev. Geophys.* **29**:141–158.

Moore, J.C., H. Narita, and N. Maeno. 1991. A continuous 770-yr of volcanic activity from east Antarctica. *J. Geophys. Res.* **96D**:17353–17359.

Mörner, N.A., and W. Karlén W., eds. 1984. Climate Changes on a Yearly to Millennial Basis. Dordrecht: Reidel.

Pecker, J.C., and S.K. Runcorn, eds. 1990. The Earth's Climate and Variability of the Sun over Recent Millennia: Geophysical, Astronomical and Archeological Aspects. *Phil. Trans. R. Soc. Lon. A* **330**:395–687.

Pfister, C. 1984. Klimageschichte der Schweiz: 1525–1860, vols. 1–2. Academia Helvetica 6. Bern: P. Haupt Verlag.

Quinn, W.H., V.T. Neal, and S.E.A. de Mayolo. 1987. El Niño occurrences over the past four and a half centuries. *J. Geophys. Res.* **92**:14449–14461.

Raisbeck, G.M., et al. 1990. ^{10}Be and δH in polar ice cores as a probe of the solar variability's influence on climate. *Phil. Trans. R. Soc. Lon. A* **330**:463–470.

Röthlisberger, F. 1986. 10,000 Jahre Gletschergeschichte der Erde. Aarau, Switzerland: Sauerländer.

Scherhag, R. 1936. Zunahme der atmosphärischen Zirkulation in den letzten 25 Jahren. *Ann. Hydrograph.* **67**:397–407.

Scherhag, R. 1939. Die gegenwärtige Milderung der Winter und ihre Ursachen. *Ann. Hydrograph.* **67**:292–303.

Schweingruber, F.H., K.R. Briffa, and P.D. Jones. 1991. Yearly maps of summer temperature in Western Europe from A.D. 1750 to 1975 and Western North America from 1600 to 1982. *Vegetatio* **92**:5–71.

Stuiver, M., T.F. Braziunas, B. Becker, and B. Kromer. 1991. Climatic, solar, oceanic, and geomagnetic influences on late glacial and holocene atmospheric ^{14}C/^{12}C changes. *Quat. Res.* **35**:1–24.

Stuiver, M., and R.D. Quay. 1980. Changes in atmospheric Carbon 14 attributed to a variable sun. *Science* **207**:11–12.

Thompson, L.G., E. Mosley-Thompson, J.F. Bolzan, and B.R. Koci. 1985. A 1500-yr record of tropical precipitation in ice cores from the Quelccaya ice cap, Peru. *Science* **229**:971–973.

Thompson, L.G., E. Mosley-Thompson, W. Dansgaard, and P.M. Grootes. 1986. The Little Ice Age as recorded in the stratigraphy of the tropical Quelccaya ice cap. *Science* **234**:361–364.

Thompson, L.G., E. Mosley-Thompson, M.E. Davis, J.F. Bolzan, J. Dai, T. Yao, N. Gundestrup, X. Wu, L. Klein, and Z. Xie. 1989. Holocene-Late Pleistocene climate ice core records from Qinghai-Tibetan plateau. *Science* **246**:474–477.

von Rudloff, H. 1967. Die Schwankungen und Pendelungen des Klimas in Europa seit Beginn der regelmässigen Instrumentenbeobachtungen (1670). Braunschweig: Viehweg.

Wang, R., et al. 1991. An approach to reconstruction of temperature on a seasonal basis using historical documents from China. *Int. J. Climatol.* **11**:381–392.

Wigley, T.M.L., M.J. Ingram, and G. Farmer, eds. 1981. Climate and History. Studies in Past Climates and Their Impact on Man. Cambridge: Cambridge Univ. Press.

Wigley, T.M.L., and P.M. Kelly. 1990. Holocene climatic change, ^{14}C wiggles and variations in solar irradiance. *Phil. Trans. R. Soc. Lon. A* **330**:547–560.

20

Can Paleoclimatic and Paleoecological Analyses Validate Future Global Climate and Ecological Change Projections?

S.H. SCHNEIDER[1]
National Center for Atmospheric Research, Boulder, CO 80307–3000, U.S.A.[2] and
Department of Biological Sciences, Stanford University, Stanford, CA 94305, U.S.A.

ABSTRACT

Heat trapping by greenhouse gases is a well-established principle. Human additions to global greenhouse gases have added some 2–3 W m^{-2} of surface-atmosphere heating over the past century. Translating such heating into precise estimates of global surface temperatures increases, however, are controversial since they involve assumptions about "feedback mechanisms" within the physical/chemical/biological components of the climate system. Most assessments assign greater than 50% change to the projection that a global warming of 1°–5°C is being built into the 21st century climate, and that this implies rates of change some 10 to 50 times faster than most sustained, global average rates of temperature change since the end of the Pleistocene. The implications of such changes for natural and managed ecosystems are potentially serious.

While a fairly strong consensus of knowledgeable climatologists holds to the view that such a range of potential warming is likely, few such scientists express confidence in the time-evolving, regionally specific projections of climate change so essential to water or agricultural planning or ecological impact analyses or biological conservation efforts. Moreover, most climatic projections typically occur over 500 × 500 km scale "grid boxes" whereas most field studies or impact assessments occur on scales orders of magnitude smaller. Static and dynamic techniques to bridge this scale mismatch in climatic models are briefly

[1] Any opinions, findings, conclusions, or recommendations expressed in this article are those of the author and do not necessarily reflect the views of the National Science Foundation. Current address: Dept. of Biological Sciences, Stanford University, Stanford, CA 94305, U.S.A.

[2] The National Center for Atmospheric Research is sponsored by the National Science Foundation.

Global Changes in the Perspective of the Past
Edited by J.A. Eddy and H. Oeschger

reviewed. I am is cautiously optimistic that some translation of global change projections from grid scale to smaller scales can be made.

Since human-induced climate changes could proceed, on a sustained globally averaged basis, at an order of magnitude faster than Pleistocene/Holocene climatic changes, dramatic and differential response of plant, insect, mammal, and bird species can be anticipated. Thus, a very large restructuring and reorientation of biological communities seems likely. Paleobotanic and other paleoecological data will need to be joined with paleoclimatic data, not to provide direct analogs for future global changes, but rather to help derive and calibrate the climate and ecological models needed for assessing the consequences of human activities, which are currently believed to be responsible for forcing global environmental changes. The unprecedented rapid rate of greenhouse-gas forcing and the need to solve a coupled, transient set of sub-models of atmosphere, oceans, sea ice, glaciers, soils, and biota make paleoclimatic reconstructions primarily useful to validate equilibrium sensitivity of the climate to sustained forcings (e.g., CO_2 doubling).

Finally, although it is a personal value judgement whether the prospect of unprecedented rapid climatic changes and their potential ecological implications justifies current actions to slow down human activities driving such global changes, I believe that cost-effective strategies to slow down the forces of global change already exist and should be implemented to buy time to study the issues further and to make it easier for both natural and managed ecosystems to deal with whatever changes do unfold.

INTRODUCTION

Global change is a problem of great significance for the Earth's future. In particular, it is critical to estimate the implications of human alteration of land characteristics, use of natural resources, or the emission of relatively active gases. These include sulfur dioxide, which can be photochemically converted to particles that might directly reflect sunlight in between or above clouds, thus cooling the climate, or which could also increase cloud albedo (e.g., Twomey et al. 1984). On the other hand, emissions of CO_2, methane, chlorofluorocarbons, and nitrous oxide (gases that can increase the surface temperature by trapping heat through enhancing the so-called "greenhouse effect") have led to a major debate over the magnitude, rate, and implications of such global greenhouse-gas changes on climate. Table 20.1 (Schneider 1992) lists a number of examples of those potential human impacts on climate. The primary methods for projecting these effects have been climatic models, since the magnitude and rate of human alteration to global biological, chemical, and physical characteristics of Earth is often unprecedented relative to those even of the paleoclimatic past. As a result, considerable uncertainty and controversy surround such projections. While it is possible that CO_2 may have been a factor of 5 to 10 larger 100 million years ago in the mid-Cretaceous (e.g., Budyko et al. 1987; Berner et al. 1983; Schneider et al. 1985), any such changes recorded in the paleochemical or paleobotanic records from which we infer paleoclimate reconstructions are for very long time averages of past climates. Such geologic strata cannot conceivably resolve the less-than-hundred-year transient time frame for which major human-induced global changes are projected to occur. Nonetheless, paleoclimate reconstructions are extremely important to help calibrate our understanding of naturally occurring climate change, an understanding

Table 20.1 Summary of the principal human activities influencing climate (Schneider 1992).

Activity	Climatic Effect	Scale and Importance of the Effect
Release of CO_2 by burning fossil fuels	Increases the atmospheric absorption and emission of terrestrial infrared radiation (greenhouse effect), resulting in warming of lower atmosphere and cooling of the stratosphere.	Global: potentially a major influence on climate and biological activity.
Release of CH_4, chlorofluoromethanes, nitrous oxide, carbon tetrachloride, carbon disulfide, etc.	Similar climatic effect as that of CO_2 since these, too, are infrared-absorbing and fairly chemically stable trace gases.	Global: potentially significant influence on climate.
Release of particles (aerosols) from industrial and agricultural practices. SO_2 is primarily concerned since it photochemically converts to sulfuric acid particles.	These sunlight-scattering and absorbing particles (especially soot) could decrease albedo over land, causing warming, and could (especially sulfate) increase albedo over water, causing cooling; they also change stability of lower atmosphere; net climatic effects still speculative, although net cooling effect seems more likely.	Largely regional since aerosols have an average lifetime of only a few days, but similar regional effects in different parts of the world could have nonnegligible net global effects; stability increase may suppress convective rainfall, but particles could affect cloud properties with more far-reaching effects.
Release of aerosols that act as condensation and freezing nuclei. Again, soot or SO_2 released by industrial activities is the primary concern.	Influences growth of cloud droplets and ice crystals; may affect amount of precipitation or albedo of clouds in either direction.	Local (at most) regional influences on quantity and quality of precipitation, but unknown and potentially important change to Earth's heat balance if cloud albedo is altered. Some calculations suggest SO_2 released between 1950 and 1980 opposed much of the northern hemispheric warming trend that otherwise would have been expected from rapid buildup of greenhouse gases during those decades.
Release of heat (thermal pollution)	Warms the surface layers directly.	Locally important now; could become significant regionally; could modify circulation.

that is essential to the credibility of the tools that we need for forecasting human-induced, time-evolving regional changes on Earth.

While this connection between past climates and future climates has long been well understood, a number of people do argue that there are direct metaphors between past "warm worlds" and anthropologically warmed future world (Kellogg and Schware 1981; Budyko 1990). It is my strong belief, however, that there are no credible, clear past warming patterns from which some regional map of paleoclimatic change can be directly applied to estimate a regional pattern of time-evolving climatic changes forced by increasing greenhouse gases (or any other "global change" forcings that are occurring over a century time scale). Once again I repeat that I do not suggest that it is improper to look at paleoclimate data as a calibration check on climate theory or models; in fact I believe quite the opposite. The key point is to use paleoclimatic information in a way that helps to provide credibility to the tools of climatic theory, not to provide a regionally specific metaphor for actual future changes (Schneider 1984).

The three basic climate projections that are needed for global change assessments are: (a) what is the equilibrium, global average sensitivity of the Earth's climate to fixed, global-scale external forcing, such as a doubling of CO_2; (b) what is the equilibrium, regional distribution of climatic anomalies (for example, where will it be wetter and dryer) resulting from equilibrium forcing; and (c) how will these regional climatic changes evolve with time. I believe that the study of the first two questions can be substantially enhanced by paleoclimatic analyses and the use of such data in validating models. However, credible insights into the last question will require a transient set of coupled atmosphere, ocean, land, biosphere, and chemical models. Although subcomponents of such coupled models can, to some extent, be validated by paleoclimate data and model analyses, because of the relatively rapid time scales of global change forcings the problem of projecting regional, time-evolving climatic changes will also require independent validations well beyond that which can be attained through paleoclimatology.

GLOBAL AVERAGE EQUILIBRIUM RESPONSE

Climate Models of the Radiation Balance

(See Schneider 1992 from which this section is adapted)

Consider the simplest possible climate model, a radiation balance for the Earth.

$$S = F \tag{1}$$

is an energy balance which requires that the amount of solar energy input to the system S is, in equilibrium, exactly balanced by the amount of terrestrial radiation energy F that leaves the system.

Translating this radiation balance into a simple climate model involves the energy balance Equation (1), the Stefan-Boltzmann law, where σ is the Stefan-Boltzmann constant (= 5.6696×10^{-8} W m^{-2} k^{-4}) and T_p is the effective black body radiative temperature of the Earth-atmosphere system in °K. Rewriting Equation (1) into the simplest possible climate model takes the form

$$Q\,(1-\alpha) = \sigma T_p^4 . \tag{2}$$

In the left-hand term, Q represents the incoming solar radiation of approximadely 340 W m^{-2} averaged over the Earth's surface (i.e., this is S_o = 1340 W m^{-2} of "solar constant" divided by 4, since the area of the Earth's surface is 4 times the area of the Earth's disk which intercepts sunlight); a is the albedo of the planet. We can define a sensitivity parameter

$$\lambda = \frac{1}{\dfrac{dF}{dT} - \dfrac{dS}{dT}}, \tag{3}$$

which represents how much temperature change the planet would undergo if either the solar input S or the long-wave radiation back to space F were to change. If one differentiates both sides, Equation (2) with respect to T and rearranges terms, one can show that for the simple climate model (2):

$$\lambda_p = \frac{T_p}{4S} . \tag{4}$$

Observations by satellite infrared sensors suggest that the effective planetary temperature for Earth infrared radiation to space is equal to that of a black body, with an absolute temperature of approximately T_p = 255°K. Therefore, λ_p from Equation (4) is approximately equal to 0.27°K W^{-1} m^2. In simpler terms this means that if the Sun were to somehow increase its energy output by 1%, the Earth's radiative temperature would eventually (i.e., in equilibrium) warm up by 0.64°K.

Climatic Feedback Mechanisms

Let us consider next what would happen if there were feedback in the system. That is, $\alpha_p = \alpha(T_p)$. Substituting $\alpha_p(T)$ into Equation (2) and differentiating both sides with respect to T, one can obtain λ_p^*, a modified expression for λ_p in the case of feedback

$$\lambda_p^* = \frac{\lambda_p}{1 + \lambda_p \left(\dfrac{1}{1-\alpha_p} \right) \dfrac{\partial \alpha_p}{\partial T_p} S}, \tag{5}$$

where λ_p is the sensitivity parameter obtained when albedo is constant (i.e., 4). It is an instructive exercise to provide some intuitive feeling for what this equation means. Suppose, for example, $\alpha_p = a - bT_p$, where a is constant and b is 0.01. What this implies is that if the Earth's temperature were to increase by 1°C, then albedo would decrease by 0.01. Thus, for $\alpha_p = 0.3$ at $T_p = 255$°K and $b = 0.01$, this implies $\lambda_p^*/\lambda_p \approx 10$, indicating that a seemingly "small" feedback ($b = 0.01$) can have dramatic impact on the sensitivity of the system!

Earth satellites have suggested that, for temperatures observed on Earth, a linear relationship between outgoing infrared radiation and surface temperature, T_s, is not a bad first approximation (e.g., Warren and Schneider 1979). Let us rewrite our simple climate model then with this linearized form assuming α is a constant:

$$Q\,(1\ \alpha) = A + BT_S\,. \tag{6}$$

Consequently,

$$\lambda_S = \frac{1}{B}\,. \tag{7}$$

Empirically, from satellite observations, B has been found to have a value of 1.83 W m^{-2} K^{-1} (Warren and Schneider 1979). If Q = 340 W m^{-2} and $\alpha = 0.3$, then $\lambda_s = 0.55$ K W^{-1} m^2. This is twice λ_p, which suggests that empirically derived values of outgoing infrared radiation as a function of surface temperature leads to amplifying feedbacks.

The most often postulated feedback that could describe this amplification is the "water vapor–greenhouse effect" feedback. It is well known that increasing surface temperature increases evaporation because evaporation increases nonlinearly with surface temperature through the Clausius-Clapeyron relationship between vapor pressure of water and temperature. In the mid-latitudes, for example, although relative humidity is fairly constant from one season to the next, absolute humidity can increase by a large factor from winter to summer due to this water vapor pressure–temperature relationship. Some recent empirical information from satellites strongly suggests that the water vapor–greenhouse feedback is indeed positive (e.g., Raval and Ramanathan 1989), and may very well account for the substantially enhanced sensitivity of the linearized semi-empirical model (6) relative to the original black body model, (2).

However, water vapor–greenhouse effect positive feedback is not the only potentially important feedback in the system. For example, consider

$$\alpha\,(T_S) = \delta + \gamma T_S\,, \tag{8}$$

Plugging (8) into (6) and using (3) yields

$$\lambda_S = \frac{1}{B + \gamma Q}\,. \tag{9}$$

What does this mean? If γ is a positive number, it means that when surface temperature increases, albedo increases. This means that whatever causes surface temperature to increase would cause an increase in the reflectivity of the planet, which would limit that original surface temperature rise, which is a negative feedback. If γ is negative, then (9) implies positive feedback since B is a positive number and the denominator of (9) would be reduced. In addition, if the absolute value of γQ is greater than B, this system would become unstable. The greatest challenge in climate modeling (demonstrated here in linearized, zero-dimensional formalism) is to determine what the sum $B + \gamma Q$ is, based upon the many processes that interact in the climate system. Moreover, these processes that affect the feedback parameter γ do not occur uniformly across the globe; rather the global average values represented by the simple expressions derived so far are the manifestations of many local or regional changes that could be larger, smaller, or even of opposite sign to the net global effect. For example, Raval and Ramanathan (1989) have shown empirically strong water vapor–greenhouse effect feedback, whereas Ramanathan and Collins (1991) have shown empirically strong negative cirrus cloud feedback in the part of the world, with tropical cumulus clouds and ocean surface temperatures greater than 303°K. Nevertheless, the exercise given by Equation (9) does show how sensitive the Earth's climate can be to seemingly small changes in these feedback processes.

Paleoclimatic Inference of Equilibrium Global Climate Sensitivity

Because forcings have changed over time, it might be possible to tease out an estimate of climate sensitivity from estimates of past average climatic temperatures and altered forcings. Unfortunately, these forcings involve simultaneous changes to the latitudinal and seasonal distribution of incoming sunlight (e.g., Berger 1988) as well as alterations in trace greenhouse gases (e.g., Oeschger et al. 1982; Lorius et al. 1990) and atmospheric aerosols (e.g., Royer et al. 1983). Moreover, lags of centuries (in oceans) to millenia (in ice sheets) will further confound any easy empirical determination of the global climatic response to these paleoclimatic forcings. Futhermore, the possibility of internal stochastic or even chaotic variability of the climate system on all time scales further complicates any easy inferences.

Nonetheless, careful and plausible decomposition of forcings and reconstruction of climatic responses have been attempted (e.g., Lorius et al. 1990 or Hoffert and Covey 1992), although these authors wisely caution that incomplete understanding of the detailed causes of past climatic change could lead to misinterpretations of any attempt to infer present climate sensitivity from past changes in climatic forcings (e.g., CO_2, CH_4, dust, ice, and insolation). Despite these appropriate cautions, Lorius et al. do obtain a value of climate sensitivity ~1°K W^{-1} m^{-2} (implying 3°–4°C equilibrium warming from a doubling of CO_2 since that would trap about 4 W m^{-2} of infrared radiative heat in the surface-atmosphere system—in the upper middle of the range of current general circulation models [GCMs]) (IPCC 1990). I suspect that additional accounting for the forcings implied by expanded ice coverage and paleodust might lower that 3°–4°C estimate into the 2°–3°C range; however, this speculation of mine needs to be confirmed.

REGIONAL DISTRIBUTION OF CLIMATE CHANGE

The most important impact assessment question surrounding the greenhouse-gas controversy is simply: What will be the regional distribution of climatic changes associated with significant increases in CO_2 and other trace greenhouse gases? (Other trace gases such as chlorofluorocarbons, methane, nitrogen oxides, or ozone are together expected to have a comparable greenhouse effect to CO_2 over the next century). To investigate such possibilities, we need a model with regional resolution. It needs to include such processes as the hydrologic cycle and the storage of moisture in the soils, since these factors are so critical to regional rainfall as well as agriculture and water supplies. To investigate plausible regional climatic change scenarios, modelers have typically run "equilibrium simulations," in which instantaneously increased values of CO_2 are imposed at an initial time and are held fixed while the model is allowed to approach a new equilibrium. Manabe et al. (1981), for example, in one of the most widely quoted results, found a summer "dry zone" in the middle of North America as well as increased moistness in some of the monsoon belts. All of this was from a quadrupling of CO_2 held fixed over time. The model was allowed to run several decades of simulated time in order to reach equilibrium. However, Manabe's team used an artificially constructed ocean that consisted of a uniform mixed layer of depth of 70 meters with no heat flow to or from the deep ocean below. While this shallow "sea" allows the seasonal cycle to be satisfactorily simulated (see Figure 20.1), a purely mixed layer ocean does not include the important processes whereby water is transported horizontally from the Tropics toward the poles or vertically between the mixed layer and the abyssal depths. The latter processes slow the approach toward thermal equilibrium of the surface waters and certainly would affect the transient evolution of the surface temperature changes in the ocean due to the actual time-evolving increase of trace greenhouse gases.

One of the most successful regional paleoclimatic simulations to date was performed by John Kutzbach and several colleagues at the University of Wisconsin in Madison. Kutzbach attempted to explain the warmest period in recent climatic history, the so-called "climatic optimum" that occurred between about five and nine thousand years ago. This was a time when summertime northern continental temperatures were probably several degrees warmer than at present and monsoon rainfall was more intense throughout Africa and Asia. Kutzbach found that the optimum could be explained simply by the fact that the tilt of the Earth's axis (its obliquity) was slightly greater then than now. Also, the orbit was such that the Earth was closer to the Sun (i.e., perihelion) in June rather than in January as it now is. These variations, these slight perturbations in the Earth's orbit, do not make substantial changes in the annual amount of solar radiation received on the Earth, but do change substantially the

Figure 20.1 A three-dimensional climate model has been used to compute the winter-to-summer temperature differences all over the globe. The model's performance can be verified against the observed data shown below. This verification exercise shows that the model quite impressively reproduces many of the features of the seasonal cycle. These seasonal temperature extremes are mostly larger than those occurring between ice ages and interglacials or for any plausible future carbon dioxide change. From Manabe and Stouffer (1980).

difference between winter and summer heating periods. About 5% more solar heat over much of the Northern Hemisphere summer and a comparable amount less in winter occurred nine thousand years ago compared to the present. This change was sufficient in Kutzbach and Guetter's (1986) simulations to substantially alter mid-continental warming in the summer months, which led to enhanced monsoonal rainfall and river runoff in the models. Their results, as seen in Figure 20.2, matched with considerable amounts of paleoclimatic evidence gathered by an international team of

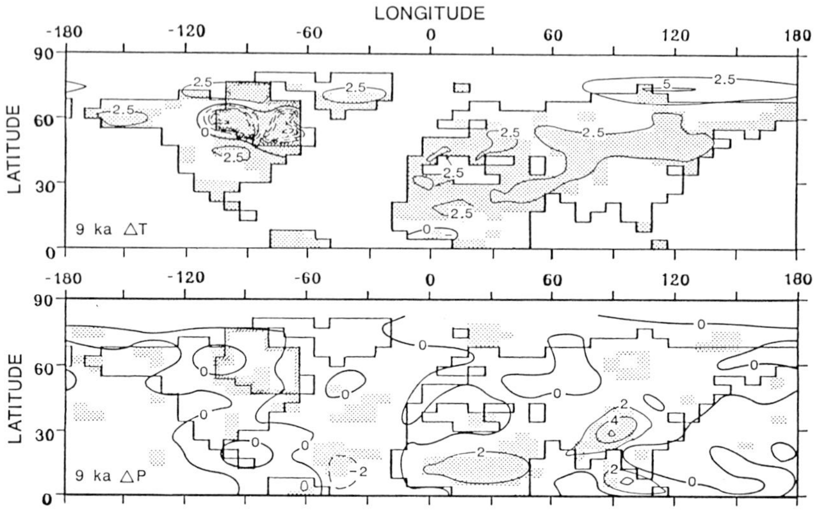

Figure 20.2 Model simulated surface temperature (top: degrees centigrade) and precipitation (bottom; mm/day) departures, 9 Ka minus control case, Northern Hemisphere, July. The shading on the departure map indicates the departures are statistically significant (two-sided t-test) at or above the 95% confidence level based upon the model's natural variability. No statistical significance is computed for ocean surface temperature, since it is prescribed. Precipitation data were smoothed with a nine point binomial filter The area covered by the North American ice sheet is outlined with a hatched line. From P. Guetter and J. Kutzbach (1990).

scientists, and are helping to explain one of the important periods in the paleoclimatic record (i.e., COHMAP 1988).

TRANSIENTS

Transient Response of Global Temperature

What if we are interested in the transient response of the system to some global change forcing (see Schneider 1992 and the volume it is contained within)? In this case, our energy balance Equation (1) is rewritten as the rate of energy storage equals solar energy input to the Earth minus infrared radiant energy out. In symbolic terms,

$$R \frac{\partial T_s}{\partial t} = Q (1 - \alpha) - F_{ir} . \tag{10}$$

where R is the "thermal inertia" of the system, i.e., the effective heat capacity of the atmosphere, oceans, land, etc. This, in turn, is proportional to the mass of each of the components of the climate system times their respective specific heats. If α is constant, or (to first order) its effects are lumped into the linear term B of (6), then Equation (10) can be rewritten

$$R\,\frac{\partial T_s}{\partial t} = Q\,(1-\alpha) - (A + BT_s) = Q\,(t) - BT_s\,. \tag{11}$$

The solution to such an ordinary differential equation is well known from classical theory, and takes the form $T_s = T_{inhomo} + T_{homo}$. The homogeneous solution, which we could call T_{pert}, would be of the form

$$T_{pert} \propto \exp\,(-t/\tau) \tag{12}$$

where

$$\tau = R/B\,. \tag{13}$$

τ is the response time of the system to a step function forcing in Q, and is simply the heat capacity of the system divided by the radiative damping. B, the radiative damping coefficient, is also intimately involved in the sensitivity of the system. Recall, from Equations (7) and (9), that the larger B is, the less sensitive the system is to changes in external forcing Q. What does this large damping mean physically? If B is a large number, then the outgoing infrared radiation to space increases by a substantial amount if surface temperature increases. That is, the input of additional heat to the system would be damped out very effectively to space if B is a large number. That would limit the sensitivity of the system to an input of energy of any kind. For example, if a few W m^{-2} of additional energy were input by an increase in CO_2 or sunlight, then a large B means that only a small temperature change is necessary to damp that extra heat back out to space. On the other hand, a small B would imply very high sensitivity to small amounts of energy input, since the temperature would have to go up a great deal to damp that extra few W m^{-2} of heating back to space.

For the response time, τ, of the system, a large B implies a short response, since rapid damping means the system would approach its reduced equilibrium response more quickly than if there were little damping associated with a small B, as Equation (13) shows. Thus, the feedback factors, which in linear form aggregate into B, not only affect inversely the overall sensitivity of the planet to forcing, but also its response time.

Equation (13) also requires the response time to be larger if the heat capacity, R, is larger. This is intuitively obvious since a more thermally massive planet (i.e., one with a large portion of oceans or with oceans mixed deeply) would take longer to respond to global thermal forcing than would one with relatively little heat capacity. The time-dependent model derived so far (i.e., 10) is for an Earth with a fixed-heat capacity R. The real world, of course, does not have a fixed heat capacity, but consists of a mix of multiple-heat capacities: capacities which, in fact, change with the climatic state (e.g., Thompson and Schneider 1979). For example, the heat capacity of the climate system over land is relatively small, consisting largely of the atmosphere itself and a few centimeters of soil. The response time is thus on the order of a month or so. The effective heat capacity of the middle of tropical oceans is largely governed by the depth of the oceanic mixed layer, which is in contact with the atmosphere. Although that mixed layer is slowly ventilated from below, the dominant term in the heat capacity is that 50–70 m mixed layer. Nevertheless, the thermal inertia of the center of tropical oceans is at least an order of magnitude or so larger than that of the center of continents. However, polar seas, such as the Norwegian or Weddell Seas, in which oceanic convection causes the mixed layer to penetrate many hundreds of meters in depth, have effective thermal inertias another order of magnitude or so larger than that of tropical oceans. Thus, to predict the transient response of the climate to global change forcings (e.g., CO_2 or other greenhouse-gas buildups) will require not simply a global average model extended to include time dependence but a model that has enough spatial and temporal resolution to capture the important nonglobal nature of the regionally heterogeneous effects of various physical processes and heat capacities (e.g., Schneider and Thompson 1981). These vertical circulations were very different at past times (e.g., Broecker et al. 1985).

The results by Kutzbach and co-workers imply that although paleoclimatic inferences can help to validate the credibility of GCMs in predicting such regional climatic changes as altered monsoons or help us to calibrate some properties of ocean models, the rapid time scale of global change forcings (10^1–10^2 yrs) implies that validation exercises in addition to paleoclimatic data-model intercomparisons will be particularly needed for evaluating the regional, time-evolving global changes that might be induced by increasing SO_2 or greenhouse gases.

GCM forecasts of regional or watershed changes in temperature, evaporation, and precipitation are most germane to estimating various consequences of greenhouse warming, such as hydrological implications. However, as suggested above, such regional forecasts are more uncertain than those at larger scales. Regional temperature changes can be much larger than global changes, and even some regions can show a negative change (e.g., Schlesinger and Mitchell 1987; IPCC 1990). For example, surface-temperature increases projected for the higher northern latitudes are up to several times larger than the projected global average response, at least in equilibrium.

Even more uncertain than regional details, but perhaps more important to long-term ecosystem responses, are estimates for such measures of climatic variability as the frequency and magnitude of severe storms, enhanced heat waves, temperature extremes, or reduced frost probabilities (Parry and Carter 1985; Mearns et al. 1984, 1990; Wigley 1985). For example, the physical principle, which says evaporation increases exponentially with surface water temperature, implies that hurricane intensities could increase with warming of the oceans (Emanuel 1987). Such issues are just now beginning to be considered and evaluated. However, these evaluations are based on so-called "equilibrium" climate-model results, in which CO_2 is doubled and held fixed over time. Because the actual greenhouse-gas increases are time-evolving exponential increases, these issues will, of course, have to be studied for realistic transient forcing cases, to be of maximum value to impact assessors.

SCALE TRANSITION

Finally, within the foreseeable future even the highest-resolution three-dimensional GCMs suitable for integrations over 50 or more years will not have a grid with nodes much less than 100 km apart; individual clouds and most ecological research (to say nothing of microclimatic or microphysiological research), for example, occur on scales far smaller than that (Kareiva and Andersen 1988). GCMs will not, therefore, be able to resolve most local biological communities or the important local and mesoscale effects of hills, coastlines, lakes, vegetation boundaries, and heterogeneous soil. (For example, at the typical grid resolution of 4° latitude by 7° longitude, a GCM cannot easily resolve the Sierras and Rockies as separate mountain chains!) For regions that have relatively uniform surface characteristics, such as a thousand-kilometer savanna or a tropical forest with little elevation change, GCM grid-scale parametric representations of surface albedo, soil type, and evapotranspiration could be used to estimate local changes. Alterations in climate predicted within one grid square would likely apply fairly uniformly across such homogeneous areas. On the other hand, steep topography or lakes smaller than GCM grids can mediate real climate, but are not resolved in the models. Therefore, even if GCM predictions were accurate at grid scale, they would not necessarily be appropriate to local conditions.

Large-scale observed climatic anomalies can be mapped to local variations. For instance, Gates (1985) analyzed the local climatic variability for the state of Oregon using a technique known as empirical orthogonal functions (Figure 20.3). In other words, a GCM of altered temperatures resolved on the western slopes of the Cascade Mountains could be applied to the maps of Oregon (Figure 20.3) to determine the climatic effects on local areas. Such a map, constructed from variations of climate observed over several years, may seem an ideal way to translate the GCM-scale-averaged grid data to the local or mesoscale. Because empirical data have been used, however, such a relation would be valid only where the causes of recent climatic variations or oscillations carry forward and include the effect of climatic changes forced by trace gases. The north-south Cascade Mountains translate a simple change in the frequency or intensity of westerly winds into a characteristic climatic signature of either wetter on the west slope and drier on the east or vice versa. Thus, the signature of climatic change from increases in trace gases may not be the same as past vacillations, many of which could have been internal oscillations within the climate system, not the result of external forcing such as changes in trace gases. For example, simply increasing the temperature on the western slopes of the Cascades will not necessarily (despite Figure 20.3) imply a decreased temperature on the eastern slopes if the temperature rise were forced by greenhouse-gas heating all across the globe, whereas eastern slope cooling, predicted from Figure 20.3, would likely hold if the

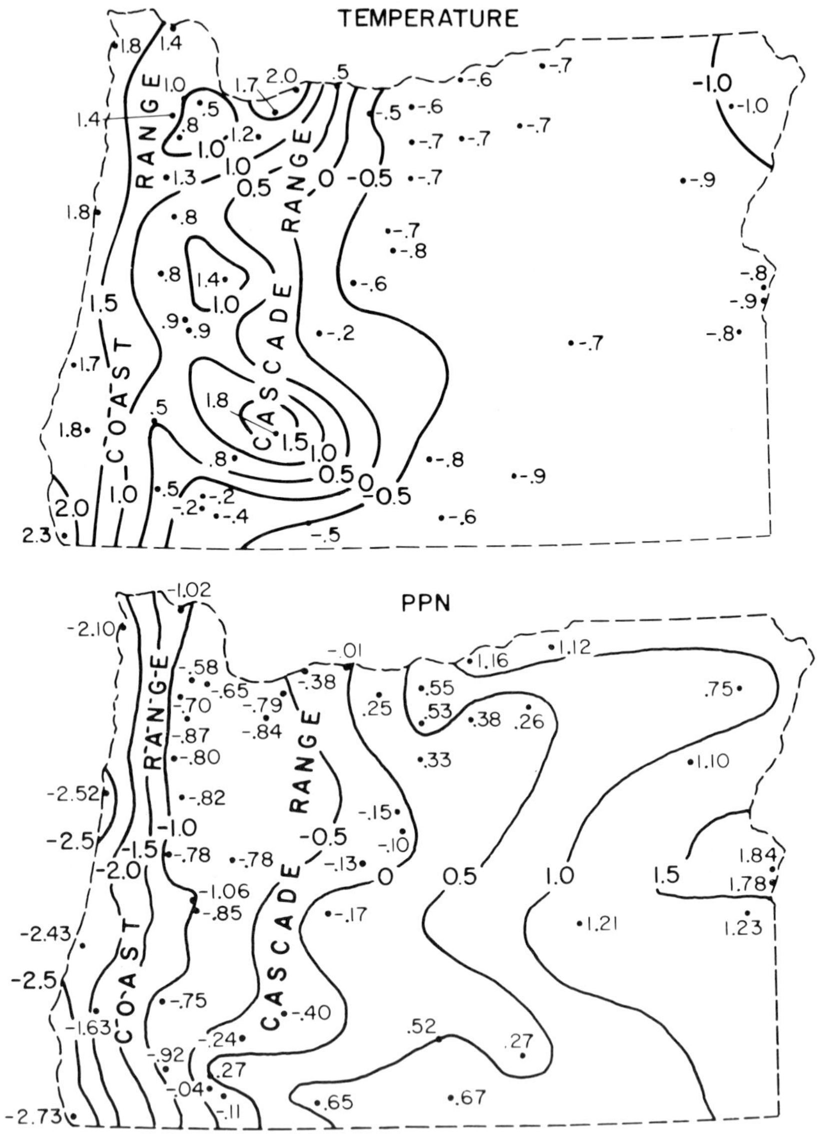

Figure 20.3 The distribution of the relationship between large-scale (area-averaged) and local variations of the monthly mean surface-air temperature (above) and precipitation (below), as given by the first empirical orthogonal function determined from thirty years of observing the monthly means at 49 stations in Oregon in comparison with the state-wide average. From Gates (1985).

western slope warming were a result of weakened westerlies. Thus, other translations of scale need to be considered to map large-scale global change projections to smaller scales that account for the causes of the large-scale changes.

One might embed a high-resolution mesoscale model within a few grid squares of a GCM, using as boundary conditions for the mesoscale model the wind, temperature, and so forth predicted by the GCM at the squares' boundaries (e.g., Dickinson et al. 1989). A mesoscale model, with grid-square scale typically 50×50 km, could then account for regional topography, soil type, and vegetation cover and could map GCM forecasts to regional topography. Figure 20.4 is an example for the western United States. For such a method to have any reasonable hope of success, however, the GCM must produce accurate grid-scale climatic statistics for the special limited grid area. To return to the Oregon case in Figure 20.3, if the climatic average of the GCM's winds in the unperturbed case (the control case) has the wrong westerly component, the climate change will probably be misrepresented in a region where topography amplifies any such error in the wind direction. A likely prerequisite for that kind of cross-scale mapping, therefore, is a sufficiently accurate grid-scale control climate for the important variables. Only then does it make sense to take the next step of imposing a scenario of trace-gas increase on the GCM to estimate how the regional- or local-scale climate might change. Although this technique has shown some promise for mid-latitudes, how embedding would work in the Tropics—where thunderstorms (smaller in scale than even the mesoscale grid squares) dominate precipitating systems—is more uncertain.

ECOLOGICAL IMPLICATIONS

Understanding the consequences of global warming on vertebrates requires at least a rudimentary understanding of the influence climate change will have on vegetation, because essential characteristics of ecosystems are shaped by vegetation (Graetz et al. 1988). Climate strongly affects the distribution of plants; many studies have shown that vegetation changes as the climate does (e.g., Woodward 1987 and references therein; COHMAP 1988; Gates 1980). Of course, there are many factors involved in range limits and abundances, such as competition and (especially in Europe) alterations from human activities (e.g., Frenzel, this volume). Nevertheless, even small changes in ambient temperature can cause dramatic effects owing to changes in the physiological demands of the plants. Lower energy demands due to warmer temperatures could easily change the delicate dominance structure that is assumed to be maintained by competition (Strain 1987).

One of the Office of Management and Budget officials of President Bush's administration once said that it would be no more difficult for individuals to adapt to a few degrees of climate warming than it is to adjust the temperature of Washington, D.C. temperatures upon disembarking from a Boston flight. Of course, what this view neglects is the basic ecology of habitats; the forests of Virginia cannot call the

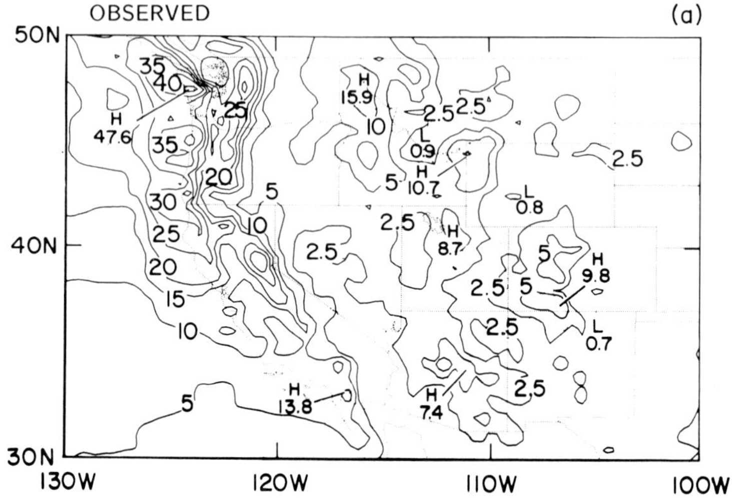

Figure 20.4a Average January total precipitation (centimeters): observations.

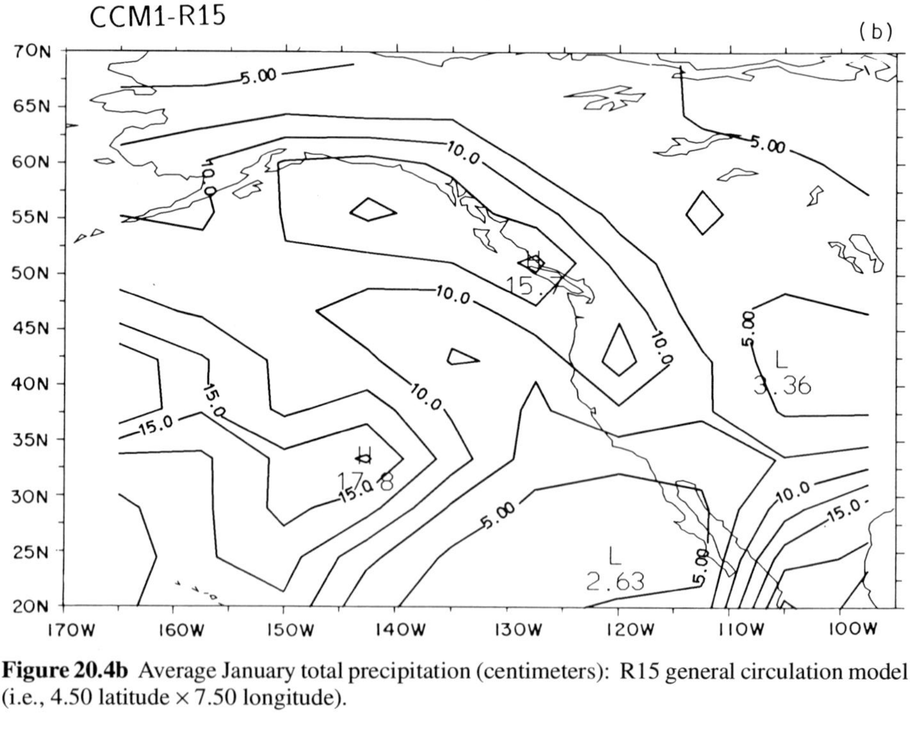

Figure 20.4b Average January total precipitation (centimeters): R15 general circulation model (i.e., 4.50 latitude × 7.50 longitude).

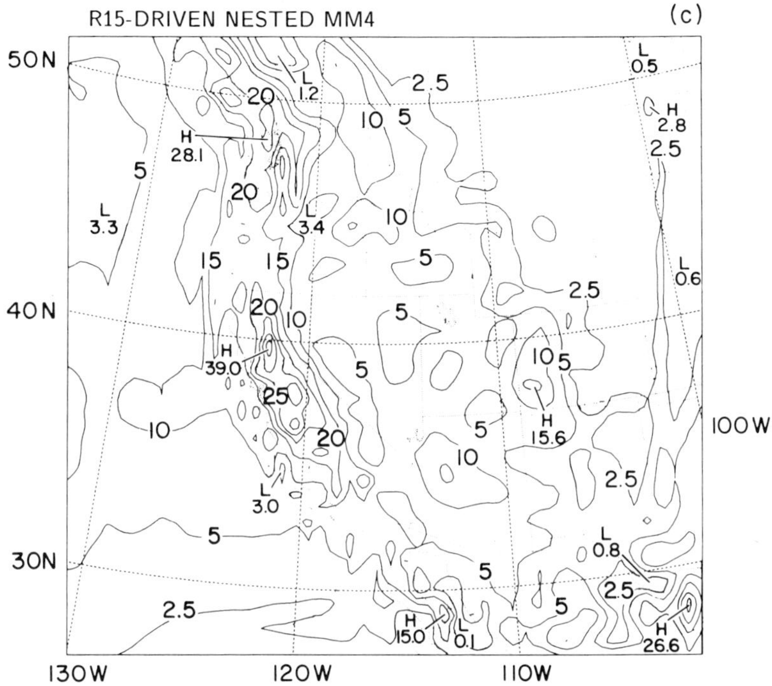

Figure 20.4c Average January total precipitation (centimeters): mesoscale model driven by output of R15 model. From Giorgi (1990).

neighborhood van lines to transplant millions of trees up to Massachusetts. Quite simply, many tree species take hundreds, even thousands of years for substantial migration (Davis 1990). The tree with the fastest known dispersal rate is spruce, which expanded its range on an average of 20 kilometers every ten years about 9000 years ago (Roberts 1988). The reason that trees move so slowly is because, unlike animals, individuals cannot move. Range expansion can only occur through seed dispersal, germination, and establishment, which, of course, takes time (Cohn 1989). Moreover, species migrate at different rates (Cohn 1989), which will result in dramatic alterations of the species composition of basically all biological communities. That is indeed what occurred when the ice ages waned about 12,000 years ago and were replaced in the following five thousand years by the relatively warm interglacial period we are now experiencing (e.g., see COHMAP 1988). At that time, a 5°C globally averaged temperature increase accompanied a 100 meter sea-level rise and dramatic vegetation changes, such as those seen in Figure 20.5. The ice-age-to-interglacial transition quite literally revamped the ecological landscape of the Earth. The time it took nature to make this adjustment was on the order of thousands of years, driven by a sustained global average-temperature change on the order of 5°C. We are currently facing the prospect of temperature changes of this magnitude in the time frame of a century, a rate that most biologists acknowledge as simply too fast for evolutionary processes, such as natural selection, to occur, thereby inhibiting species' ability to adapt and threatening some with extinction. Indeed, the importance in the difference between evolutionary and ecological time scale was noted by Davis (1990): "The fossil record shows that most forest trees were able to disperse rapidly enough to keep up with most of the climatic changes that took place in recent millennia. These changes were much more gradual than the climatic changes projected for the future. Even so, there were occasional periods of disequilibrium between plant distributions or abundances, soils, and climate that lasted a century or more. The most rapid dispersal rates known from the fossil record, however, are an order of magnitude too slow to keep up with the temperature rise expected in the coming century." Additionally, land use by humans has fragmented the natural environment in such a way, in recent history, that even if the climatic changes were slow enough, dispersal may not occur for many species without substantial human intervention (Main 1988; Arnold 1988).

Climatic factors can strongly dictate the biogeographic patterns of many species. Frequently animals are most abundant in areas where the least amount of energy is needed for survival and, range boundaries occur where metabolic energy needed for survival (i.e., to maintain thermal homeostasis) becomes too high. Root (1988) has shown a very high degree of statistical correlation between the distribution and abundance of many bird species and large-scale environmental factors, such as temperature and precipitation.

What this type of analysis implies is that as climate changes, the physiological tolerance of some birds could cause them—habitat permitting—to change their ranges as rapidly as the climate changes. The range changes in those species for which the habitat is not permitting will probably not be as extreme or as rapid, and those species with ranges not associated with ambient temperature will probably not exhibit an immediate change in range. Such differential movements of species will certainly cause a tearing apart of communities, thereby forcing potentially dramatic restructurings and reorganizations. Of course, that is what nature did during the Pleistocene–Holocene transition: a climatic change that was, on a sustained global basis, an order of magnitude slower than that projected for the century ahead.

Climate change is not necessarily a threat to the viability of all climate-sensitive species. However, the transient nature of most projected human-induced significant climate change is on a scale of decades while the adaptability of many species is on a time scale of centuries. This implies that substantial disequilibrium within ecosystems could be created owing to maladaptions, significant shifts in species abundances or ranges, and inevitable extinctions (e.g., McDonald and Brown 1991). Consequently, the only outcome that can be predicted with virtual certainty is major surprises. The only forecast that seems certain is that *the more rapidly the climate changes, the higher the probability of substantial disruption and surprise within natural systems.*

Paleoecological research is essential to the understanding of the connections among vegetation systems, fauna, biogeochemical cycling, and climate change. However, as in the case of paleoclimatic reconstructions and transient global climate changes, such paleoanalogies are only of partial use, owing to the very rapid nature of global changes from human interaction.

CONCLUSIONS

In summary, then, paleoclimatic reconstructions require substantial understanding of the biogeochemical cycles and of paleoecological conditions. The latter help to calibrate ecosystem models which, in turn, are needed to project the potential future ecological implications of global climate change. Thus, not only must we understand the vegetation and faunal interactions with climate in order to reconstruct the paleoclimate itself, we need to use that knowledge in order to derive and validate models of ecosystem–climate interactions. These, in turn, will be needed in order to provide assessments of how ecosystems might change over time.

I believe that the global change time scales (10^1–10^2 yrs) pose a particular challenge, since differential rates of response for large-scale vegetation, mammals, insects, and birds are all different, and that the global change time scale may very well see community responses very different than individual species responses (e.g., Root 1992). While there is some paleoecological evidence for this view, we must bear in mind that paleoecological metaphors can only be pushed so far as analogies to global warming.

I also conclude that paleoclimatic analyses and assumptions of paleoclimatic forcings and responses suggest that climate sensitivity based on paleoclimatic reconstruction is (certainly to an order of magnitude and probably to a factor of 2 or 3) consistent with the standard range of equilibrium climate model sensitivity. While forcing and responses are not precisely known well enough to pin down the global average climate sensitivity much better than a factor of 2 or so, this is indeed encouraging! Moreover, regional paleoclimatic inferences, such as the enhanced monsoonality associated with orbital element variations, can provide some credibility and validation testing for GCMs' projections of regional response to global change forcing. While this is also very encouraging, it does not, unfortunately, help us validate how transient climatic change would evolve, since this involves turnover time scales in the oceans of 10 to 100 years, and such a time-evolving, regional transient forecast will simply require coupled high resolution models of all the climate subsystems driven by realistic, transient trace-gas forcings. While that requirement precludes the possibility that paleoclimatic regional patterns could be applied as direct metaphors to the time-evolving climate change in the decades ahead, nonetheless the validations given and the credibility thereby obtained through an intercomparison of climate modeling studies with paleoclimatic analyses is invaluable.

Finally, I believe that it will be decades, at least, before highly credible, validated, regional, time-evolving scenarios of climate change will be possible—let alone credible projections of ecosystem responses. Unfortunately, that is the time scale over

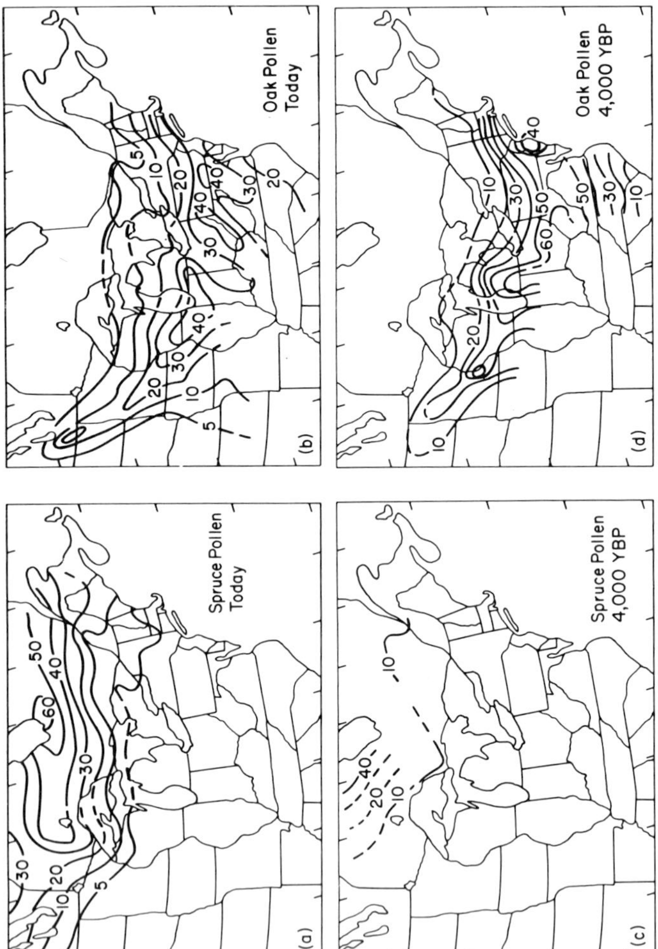
Spruce Pollen Today
(a)
Oak Pollen Today
(b)
Spruce Pollen 4,000 YBP
(c)
Oak Pollen 4,000 YBP
(d)

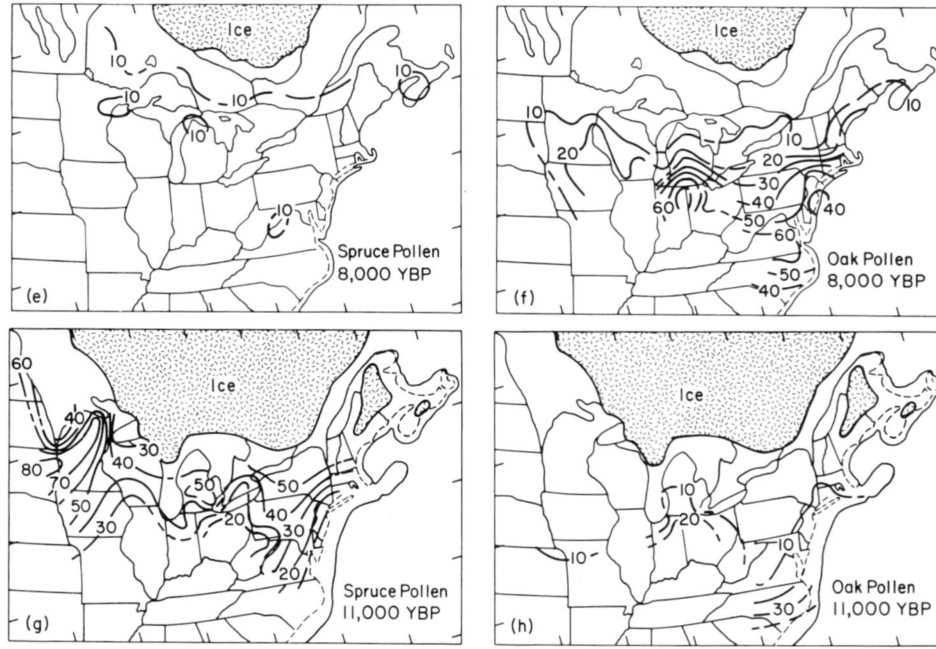

Figure 20.5 The relative fraction (in percentage) of fossil pollen from various species of trees is shown on the maps. These pollen patterns can be correlated to climatic conditions. These maps show, for example, that at the end of the last Pleistocene Ice Age (11,000 years ago) much of the northeastern United States was covered with spruce forests, which are found today in central Canada near Hudson Bay. As climatic conditions ameliorated, the spruce "chased" the retreating ice sheet northward into Canada, whereas more temperate-climate trees like oak moved outward from their refuge in the Carolinas into the middle Atlantic States and the midwest. However, as COHMAP (1988) has shown, the nature of the ecosystem was dramatically altered from a pre-Holocene spruce parkland habitat to a post-Holocene Boreal forest canopy. Around 4000 years ago, the distribution of these temperate-climate trees stabilized in what is roughly their present configuration. From Bernabo and Webb (1977).

which these global changes will likely evolve. Therefore, any human policy response to deal with the advent or prospect of global changes will have to recognize that definitive scientific answers are not likely before the system itself "performs the experiment" of determining what those global changes and their impacts will be. Whether or not such changes should be resisted now as an urgent matter of public policy is a value judgment that is not treated here, but one which I have been personally concerned with for many years (e.g., see Schneider 1990). When the fate of the Earth is in the balance, my values dictate that we err on the side of caution and look for cost-effective ways to slow down the rate of emissions that force global changes.

REFERENCES

Arnold, G.W. 1988. Possible effects of climatic change on wildlife in Western Australia. In: Greenhouse: Planning for Climate Change, ed. G.I. Pearman. Melbourne: CSIRO.

Berger, A. 1988. Milankovitch theory and climate. *Rev. Geophys.* **26**:624–657.

Bernabo, J.C., and T. Webb III. 1977. Changing patterns in the Holocene pollen record of northeastern North America: A mapped summary. *Quat. Res.* **8**:64-96.

Berner, R.A., A.C. Lasaga, and R.M. Garrels. 1983. The carbonate-silicate geochemical cycle and its effect on atmospheric carbon dioxide over the past 100 million years. *Amer. J. Sci.* **283**:641–683.

Broecker, W.S., D.M. Peteet, and D. Rind. 1985. Does the ocean-atmosphere system have more than one stable mode of operation? *Nature* **315**:21–26.

Budyko, M.I. 1990. Climate prediction based on past and current analogues. In: Climate Change: Science, Impacts and Policy, Proc. of the Second World Climate Conference, ed. J. Jager and H.L. Ferguson. Cambridge: Cambridge Univ. Press.

Budyko, M.I., A.B. Ronov, and A.L. Yanshin. 1987. History of the Earth's Atmosphere. Berlin: Springer.

COHMAP. 1988. Climatic changes of the last 18,000 years: observations and model simulations. *Science* **241**:1043.

Cohn, J.P. 1989. Gauging the biological impacts of the greenhouse effect. *BioScience* **39**:142–146.

Davis, M.B., 1990. Climatic change and the survival of forest species. In: The Earth in Transition: Patterns and Processes of Biotic Impoverishment, ed. G.M. Woodwell. Cambridge: Cambridge Univ. Press.

Dickinson, R.E., R.M. Errico, F. Giorgi, and G.T. Bates. 1989. A regional climate model for the western United States. *Clim. Change* **15**:383–422.

Emanuel, K.A. 1987. The dependence of hurricane intensity on climate. *Nature* **326**:483–485.

Gates, D.M. 1980. Biophysical Ecology. Berlin: Springer.

Gates, W.L. 1985. The use of general circulation models in the analysis of the ecosystem impacts of climatic change. *Clim. Change* **7**:267–284.

Giorgi, F. 1990. Simulation of regional climate using a limited area model nested in a general circulation model. *J. Climate* **3**:941-963.

Graetz, R.D., B.H. Walker, and P.A. Walker. 1988. The consequences of climatic change for seventy percent of Australia. In: Greenhouse: Planning for Climate Change, ed. G.I. Pearman. Melbourne: CSIRO.

Guetter, P., and J. Kutzbach. 1990. A modified Koppen classification applied to model simulations. *Climat. Change* **16**:193–216.

Hoffert, M.I., and C. Covey. 1992. Deriving global climate sensitivity from paleoclimate reconstructions. *Nature* **360:**573–576.

IPCC (Intergovernmental Panel on Climate Change). 1990. Scientific Assessment of Climate Change. Geneva: World Meteorological Organization.

Kareiva, P., and M. Andersen. 1988. Spatial aspects of species interactions. In: Advances in Theoretical Ecology, ed. A. Hastings. New York: Springer.

Kellogg, W.W., and R. Schware. 1981. Climatic Change and Society: Consequences of Increasing Atmospheric Carbon Dioxide. Boulder, CO: Westview.

Kutzbach, J.E., and P.J. Guetter. 1986. The influcence of changing orbital parameters and surface boundary conditions on climate simulations for the past 18,000 years. *J. Atmos. Sci.* **43**:1726–1759.

Lorius, C., J. Jouzel, D. Raynaud, J. Hansen, and H. Le Treut. 1990. The ice-core record: climate sensitivity and future greenhouse warming. *Nature* **347**:139–145.

Main, A.R. 1988. Climatic change and its impact on nature conservation in Australia. In: Greenhouse: Planning for Climate Change, ed. G.I. Pearman. Melbourne: CSIRO.

Manabe, S., and R.J. Stouffer. 1980. Sensitivity of a global climate model to an increase of CO_2 concentration in the atmosphere. *J. Geophys. Res.* **85**:5529-5554

McDonald, K.A., and J.H. Brown. 1992. Using montane mammals to model extinctions due to global change. *Conserv. Biol.* **6:**409–415.

Mearns, L.O., R.W. Katz, and S.H. Schneider. 1984. Extreme high temperature events: changes in their probabilities and changes in mean temperature. *J. Clim. App. Meteor.* **23**:1601–1613.

Mearns, L.O., S.H. Schneider, S.L. Thompson, and L. McDaniel. 1990. Analysis of climate variability in general circulation models: compared with observations and changes in variability in $2\times CO_2$. *Geophys. Res.* **95**:20469.

Oeschger, H., B. Stauffer, A. Neftel, J. Schwander, and R. Zumbrunn. 1982. Atmospheric CO_2 content in the past deduced from ice core analyses. *Ann. Glaciol.* **3**:227–232.

Parry, M.L., and T.R. Carter. 1985. The effect of climatic variations on agricultural risk. *Clim. Change* **7**:95–110.

Ramanathan, V., and W. Collins. 1991. Thermodynamic regulation of ocean warming by cirrus clouds deduced from observations of the 1987 El Niño. *Nature* **351**:27–32.

Raval, A., and V. Ramanathan. 1989. Observational determination of the greenhouse effect. *Nature* **342**:758.

Roberts, L. 1988. Is there life after climate change? *Science* **242**:1010–1012.

Root, T.L. 1988. Environmental factors associated with avian distributional boundaries. *J. Biogeog.* **15**:489–505.

Root, T.L. 1993. Effects of global climate change on North American birds and their communities. In: Biotic Interactions and Global Change, ed. J. Kingsolver, P. Kareiva, and R. Huey, pp. 280–292. Sunderland, MA: Sinauer.

Royer, A., M. DeAngelis and J.R. Petit. 1983. A 30,000 year record of physical and optical properties of microparticles from an east antarctic ice core and implications for paleoclimate reconstruction models. *Clim. Change* **5**:381–412.

Schlesinger, M.E., and J.F.B. Mitchell. 1987. Climate model simulations of the equilibrium climatic response to increased carbon dioxide. *Rev. Geophys.* **25**:760–798.

Schneider, S.H. 1984. On the empirical verification of model-predicted CO_2-induced climatic effects. In: Climate Processes and Climate Sensitivity, ed. J. Hansen and T. Takahashi. *Geophys. Monog.* **29**:187–201. Washington, D.C.: American Geophysical Union.

Schneider, S.H. 1990. Global Warming: Are We Entering the Greenhouse Century? New York: Vintage.

Schneider, S.H. 1992. Introduction to climate modeling. In: Climate Systems Modeling, ed. K.E. Trenberth, pp. 3–26. Cambridge: Cambridge Univ. Press.

Schneider, S.H., and S.L. Thompson. 1981. Atmospheric CO_2 and climate: importance of the transient response. *J. Geophys. Res.* **86**:3135–3147.

Schneider, S.H., S.L. Thompson, and E. Barron 1985. Mid-Cretaceous continental surface temperatures: are high CO_2 concentrations needed to simulate above-freezing winter conditions? In: The Carbon Cycle and Atmospheric CO_2: Natural Variations Archean to Present, ed. E.T. Sundquist and W.S. Broecker. *Geophys. Monog.* **32**:554–559. Washington, D.C.: American Geophysical Union.

Strain, B.R. 1987. Direct effects of increasing Atmospheric CO_2 on plants and ecosystems. *Trends Ecol. Evol.* **2**:18–21.

Thompson, S.L., and S.H. Schneider. 1979. A seasonal zonal energy balance climate model with an interactive lower layer. *J. Geophys. Res.* **84**:2401–2414.

Twomey, S., A.M. Piepgrass, and T.L. Wolfe. 1984. An assessment of the impact of pollution on global cloud albedo. *Tellus* **36B**:356–366.

Warren, S.G., and S.H. Schneider. 1979. Seasonal simulation as a test for uncertainties in the parameterizations of a Budyko-Sellers zonal climate model. *J. Atmos. Sci.* **36**:1377–1391.

Wetherald, R.T., and R.J. Stouffer. 1981. Summer dryness due to an increase of atmospheric CO_2 concentration. *Clim. Change* **3**:347–385.

Wigley, T.M.L. 1985. Impact of extreme events. *Nature* **316**:106–107.

Woodward, F.I. 1987. Climate and Plant Distribution. Cambridge: Cambridge Univ. Press.

Standing, left to right:
B.D. Santer, F.H. Schweingruber, H. Flohn, T. Litt, A. Berger

Seated, left to right:
M. Stuiver, J.A. Eddy, J. Imbrie, S.H. Schneider

21

Group Report: How Can Paleodata Be Used to Evaluate the Forcing Mechanisms Responsible for Past Climate Changes?

B.D. SANTER, Rapporteur

A. BERGER, J.A. EDDY, H. FLOHN, J. IMBRIE, T. LITT, S.H. SCHNEIDER, F.H. SCHWEINGRUBER, M. STUIVER

INTRODUCTION

The general issue addressed by our group was how paleodata could be used to identify and evaluate the forces that have brought about significant climate changes in the past. An obvious example involves the major ice ages of the Quaternary, now thought to be paced by orbitally induced changes in the distribution of sunlight over the surface of the Earth. Indeed, evidence for the important role of orbital forcing comes entirely from paleodata—principally from deep-sea cores that tell, indirectly, of past changes in the terrestrial environment (including the volume of ice on land) and, directly, of changes in the surface and deep ocean. What we now ask from these and other data is a more sophisticated question: What are the feedbacks through which the climate system responds to Milankovitch forcing? Answers can come in part by using paleodata to test and ultimately validate analytical models. These tests also point to critical gaps in existing data coverage.

A second example deals with a more recent period of time and concerns the role of direct solar forcing of past climate. The question of the possible influence of solar output variations in dictating the climate of the last 1,000 years is of more than academic interest because of the possible effect of solar variations in enhancing or ameliorating past and future effects of global warming by greenhouse gases. Here paleodata from ice cores and tree rings can tell us more about the past behavior of

Global Changes in the Perspective of the Past
Edited by J.A. Eddy and H. Oeschger

solar activity. By combining this proxy history with modern direct measurements of solar output variations, we can hope to recover the more relevant history of changes in the flow of energy from the Sun over the last several hundred thousand years.

A number of other important unknowns regarding forcing factors can be illuminated by the expanded study of paleodata, including the dimensionality of climate change on different time scales and the possible use of paleodata to establish the "noise" level of internally generated natural variability in the climate system, particularly on the 10- to 100-year time scale appropriate to the detection of a slowly evolving enhanced greenhouse effect signal. One can also apply various types of available paleodata to illuminate the causes of the abrupt changes in the pacing and nature of major ice ages that took place about a million years ago, and the significant differences between different glacial (or interglacial) episodes of the past.

In each of these cases, set out more precisely in the form of six questions below, we consider paleodata as a source of information on the more advanced question of climate forcing: not only *what* happened in the past, but *why*. By understanding the processes that drive and govern past climates, we should be able to narrow the confidence limits that must be placed around predictions of the course of future climates.

WHAT ARE THE IMPORTANT DIMENSIONS AND RESPONSE TIMES OF CLIMATE CHANGE ON THE 100- TO 100,000-YR TIME SCALES?

The climate system consists of coupled interactions involving the atmosphere, oceans, ice sheets, land surface, and biota. There is, therefore, a rich menu of variables which must be considered when describing the dimensions of the climate state. The most aggregated variables include, e.g., globally averaged surface air temperature, ice volume, CO_2 and aerosol concentrations, sea-ice area, or masses of the terrestrial or marine biosphere. However, these zero-dimensional variables need to be supplemented with one-, two-, and three-dimensional fields of temperature, moisture, motions (winds and ocean currents, upwelling and downwelling), chemical constituents, distribution and abundance of vegetation, soil moisture, cloudiness, radiation fields, and river runoff. Daily, monthly, and interannual variability of these annual cycle variables are also important dimensions of the climate state inasmuch as living systems are nonlinearly affected by the probability of certain threshold-crossing events (e.g., freezing, droughts, or a short period of heat and drought).

On longer time scales, measures of changing vegetation patterns, oceanic circulation, ice fields, and chemical composition are used to provide estimates of the characteristic times of important climatic processes. For example, instrumental observations, paleoclimatic reconstructions, and paleoecological data have suggested characteristic time scales for various climatic variations. Yet observations alone are not sufficient, since what is observed includes both the internal response characteristics of a given system and the time scales of external factors that may be forcing the system to change. A continuous interplay, therefore, among instrumental observations, paleodata, models, and theory is needed to separate internal response times from external forcing scales.

For example, we know that the atmosphere itself has a characteristic radiative relaxation time of weeks, the mixed-layer of the ocean responds in years, the response of deep oceans and forests is centuries, soils form and decay over 10^3 years, ice shelves vary on 500- to 5,000-yr scales, and continental ice sheets wax over 10^4- to 10^5-yr scales but appear to wane on < 10^3- to 10^4-yr scales (depending on how the ice sheet is grounded). The response of the crust to ice-sheet loading has time constants on the order of 2,000 to 3,000 years. Biogeochemical cycling of carbon occurs on 10-yr scales for CH_4, 100-yr scales for anthropogenic, atmospheric and biospheric CO_2 changes; 10^2- to 10^4-yr scales for oceanic CO_2 fluctuations; and 10^6- to 10^8-yr scales for weathering/tectonic influences on CO_2.

The distinction between processes and response times operating on different time scales, although useful, should not obscure the fact that short- (interannual, decadal, century) and long-period (10^3 to 10^6 years) phenomena coexist in the climate system. For example, the transition between glacials and interglacials is characterized by abrupt events (e.g., the Younger Dryas event at ca. 10.7 Ka B.P.). Other abrupt events exist in the recent instrumental record (at least at the regional scale; Flohn 1986), such as the decadal-scale cooling in the North Atlantic, which commenced in the 1950s. The interaction between processes with very different time scales may also be important at the initiation of glaciation after a long interglacial period. If a large, high elevation plateau (e.g., Baffin Island, with an area on the order of 10^3 km^2) has a climate close to the glacial–interglacial transition (humid with cold, long winters and short, warm summers), a single anomalously cool summer may result in the survival of the snow cover until the next winter, thus increasing the probability of snow cover survival from year to year until a fairly stable permanent snow field is formed. Studies by Dansgaard et al. (1989) and Thompson et al. (1989) indicate that such abrupt climate changes have occurred in the past.

The changes now taking place—tropical warming on the order of 1°C over 25 years in the mid-troposphere, increase of evaporation and water vapor content in the Tropics by 15–20% over 40 years, simultaneous intensification of the atmospheric circulation above the oceans by up to 50% in 25 years—are of the same time scale and roughly the same magnitude. Many studies have attempted to link such recent events to anthropogenically induced increases in the atmospheric concentrations of CO_2, CH_4, and other trace gases. However, we must also address the possibility that the rapid climate changes seen in recent decades may be due to nonanthropogenic factors. To this end, a thorough interrogation of paleodata may enable us to learn more about the magnitude, spatial pattern, and possible causes of past abrupt changes, and guide us towards a better understanding of the mechanisms involved in more recent changes.

As in the dimension of time, we also have to distinguish in space between large- and small-scale processes. The range of spatial scales involved in climate spans many orders of magnitude. Even a modeler with the aim of simulating climate on the spatial scale of a 500 × 500 km grid-box attempts to incorporate the subgrid-scale variation of important processes, such as the effects of clouds and the fluxes of heat and momentum at the air–sea interface. Therefore, in simulating climate over the next 100 years, we should not lose sight of the fact that we are working in a four-dimensional continuum, and that time and space scales spanning many orders of magnitude coexist and interact in a complex way.

HOW IMPORTANT AN AGENT IS SOLAR FORCING ON CENTURY TIME SCALES? WHAT OCEANIC AND OTHER MECHANISMS ARE INVOLVED?

It is possible that variations in solar radiative output may have contributed to the record of past climatic variations. Although direct measurements of solar irradiance (i.e., total radiative energy) suggest only a tenth of one percent change over an 11-yr solar cycle (too small to have caused more than a tenth of a degree C global climate change), such measurements have only been reliably made over the past decade or so. Some scientists think it is possible that the Sun could vary its output by as much as several tenths of one percent over century time scales. Such forcing could not compete with hypothesized greenhouse-gas increases projected into the 21st century (i.e., warming of 1°–5°C); nonetheless it is critical to know what the contribution, if any, of solar forcing has been to the climate record of the past few centuries. Only if such a solar component of climate change were subtracted out from the record, could we infer empirically how greenhouse gases, dust veils, or other mechanisms may have contributed to past climate variations.

The central challenge we face in identifying a possible contribution of solar forcing to future climate change is that there are no known natural archives that hold information on the past behavior of what we most want to know, namely, what the solar radiation inputs to the Earth have been.

For the period before about A.D. 1900, and for most purposes before A.D. 1980, what we know about solar behavior, directly or indirectly, is the solar activity conventionally described in terms of the sunspot number. However, the connection between sunspot number and solar radiative inputs to the Earth is known only for the shortest-period changes (i.e., the change in the solar constant during the last 11-yr sunspot cycle). In terms of what we know of the physical Sun, there is every reason to expect that longer-term changes in the solar radiative output could well exceed in magnitude the 0.06% peak-to-peak change that characterized the last 11-yr sunspot cycle.

The most promising source of data on the past behavior of solar activity are tree-ring ^{14}C and ice-borne ^{10}Be. Clear quasi-periodic behavior on a time scale of ca. 200 years has been observed in both the long-term ^{14}C and ^{10}Be records (Stuiver et al. 1991; see Figures 21.1 and 21.2).

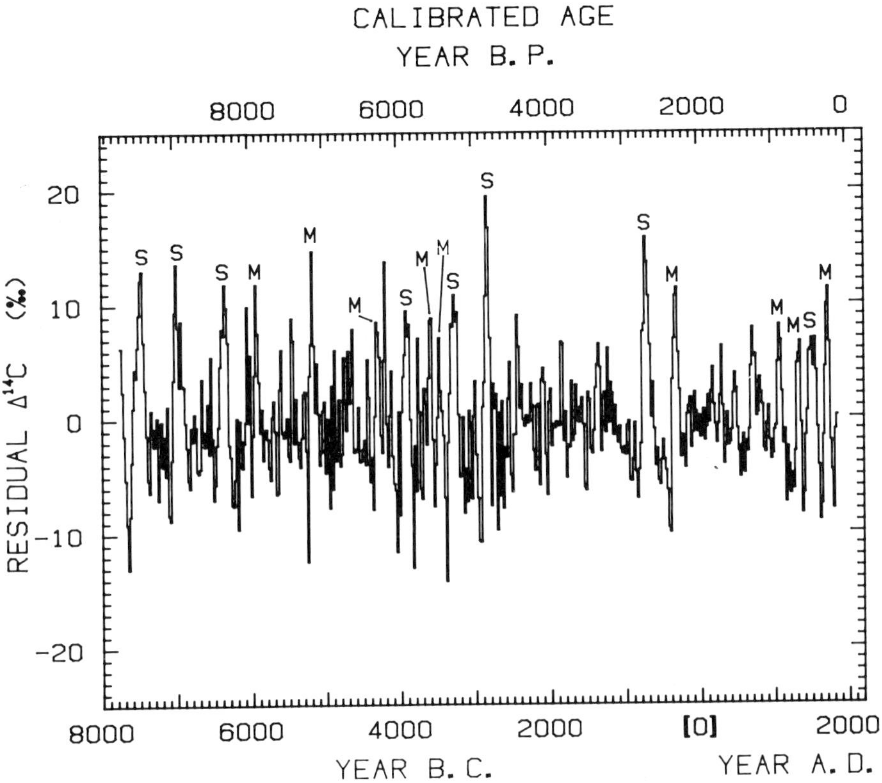

Figure 21.1 Residual $\Delta^{14}C$ in per mil, obtained after removing a spline approximating a 400-year moving average from the atmospheric $\Delta^{14}C$ record. The M and S nomenclature identify Maunder- and Spörer-type perturbations. From Stuiver et al. (1991).

A prediction of climate change over the next 100 years cannot be made without improving our knowledge about ocean circulation changes on a 10- to 100-year time scale. Variability in regional deep-water formation has been demonstrated on a 10-year time scale. Related changes in upwelling rate on a 10- to 100-year time scale may have a substantial impact on coastal climates and related tradewind variability.

Recommendations

1. Our primary need is to collect and analyze more high-temporal-resolution ^{10}Be data on the full range of the past for which good chronologies are available. This will increase our general understanding of the range of behavior of solar activity (and only indirectly of the range of possible behavior of solar radiative output).

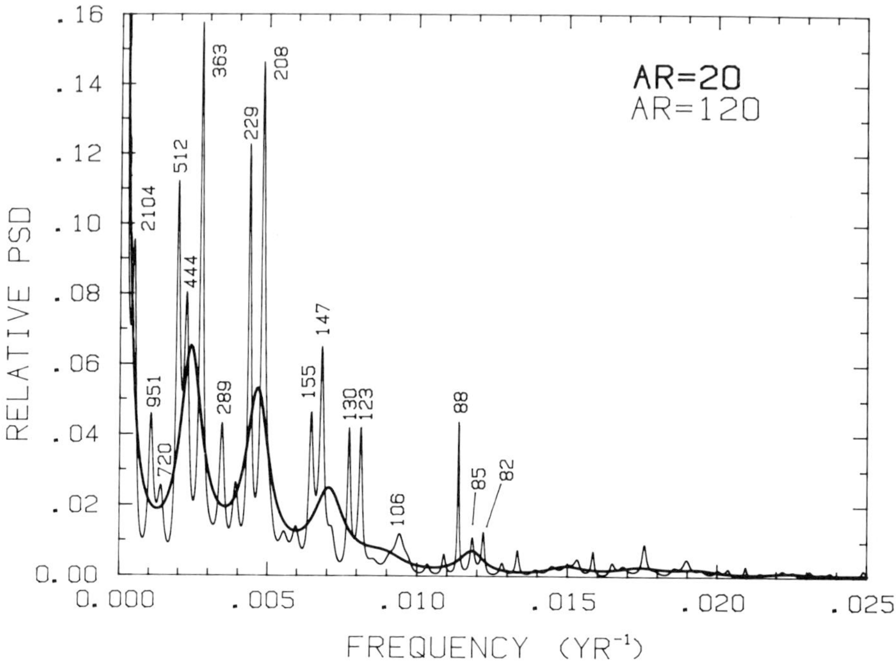

Figure 21.2 Power spectral density (PSD) normalized on the largest peak, obtained from the Maximum Entropy Method (MEM) analysis of the $\Delta^{14}C$ record. Autoregressive (AR) orders are 20 and 120. The periods are given in years for each individual peak. The peaks present at AR = 20 are 414, 216, 142 and 85 years. From Stuiver et al. (1991).

2. The interpretation of the atmospheric ^{14}C record (solar-induced production rate changes versus climate-related perturbations of the carbon reservoirs) can be improved by focusing on high-resolution properties of sediments in regions of oceanic upwelling.
3. We need to advance, by whatever means possible, the description of global and regional climate for the period of time in which good records of past solar activity exist. This will allow more meaningful comparisons of solar changes and climate changes on time scales of tens to thousands of years. We particularly need improved global descriptions of the Little Ice Age, the so-called Medieval Warm Period, and the Holocene Maximum. Examples of the type of high temporal resolution data available include:
   a. *Historical data*. The quality of these data is good if records from different observers can be compared; however, time control may be complicated by calendar changes, and the spatial coverage is limited to densely populated regions. Continuous local records are available back to about A.D. 1400 in Europe, to about A.D. 1800 in North America, and back to 200 B.C. in China.

b. *Laminated ice.* Time control is good in the upper levels (e.g., back to about 10,000 B.P.) but is more problematic in lower levels. Spatial coverage includes the ice sheets on Antarctica and Greenland, ice caps on Iceland, Baffin Island, Peru, and Tibet and mountain glaciers in widely distributed areas.
c. *Laminated sediments.* Although the potential for extracting important climatic information back to about 10,000 years from these records is high, this potential has yet to be realized. Much work needs to be done to cross-correlate between different localities.
d. *Tree rings.* Many records 300 years long are available; in some areas records extend back to 1,000 or even 10,000 years (e.g., Briffa et al. 1990; see Figure 21.3). A wealth of temperature information can be obtained from trees in the boreal and subalpine zones. Information on precipitation can be obtained from trees in arid and semi-arid zones. The climatic information from trees in temperate zones is more uncertain. Another possible difficulty is the effect of CO_2 fertilization. As in the case of laminated sediments, correlation between localities is essential (e.g., Schweingruber et al. 1991). With regard to spatial coverage, observational networks already exist in the boreal zones and some mountain regions, in the arid zone of the U.S. southwest, and in North Africa. The potential for developing networks in Siberia is enormous.

4. A substantial time range is needed because comparisons of sunspot numbers and global temperature change over intervals of a few hundred years are of limited statistical value when demonstrating a solar influence on climate. Geographic regions with potentially stronger solar influence should be identified by means of appropriate climate model studies.

In summary, we need to (a) evaluate the present archives, (b) obtain more cores in alpine regions, (c) evaluate the time control in laminated sediments, (d) expand and cross-check tree-ring networks, and (e) evaluate the climatological inferences drawn from material in temperate zones.

CAN WE USE PALEODATA TO VALIDATE THE DETAILS OF THE 10- TO 100-YR NOISE OBTAINED FROM MODELS? WHAT DATA ARE NEEDED AND ARE POSSIBLE?

Noise Estimates from Stochastic Forcing Experiments

A frequently encountered problem is that of determining whether the trend in an experiment with time-dependent forcing (e.g., the 100-year trend in surface air temperature from an experiment with a coupled ocean-atmosphere general circulation model [GCM]) is large relative to the trends that could occur due to internally generated natural

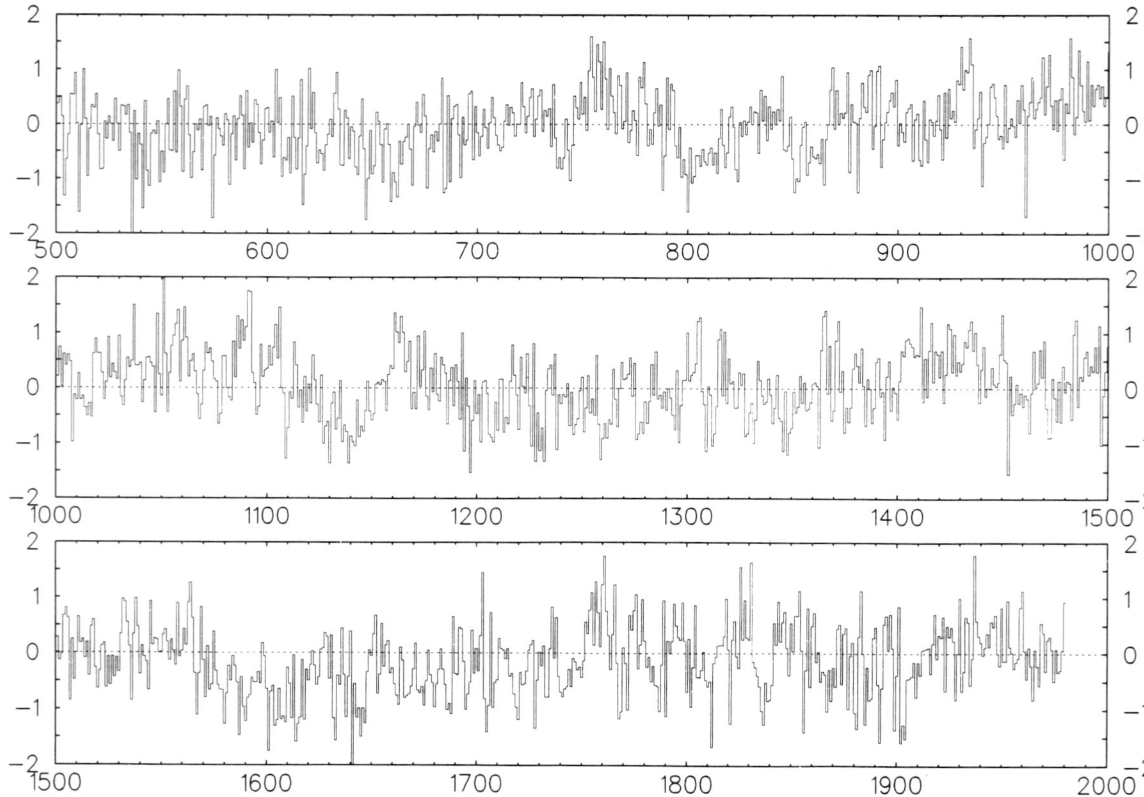

Figure 21.3 Reconstructed northern Fennoscandian temperature anomalies (°C, April–August mean) relative to 1951–1970. The dramatic temperature decrease from A.D. 1105 to A.D. 1156 is evident. The most "normal" period starts around A.D. 500 and ends at approximately A.D. 730. From Briffa et al. (1990).

variability alone (Mikolajewicz et al. 1991; Santer et al. 1993). Since any climate signal due to an enhanced greenhouse effect evolves on a decade-to-century time scale, we are interested in analyzing noise on similar time scales.

One of the difficulties is that state-of-the-art coupled models have generally been integrated for only ≤ 100 years (e.g., Manabe et al. 1991; Cubasch et al. 1992). From a single 100-yr control integration, we cannot derive a reliable estimate of the noise due to century time-scale, internally generated natural variability. For the moment, therefore, we have to rely on long stochastic experiments (> 1,000 years) with ocean GCMs in order to obtain details of the noise on century time scales (Mikolajewicz and Maier-Reimer 1990). However, we do not know whether the low-frequency variance information derived from such "ocean only" stochastic experiments (Figure 21.4) is representative of how variance is partitioned on different time scales in the real-world ocean. Furthermore, we do not know whether such model-based noise estimates are resolution dependent or differ strongly from model to model.

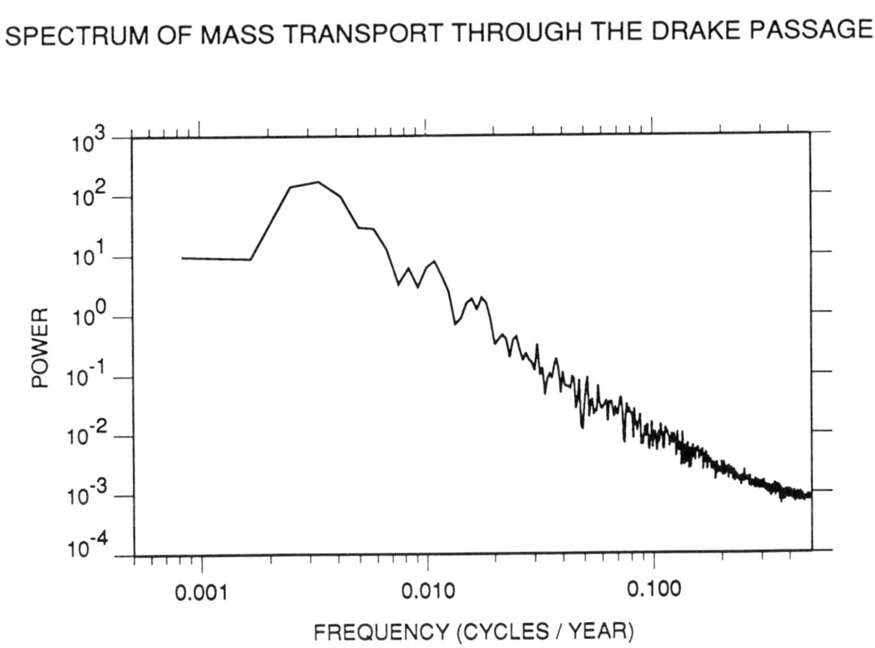

Figure 21.4 Spectrum of mass transport through the Drake Passage from the 3,800-year "ocean only" experiment performed by Mikolajewicz and Maier-Reimer (1990). Results are for a chunk length of 1,200 years and are averaged over three non-overlapping chunks. Input time series were normalized. Note that the spectrum has maximum power at a period of approximately 320 years. The mode of variability associated with this spectral peak describes the movement of large-scale salinity anomalies through the Atlantic.

Can we use paleodata as "ground truth" to verify the noise behavior simulated in the stochastic forcing experiments? Hopefully, there are opportunities for obtaining paleodata of sufficient length and high temporal resolution, which would enable us to test whether there is good empirical evidence for these simulated modes of variability.

Using model output as a guide, we should, in principle, be able to identify particular regions, variables, and time scales to test the validity of the model noise. For example, can we find characteristic phase relationships, e.g., for ocean diagnostics in the Atlantic versus Pacific or the Northern Hemisphere versus Southern Hemisphere?

Availability of High Temporal Resolution Paleoclimate Data from the Ocean

Annually dated or somewhat less well-dated paleoenvironmental records are more common for low-latitude surface ocean conditions than for higher latitudes. A few of these low-latitude records already exist, many are in the process of being developed, and many more could be acquired (e.g., see Overpeck and Cole 1990).

In the Caribbean region, coral records provide some information regarding change near Florida over the past 350 years (Druffel 1982). The development of short (< 500 years) coral records of Orinoco River outflow (Swart, in progress) and North Atlantic tradewind variability (Cole, in progress) has begun. A longer (13,000 year) marine sediment record of North Atlantic tradewind variability exists (Overpeck et al. 1989; Peterson et al. 1991) and will potentially provide an interannual (varve-based) record in the future. Barbados corals also may provide information on surface water conditions for the late Quaternary (Fairbanks 1989), as could other long drill cores from submerged reefs throughout the Tropics.

The long-term variability of the equatorial Pacific is being examined in a number of coral-based studies (Cole et al. 1992). Coral records extending beyond the length of the instrumental records have been, or are being generated near Panama (Linsley et al. 1991), Galapagos (Dunbar et al. 1991), and Australia (Isdale 1984).

Paleoclimate records from the ocean provide the most logical starting point for validating the variance statistics simulated by ocean models. However, certain land-based records may also prove useful, particularly those from islands or close to coastal locations (see above: HOW IMPORTANT AN AGENT IS SOLAR FORCING ON CENTURY TIME SCALES?). For example, certain records from tree rings, ice cores, lake sediments, and instrumental data may provide insights regarding long-term variability in specific regions of the world ocean (Diaz and Markgraf 1992). Many of these records are annually dated and span the past several centuries whereas some are less well-dated and extend back into the Holocene.

High-resolution firn stratigraphies from Antarctica and Greenland and seasonal meltwater signals in ice-core crystallographic structure may also constitute promising indicators of ocean circulation changes over the last several hundred years (see, e.g., Oeschger and Langway 1989). There is considerable potential for extending such records back in time.

WHAT ARE THE POSITIVE AND NEGATIVE FEEDBACKS BY WHICH THE CLIMATE SYSTEM RESPONDS TO MILANKOVITCH FORCING DURING THE LATE PLEISTOCENE?

Two basic modeling approaches have been used to investigate climatic feedback mechanisms with climate models. The first is to use highly resolved three-dimensional GCMs, in which grid resolution (i.e., 500×500 km) fields of motion, snowfall, cloudiness, vertical sensible heat transfer, radiative fluxes, and snow melt are all calculated. However, because such models are computer intensive, they can only be run over decades of simulated time. Thus, they can investigate detailed processes of glacier inception or ablation but are, strictly speaking, only snapshots of a dynamical process in which mass, momentum, and energy are exchanged across the atmosphere, ocean, ice, land, and biota interfaces.

The second category of models accounts for these exchanges across interfaces; however, in view of the practical computational limitations involved in simulating hundreds of thousands of years, such coupled dynamical models make major simplifications to physical comprehensiveness or spatial resolution. A mix of research using both types of model, cross-checking their validity against both instrumental and proxy data, is the best research strategy. For example, GCMs could be used to help parameterize two-dimensional models, which then could be run over longer time periods. Even though the GCMs only resolve "fast physics," these processes could have residual effects over ice-age cycle time scales, and thus both approaches are needed.

In our discussion group, we only had time and resources to examine a few specific models in detail. These are reported here because we believe that they capture the nature of many of the important climate feedback issues. Given our constraints, we did not attempt to provide a comprehensive summary of all relevant modeling efforts in snapshot (Kutzbach and Street-Perrott 1985; Rind et al. 1989) or dynamical modeling (Le Treut and Ghil 1983; Pollard 1983; Neeman et al. 1988; Berger et al. 1990; DeBlonde and Peltier 1991; Herterich, this volume). We hope the discussion here, which focuses on only a limited subset of available studies, will usefully illustrate many of the issues involved in modeling climatic feedbacks on the glacial–interglacial cycle time scale.

In this section we first examine the feedback loops contained in a 2.5-dimensional climate model (LLN) recently introduced by Berger et al. (1990). We then discuss different feedbacks, which have been incorporated in other models, and point out the difficulty of choosing among alternate formulations and determining which, if any, are correct.

Feedbacks Mechanisms in the LLN Model

A two-dimensional (latitude–altitude) climate model has been validated over the present-day climate (Gallée et al. 1991) and asynchronously coupled to a model of the three main Northern Hemisphere ice sheets and their underlying bedrock (Gallée et al. 1992). The aim was to assess the influence of a number of important processes and feedbacks in an ice-age simulation using astronomically derived insolation (Imbrie et al., this volume) and CO_2 data from the Vostok ice core (Lorius et al. 1990).

One important feature of this model is that it simulates both the glaciation and deglaciation of the last climatic cycle (see Figure 21.5). This is due to a number of feedbacks introduced in the model. Although these represent only a subset of all the feedbacks acting in the climate system, it is useful to analyze their relative contributions.

Entering the Glaciation

In high northern latitudes, summer insolation started to decrease roughly 123 Ka B.P., with a minimum reached 11 to 12 Ka later. For June, this represents a change of about 20%. Each time this type of decrease occurs, positive feedbacks amplify the response of the climatic system to such external forcing, not only until the minimum of insolation is reached, but for several thousand years later due to the inertia of the slow-response part of the system. During the same time, negative feedbacks progressively slow the buildup of ice sheets up to the beginning of their melting, when the ice maximum is reached.

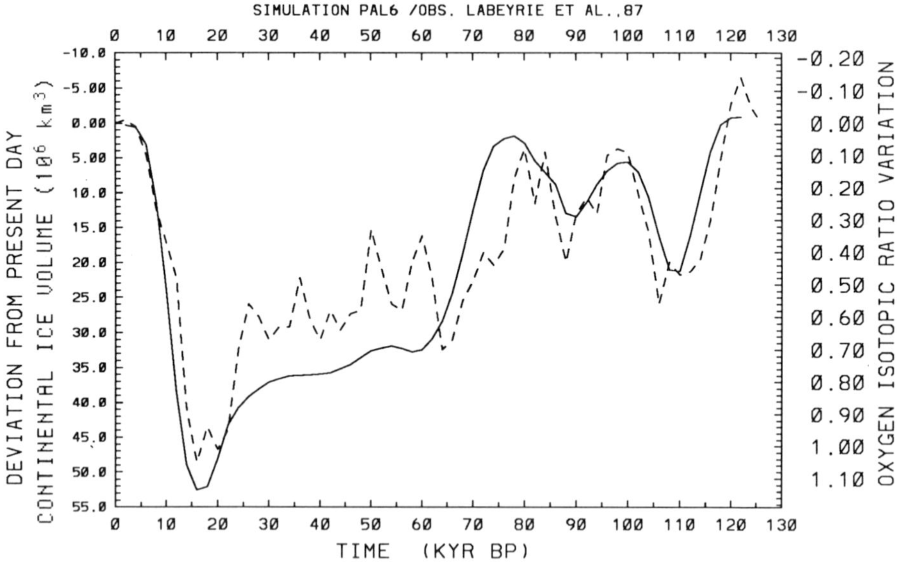

Figure 21.5 Variations over the last glacial-interglacial cycle of the deviation from the observed present-day value (30.5×10^6 km^3) of the simulated total continental ice volume over the Earth (Gallée et al. 1992; solid line) and the variation of the global sea-water oxygen ratio given by Labeyrie et al. (1987; dashed line).

When such insolation decrease occurs, temperature (T) also decreases, which increases the portion of precipitation falling as snow in high latitudes and consequently enlarges the extent of the snow field. At the same time, the taiga is replaced by tundra, which directly affects the albedo of the surface, increasing the albedo of the vegetated surface covered by snow. Therefore, both the enlarged snow field and the change in vegetation type lead to an increase in the surface albedo, creating a positive feedback. This is reinforced by the subsequent reduction of the water vapor content of the atmosphere, resulting in a decrease in the downward-directed infrared radiation. An important amplification is also due to the parallel decrease in the atmospheric CO_2. This cooling, however, is also responsible for a decrease of the upward infrared radiation, latent heat, and the sensible heat fluxes at the surface, which feedback negatively on T.

After this initiation phase lasting roughly 4 Ka, the zonal mean temperature decrease at the surface leads to atmospheric cooling, which directly increases snowfall in the regions where the ice sheets start to form. The radiative budget there becomes negative, which results in a net accumulation at the surface and suppressed ablation. As a consequence, the ice sheets grow and depress the lithosphere below. As the altitude of the ice sheets increases, their surface temperature decreases (a positive feedback), which progressively decreases the snowfall (a negative feedback) and finally stops the latitudinal growth of the ice sheet. At the same time, the ice sheets extend to the south. Because of the continentality effect, snowfall decreases at the top of the ice sheets in the continental interior (a negative feedback) and slows down the southward motion of the southern front of the ice sheets. Finally, the maximum ice volume is reached at 110 Ka B.P., 4 Ka later than the insolation minimum.

The Deglaciation Process

The effect of the insolation increase from 114 Ka B.P. is first manifested at the southern edge of the ice sheets, which begin to melt. At the same time, the absence of snowfall triggers the "snow aging" process in regions of the ice sheets where the temperature is above –10°C. This significantly decreases the albedo of both nonmelting and melting snow cover, mainly towards the south of the ice sheets. Concurrently, snow accumulation remains positive over the northern part of the ice sheets, thus creating a N–S flux of ice within the ice sheets. At the southern edge, melting is faster than ice flow, and the isostatic rebound is not rapid enough to level the ice sheet, which could prevent future net ablation in the south. This is not the case at the northern edge of the ice sheets, where isostatic rebound maintains a high ice-sheet elevation, allowing the N–S flux of ice to continue efficiently. This causes a decrease in the volume and height of the ice sheets, accompanied by an increase in their surface temperature and hence the global ablation. The result is a decrease of the zonal surface albedo and a further replacement of the tundra by taiga, which feeds back positively on the albedo decrease, both directly and by reducing the albedo of the continental surfaces covered by snow. This finally leads to an ice volume minimum about 5 Ka after the insolation maximum.

Conclusions

Sensitivity experiments with this 2.5-D model have shown that ablation is more important to the ice-sheet response than snow precipitation variations (as in the Herterich model, this volume), and that the ice-sheet surface energy balance and, in particular, the absorbed solar radiation, are of crucial importance. A key mechanism in the deglaciation after the last glacial maximum appears to be the "aging" of snow, which decreases its albedo. The other factors that play an important role are the ice-albedo feedback, the ice-sheet altitude ("desert effect"), insolation, the isostatic rebound, changes in vegetation type, the ice-sheet extent ("continentality effect"), CO_2 changes, and the temperature-water vapor feedback (Gallée et al. 1992). In the transient experiment, 50% of the cooling at the last glacial maximum and 30% of the Northern Hemisphere ice volume change are attributable to the CO_2 change, while 40% of the climate changes related to this CO_2 change are due to the water vapor feedback.

These processes and feedbacks are only those which play a major role in this specific model. Other important feedbacks are not included, e.g., those involving the deep ocean circulation, clouds (particularly over the ice sheets), and dust in the atmosphere.

In view of the relatively good simulations obtained by the LLN model, there are many important implications that need to be explored. First, were the results fortuitous or, rather, were most of the important processes captured in two dimensions by neglecting such factors as changes in deep-water formation? Second, in view of the critical importance of snow–albedo–temperature–vegetation feedbacks on the initiation and deglaciation processes, can the parameterizations of these feedbacks be improved?

Alternate Choices of Feedback Mechanism in Low-dimensional Models

The model of DeBlonde and Peltier (1991) for example, differs substantially from the LLN model of Gallée et al. (1991) in the mix of ingredients that are proposed to account for the same phenomenon. The former is a global one-level seasonal energy-balance model, which differentiates land from sea in a geographically realistic way and in which multiple ice sheets evolve in specific geographic subdomains in response to the applied insolation anomalies. The model incorporates the influence of glacial isostatic adjustment, of elevation-desert feedback, and a realistic surface topography. The hydrological cycle is driven by a precipitation field that matches present-day precipitation within each of the cryospheric subdomains, and the ice-sheet mass balances are computed as monthly averages on the basis of a geographically specific annual cycle that is diagnostically determined from the ice sheets. By itself, the mix of physical ingredients explains rather nicely the observed locations of the main Northern Hemisphere ice sheets that formed during the last glacial cycle, and also the thicknesses that these ice sheets attained. Without additional feedback loops, however, the model is unable to explain fully the deglaciation process.

A feedback that might achieve the desired effect in this regard is that associated with atmospheric dust. Ice cores and deep-sea sedimentary cores show that late glacial episodes are characterized by dry and dusty conditions for many thousands of years prior to major deglaciation events. When the albedo of the ablation zones in the DeBlonde–Peltier model is modulated in a way implied by the dustiness records, complete deglaciations are then predicted to occur. In the LLN model, an impact of "ice-aging" is invoked to effect a similar modulation of albedo, and this, too, effects the desired collapse of the ice sheet in response to the Milankovitch anomalies.

How are we to distinguish which, if either, of these mechanisms are correct? It would appear that the only acceptable course is to further disaggregate the model and describe more of the physics explicitly.

Feedbacks in General Circulation Models

GCMs suffer from many well-known deficiencies with respect to their ability to represent hydrogeological processes accurately. Probably the best evidence of this is provided by the difficulties that GCM groups have experienced in attempting to nucleate ice sheets in response to applied orbital insolation variations, commencing, for example, at 124 Ka B.P. (e.g., Rind et al. 1989). It does, therefore, seem that there are serious flaws in the sensitivities of the hydrologic cycles in all conventional atmosphere–ocean GCMs. In terms of the fundamental issues posed by paleoclimate observations, it would appear crucial that these problems be recognized and solved before we will be able to proceed to employ fully disaggregated models (atmosphere–ocean GCMs) to address the issues of ice-age climate. A satisfactory solution to these problems would also enhance the credibility of predictions of future climates that are made by the same models.

Towards this end, it is useful to compare the feedback parameterizations in the LLN model with those in a typical GCM. The positive feedback associated with the appearance of ice sheets is its effect on increasing the surface albedo. This occurs in both the GCM as well as in the LLN model, associated with the reduction of atmospheric mass above a taller ice sheet, and therefore reduced atmospheric greenhouse capacity. Negative feedbacks in both models include reduction in precipitation, associated with the colder climate, and a reduction in thermal radiation loss from the colder surface. In the LLN model, a primary negative feedback, which is ultimately responsible for melting ice after the 18 Ka maximum, is the reduction in albedo caused by aging snow. This feedback is also in the GCM; however, it is overshadowed by the negative feedback associated with reductions in sensible and latent heat loss from the cold ice surface. The presence of snow and ice forces the ground temperature below that of the surface air, reducing turbulence and associated fluxes. It is this effect that helps the GCM remove ice and snow.

In order for a GCM to grow ice at approximately 115 Ka B.P., it must calculate a positive mass balance, i.e., precipitation as snow must exceed melting plus runoff. The problem is that the precipitation does not exceed the melting. Either more snow must fall in the seasons with temperatures below freezing or the net heat at the surface must be reduced. Winter snow is limited by the cold air temperatures above the ice; in addition, it has the tendency of generating high pressure above ice sheets (the "desert effect"). To reduce net heating will require increased or thicker low clouds above the ice sheet (by virtue of the ice-sheet melting, the GCM already has an extensive low cloud cover), greater reduction in incident solar radiation, or, by whatever means, colder atmospheric temperatures during summer.

This last effect would solve all the problems: it would reduce the change in sensible and latent heat fluxes, which limit ice duration, and it would reduce the amount of precipitation falling as rain, changing it to snow. (For example, when the GISS group raise the ice albedo to 100%, spectrally integrated, the ice could be maintained. Yet this value is unrealistic.) The solar radiation reduction in summer does not provide enough cooling; nor does additional reduction of CO_2. If the global temperature was significantly reduced, including the North Pacific—implying a greater global sensitivity, with increased sea ice, altered clouds, and reduced water vapor–summer temperatures would cool.

WHAT DRIVES THE 100 Ka ICE-VOLUME CYCLE AND WHY DID IT START ABOUT 1 MILLION YEARS AGO?

For at least the past 3.5 million years, much of the variance in the benthic marine record of $\delta^{18}O$ has been concentrated in three narrow-band cycles, having periods near 23 Ka, 41 Ka, and 100 Ka. These organized oscillations are superimposed on a long-term trend toward more enriched values that reflects the combined effect of a cooling of deep water and the storage of ice on land. Exactly how much of the change over any given interval is due to ice and how much to temperature is not clear; however, there is little doubt that the record of the past 2.5 Ma is dominated by the ice-volume effect. Several features of this record—including the origin of the 23 Ka, 41 Ka, and 100 Ka cycles—cry out for a satisfactory explanation. First, we want to know why, between about 3.5 and 2.5 Ma, a conspicuous downward step (toward cooling or ice growth) occurred, and why a second downward step (that almost certainly reflects a major increase in ice volume) occurred about 1 Ma. In addition, we want to know why the spectrum of climate as recorded in this isotopic index has changed so dramatically during the mid-Pleistocene transition at about 1 Ma, from an earlier regime dominated by a combination of the 23 Ka and 41 Ka cycles, to a later regime in which 70% of the variance (and 50% of the amplitude of the observed oscillation) occurs as a 100 Ka cycle, where fast deglaciation events called terminations mark the end of major glaciations.

All of these problems have been widely recognized for many years and have received serious study by modellers and observationalists. Yet none of them have been satisfactorily solved. From the perspective of this report, which is directed at the problem of identifying the forcing of climatic changes, it might be argued that the problem of the origin of the 23 Ka and 41 Ka cycles is nearest to a satisfactory solution. In part this is because we have clear evidence (Imbrie et al. 1984; see also Figure 21.6) that these climatic cycles are linearly related to external, Milankovitch-forcing functions (variations in incoming radiation that are driven by the 23 Ka cycle of equinoctial

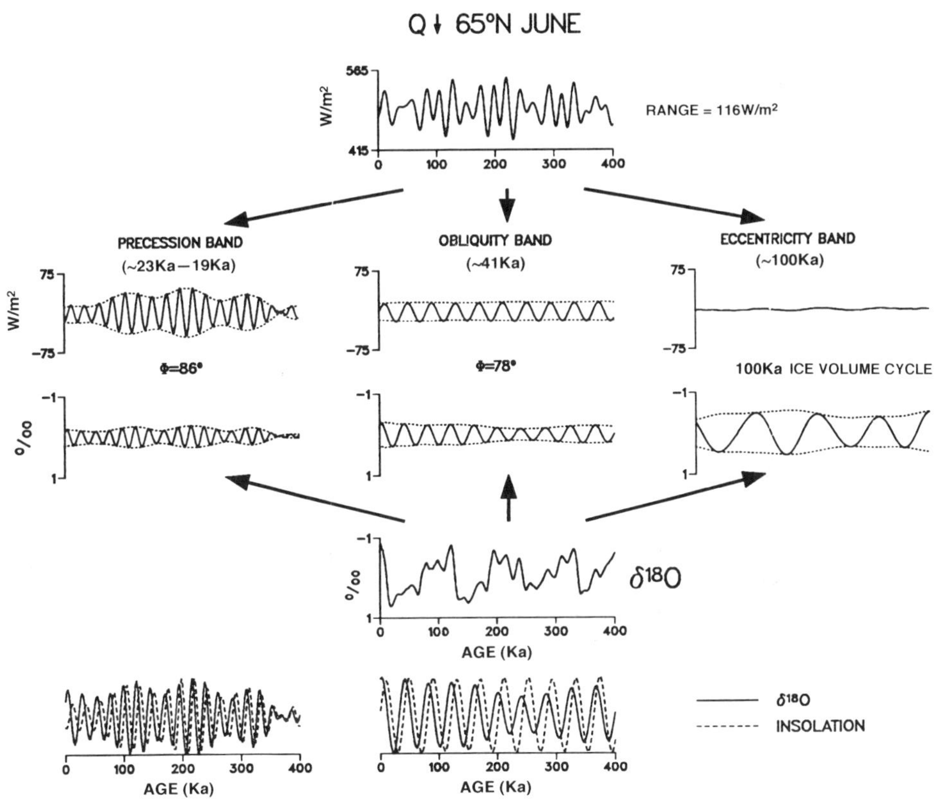

Figure 21.6 Evidence for Milankovitch forcing. Using bandpass filters, radiation time series and $\delta^{18}O$ time series of the SPECMAP stack are partitioned (arrows) into frequency components in the precession, obliquity, and 100-Ka (eccentricity) bands. In the precession and obliquity bands, the strong coherency and consistent phase relationship between the radiation and climate curves argue that the system has a strong, nearly linear response to radiation. However, in the 100-Ka band, where the amplitude of the climatic oscillation is the largest, the amplitude of the radiation curve is very small. The Milankovitch forcing for the 100-Ka climate cycle, if it exists, must be strongly nonlinear. Data and procedures are referenced in Imbrie et al. (1984).

precession and the 41 Ka cycle of obliquity), and, in part, because numerous modeling studies have shown that several parts of the climate system are sensitive to the rather large changes in irradiation (~120 W/m^2) that occur in these frequency bands (Prell and Kutzbach 1987; Kutzbach and Gallimore 1988).

Our focus of attention here is therefore on two problems where substantial doubt exists about the identity of the forcing function: What drives the 100 Ka cycle? Why did this cycle start about 1 million years ago?

What Drives the 100 Ka Cycle?

Five classes of explanation have been proposed (Imbrie et al. 1989):

1. One group of theories explains the cycle as a linear response to Milankovitch forcing, i.e., a response to variations in insolation that are driven by changes in orbital geometry. Here the main difficulty is that the range of the eccentricity-related variation in radiation that occurs at periods near 100 Ka is very small (~4 W/m^2) compared to the range of radiation (~116 W/m^2) that is associated with the much larger climatic response in the precession and obliquity bands. This class of theories invoke a linear resonance.
2. Another group of theories explains the cycle as a nonlinear response to Milankovitch forcing, either by a stochastic resonance in the system that shifts power from the precession band to the 100 Ka band (Ghil and LeTreut 1981) or by climatic mechanisms, which may be described as acting either like a rectifier or a band-pass filter (Calder 1974; Imbrie and Imbrie 1980; DeBlonde and Peltier 1991; Herterich, this volume). The rectifier-type of models so far proposed are unsatisfactory in that they generate power not only at 100 Ka, but also at the 413 Ka period which dominated the Late Pleistocene eccentricity spectrum (but which is either absent or a very minor feature of the Late Pleistocene $\delta^{18}O$ spectrum). A model proposed by Berger (1990), containing a set of climatic feedbacks that act as a band-pass filter, avoids this problem.
3. Another group of theories explains the cycle as an internal oscillation driven by a combination of positive and negative feedbacks with different time constants. The oscillation occurs in the absence of any external forcing. In the model of Maasch and Saltzman (1990), for example, the internally driven 100 Ka oscillation interacts with the externally forced oscillations at 23 Ka and 41 Ka to set the phase of the longer cycle to occur as it does in the data, i.e., nearly in phase with eccentricity.

   For the past million years, the low-frequency component of the $\delta^{18}O$ record from ODP site 677 (Shackleton et al. 1990) remains approximately in phase with eccentricity. In the older record, the low-frequency component is less clear; however, some of the strongest "cycles" appear to be in phase with eccentricity.

4. External forcing functions other than Milankovitch have also been suggested as a driving mechanism. Possible candidates include appropriately spaced changes in intergalactic dust concentrations and changes in solar irradiance. Yet, arguments can be developed to shed serious doubt on this solar mechanism. First, there is no theoretical reason suggesting that solar output *should* show power at a period of 100 Ka. (At the same time, there is no reason why there should not be internal solar variations at that frequency.) Second, the gravitational forces that might be invoked as a mechanism to initiate changes in solar radiation (based on the position of planets in their orbit) are negligible. In the extreme case where all the planets align perfectly on one side of the Sun, the effect would be to raise a tide in the solar atmosphere with an amplitude of only 1 cm.
5. Finally, it has been suggested that the 100 Ka cycle, and its onset, may be viewed as an expression of deterministic chaos. However, even if the most convenient mathematical model of the 100 Ka cycle was of a system of this type, the geologists and climate modellers who deal with the problem are not relieved of the obligation of discovering and modeling the physical mechanisms by which the cycle is driven.

Why Did the 100 Ka Cycle Start about 1 Million Years Ago?

Two general types of explanation have been proposed to explain the onset of the 100 Ka regime about 1 million years ago.

The first possibility is that a change in the orbital geometry occurring at this time might trigger the onset of the 100 Ka regime. However, as shown by Imbrie et al. (this volume), the evolution of the 100 Ka eccentricity cycle provides a strong argument against any simple formulation of this hypothesis. This is because the amplitude of the 100 Ka oscillation in eccentricity *decreases* by about 50% at approximately the time of the *onset* of the 100 Ka climate cycle.

Second, most students of the problem have suggested that the time of onset reflects a threshold value in some slowly changing boundary condition (such as continent-ocean geography, continental elevation, or atmospheric chemistry). In simplest terms, the increase in mean ice volume and the onset of the 100 Ka regime which occurs at about the same time, might be viewed as a threshold response of the system to the long-term global cooling trend of the last 100 Ma. Alternatively, the cooling event could be an indication of accelerated change in the forcing. In any case, the cooling could shift the mean state of the climate system, enhance positive snow-ice/albedo feedbacks favorable to additional high-latitude cooling, and thereby favor larger glaciations.

Many hypotheses look to tectonic sill depths as the cause of Tertiary cooling. Sill depths can affect ocean circulation and thereby CO_2. By 1 Ma, the Panama Isthmus had already closed and so cannot be the cause; however, deepening of sills in the Norwegian Sea remains a possibility. Reduced input of CO_2 from volcanoes due to seafloor spreading can be rejected as a cause over this interval because global mean spreading rates were constant or increasing slightly.

Uplift of high plateaus in Asia and other continents is a possible source of tectonic forcing. Plateau uplift appears to account for numerous regional Tertiary climatic trends (wetter, drier) at middle latitudes. This results from several major effects on atmospheric circulation, including creation of vigorous monsoons. A plausible explanation of Tertiary cooling is increased chemical weathering of silicate rocks freshly exposed on plateau margins by faulting and folding and subjected to intense monsoonal rainfall and rapid runoff.

Climate simulations suggest that changes in land–sea distribution over the past 100 Ma may have caused a 15°C decrease in summer temperatures over Greenland (Hyde et al. 1989), possibly causing threshold formation of the Greenland ice sheet. In any case, the cooling could shift the mean state of the climate system.

Recommendations

1. Efforts to model the 100 Ka cycle should be designed to capture and predict the behavior of one or more elements of the climate system in addition to that of global ice volume. These "side predictions" can then be used to test the model performance against data.
2. Experiments with atmospheric GCMs and fully coupled atmosphere–ocean GCMs should be conducted to develop and test promising ideas advanced to explain the onset of the 100 Ka regime.
3. Time series that capture the behavior of many different elements of the system's evolution should be developed and compared with the output of time-dependent models, and with theories which attempt to account for the onset of 100 Ka regime.

HOW SIMILAR ARE DIFFERENT "GLACIALS" AND "INTERGLACIALS"?

Given the complexity of the climate system and the complexity of the narrative of past climate, it is often necessary to develop simple models of climate change that can treat the transition between an interglacial state (like the Holocene) and a glacial state (like the last glacial maximum), or visa versa. However, our multidimensional record of late Pleistocene climate is now sufficiently detailed to indicate that any given interglacial interval differs from another, and that any given glacial interval differs from another, in climatically significant ways. For example, the amount of carbon storage in the deep ocean during glacial Stage 8 is significantly greater than that during Stage 2, as indicated by the $\delta^{13}C$ record of benthic forams (Shackleton and Pisias 1985). The vertical gradient of $\delta^{13}C$ at several places in the ocean is significantly greater during interglacial Stage 5e than it is during Stage 1 (Curry and Crowley 1987). The

$\Delta\delta^{13}C$ proxy for North Atlantic Deep Water flux at site 607 in the North Atlantic is slightly more enriched during Stage 3 that it is during Stage 1 (Raymo et al. 1990). In addition, the poleward extent of warm water during Stages 1 and 5e is significantly greater in the Norwegian Sea and in the Antarctic Ocean than it was during preceding interglacials (Kellogg 1977; Kellogg et al. 1978; Howard and Prell 1992).

That such differences between one glacial or interglacial state and another should occur can be regarded as a natural consequence of the structure of the Milankovitch-forcing functions. These take the form of sinusoidal waves which—not being harmonics—must combine to produce a complex time history of radiation fields.

Recommendations

1. Time-dependent models should be developed which predict the behavior of several different elements of the climate system.
2. Proxy data sets should be developed on land and at various depths in the ocean to allow the spatial pattern of climate to be sampled at different places in the climate system and at different times in the climate narrative.

SUMMARY OF RECOMMENDATIONS

Below we summarize the principal recommendations developed in the preceding sections of our working group report, separated into two groups according to the time scale of the phenomenon of interest.

Time Scale of the Quaternary Glacial Cycles

1. Determine the places of origin of the major ice sheets. Where ice sheets began growing is a sensitive indicator of the processes that brought them into being. The integration of geological evidence with modeling experiments provides a way to accomplish this objective.
2. Correlate marine and terrestial records of past environmental changes with a view towards documenting spatial patterns that distinguish different episodes of glacial and interglacial climate.
3. Improve our knowledge of the onset and driving of the 100 Ka cycle by constructing models that predict changes in parts of the atmosphere and ocean which are linked to ice volume; these predictions should then be tested against time series or vertical sections of appropriate proxy indicators. The relative phasing of responses of different parameters can be used to distinguish between different causal mechanisms.
4. Improve the ability of GCMs to simulate the growth and decay of major ice sheets. More realistic models will be needed to test processes and to make more profitable use of the data called for in Pt. (1) above.

5. Improve parameterizations of processes crucial to the formation and removal of ice sheets, including modeling of the planetary boundary layer, processes involving albedo, and changes in surface albedo with aging. Paleoclimate models now suggest that the formation and decay of the major ice sheets play significant roles in glacial–interglacial cycles.

Time Scales of 10–1,000 Years

1. Utilize marine paleodata of high temporal resolution to verify the noise behavior in stochastic experiments performed with ocean models, with particular emphasis on the 10- to 100-yr time scale. The purpose is to separate the natural high-frequency noise in the coupled climate system from indications of significant driving forces.
2. Improve, by every means possible, the spatial coverage and time controls of high-resolution climatic records of the past 1,000 years. Limitations in our current knowledge of the geographic extent, temporal extent, and significant details of purported climatic features of this period severely constrain efforts to identify their causes. One important use of these data will be to identify the regions, if any, in which climate change can be related to solar forcing.
3. Collect and analyze ^{10}Be data from high-resolution ice cores and compare these data, where possible, with historic records of solar activity, the tree-ring record of ^{14}C, and the geomagnetic record. The aim is to increase our knowledge of year-to-year, decadal, and century-scale changes in solar behavior.
4. Extend present efforts to model solar activity changes, from the 11-year cycle to longer, less-regular features of significant solar change, such as the Maunder and Spörer Minima. More reliable models of solar activity, based upon the most accurate models of the structure and dynamics of the solar interior, offer our best hope of predicting decade-to-century changes in the radiative outputs of the Sun.

ACKNOWLEDGEMENTS

A number of conference participants outside of our group provided valuable input to this report. We would like to acknowledge contributions made by Tom Crowley, Jean Jouzel, Jonathan Overpeck, Dick Peltier, Dominique Raynaud, David Rind, Bill Ruddiman, and Nick Shackleton.

REFERENCES

Berger, A. 1990. Testing the astronomical theory with a coupled climate-ice sheet model. *Glob. Planet. Change* **3**:125–141.

Briffa, K.R., T.S. Bartholin, D. Eckstein, P.D. Jones, W. Karlén, F.H. Schweingruber, and P. Zetterberg. 1990. A 1,400-year tree-ring record of summer temperatures in Scandinavia. *Nature* **346**:434–439.

Calder, N. 1974. Arithmetic of ice ages. *Nature* **252**:216–218.

Cole, J.E., G.T. Shen, R.G. Fairbanks, and M. Moore. 1992. Coral monitors of El Niño–Southern Oscillation dynamics across the equatorial Pacific. In: Paleoclimatic Aspects of El Niño–Southern Oscillation, ed. H.F. Diaz and V. Markgraf. Cambridge: Cambridge Univ. Press, in press.

Cubasch, U., K. Hasselmann, H. Höck, E. Maier-Reimer, U. Mikolajewicz, B.D. Santer, and R. Sausen. 1992. Time-dependent greenhouse warming computations with a coupled ocean–atmosphere model. *Clim. Dynam.* **8:**55–69.

Curry, W.B., and T.J. Crowley. 1987. The $\delta^{13}C$ of equatorial Atlantic surface waters: Implications for ice age pCO_2 levels. *Paleoceanog.* **2**:489–517.

Dansgaard, W., J.W.C. White, and S.J. Johnsen. 1989. The abrupt termination of the Younger Dryas climate event. *Nature* **339**:532–534.

DeBlonde, G., and W.R. Peltier. 1991. A one-dimensional model of continental ice volume fluctuations through the Pleistocene: implications for the origin of the Mid-Pleistocene climate transition. *J. Climate* **4**:318–344.

Diaz, H.F., and V. Markgraf, eds. 1992. Paleoclimatic Aspects of El Niño–Southern Oscillation. Cambridge: Cambridge Univ. Press, in press.

Druffel, E.R.M. 1982. Banded corals: changes in oceanic Carbon-14 during the Little Ice Age. *Science* **218**:13–20.

Dunbar, R.B., G.M. Wellington, M.W. Colgan, and P.W. Glynn. 1991. Eastern tropical Pacific corals monitor low-latitude climate of the past 400 years. In: Proceedings of the Seventh Annual Pacific Climate (PACLIM) Workshop, ed. J.L. Betancourt, pp. 183–198. Sacramento: California Dept. of Water Resources.

Fairbanks, R.G. 1989. A 17,000-year glacio-eustatic sea level record: influence of glacial melting rates on Younger Dryas event and deep-ocean circulation. *Nature* **342**:637–642.

Flohn, H. 1986. Singular events and catastrophes now and in climatic history. *Naturwissen.* **73**:136–149.

Gallée, H., J.P. van Ypersele, T. Fichefet, C. Tricot, and A. Berger. 1991. Simulation of the last glacial cycle by a coupled sectorially averaged climate-ice-sheet model. I. The climate model. *J. Geophys. Res.* **96**:13139–13161.

Gallée, H., J.P. van Ypersele, T. Fichefet, C. Tricot, and A. Berger. 1992. Simulation of the last glacial cycle by a coupled sectorially averaged climate-ice-sheet model. II. Response to insolation and CO_2 variation. *J. Geophys. Res.* **97:**15713–15740.

Ghil, M., and H. Le Treut. 1981. A climate model with cryodynamics and geodynamics. *J. Geophys. Res.* **86**:5262–5270.

Howard, W.R., and W.L. Prell. 1992. Late Quaternary surface circulation of the Southern Indian Ocean and its relationship to orbital variations. *Paleoceanog.*, in press.

Hyde, W.T., T.J. Crowley, K.-Y. Kim, and G.R. North. 1989. Comparison of GCM and energy balance model simulations of seasonal temperature changes over the past 18,000 years. *J. Climate* **2**:864–887.

Imbrie, J., J.D. Hays, D.G. Martinson, A. McIntyre, A.C. Mix, J.J. Morley, N.G. Pisias, W.L. Prell, and N.J. Shackleton. 1984. The orbital theory of Pleistocene climate: support from a revised chronology of the marine $\delta^{18}O$ record. In: Milankovitch and Climate, Part 1, ed. A. Berger, J. Imbrie, J. Hays, G. Kukla, and B. Saltzman, pp. 269–305. Dordrecht: Reidel.

Imbrie, J., and J.Z. Imbrie. 1980. Modeling the climatic response to orbital variations. *Science* **207**:943–953.

Imbrie, J., A. McIntyre, and A. Mix. 1989. Oceanic response to orbital forcing in the late Quaternary: observational and experimental strategies. In: Climate and Geosciences, ed. A. Berger, S.H. Schneider, and J.C. Duplessy, pp. 121–164. Boston: Kluwer.

Isdale, P. 1984. Fluorescent bands in massive corals record centuries of coastal rainfall. *Nature* **310**:578–579.

Kellogg, T.B. 1977. Paleoclimatology and paleo-oceanography of the Norwegian and Greenland Seas: The last 450,000 years. *Marine. Micropal.* **2**:235–249.

Kellogg, T.B., J.C. Duplessy, and N.J. Shackleton. 1978. Planktonic foraminiferal and oxygen isotopic stratigraphy and paleoclimatology of Norwegian deep-sea cores. *Boreas* **7**:61–73.

Kutzbach, J.E., and R.G. Gallimore. 1988. Sensitivity of a coupled atmosphere–mixed-layer ocean model to changes in orbital forcing at 9,000 years B.P. *J. Geophys. Res.* **93**:803–821.

Kutzbach, J.E., and F.A. Street-Perrott. 1985. Milankovitch forcing of fluctuations in the level of tropical lakes from 18 to 0 Ka BP. *Nature* **317**:130–134.

Labeyrie, L.D., J.C. Duplessy, and P.L. Blanc. 1987. Deep-water formation and temperature variations over the last 125,000 years. *Nature* **327**:477–482.

Le Treut, H., and M. Ghil. 1983. Orbital forcing, climatic interactions, and glaciation cycles. *J. Geophys. Res.* **88**:5167–5190.

Linsley, B.K., R.B. Dunbar, D.A. Mucciarone, and G.M. Wellington. 1991. Seasonal to decadal scale oceanographic variability recorded by eastern Pacific corals over the last 200+ years. AGU 1991 fall meeting programs and abstracts. *EOS* **72**:66.

Lorius, C., J. Jouzel, D. Raynaud, J. Hansen, and H. Le Treut. 1990. The past and future in ice cores. *Nature* **347**:139–145.

Maasch, K.A., and B. Saltzman. 1990. A low-order dynamical model of global climatic variability over the full Pleistocene. *J. Geophys. Res.* **95**:1955–1963.

Manabe, S., R.J. Stouffer, M.J. Spelman, and K. Bryan. 1991. Transient responses of a coupled ocean-atmosphere model to gradual changes of atmospheric CO_2. I. Annual mean response. *J. Climate* **4**:785–818.

Mikolajewicz, U., and E. Maier-Reimer. 1990. Internal secular variability in an ocean general circulation model. *Clim. Dynam.* **4**:145–156.

Mikolajewicz, U., E. Maier-Reimer, and T.P. Barnett. 1991. Acoustic detection of greenhouse-induced climate changes in the presence of natural variability. Max-Planck-Institut für Meteorologie Report No. 69, Hamburg.

Neeman, B.U., G. Ohring, and J.H. Joseph. 1988. The Milankovitch theory and climate sensitivity. II. Interaction between the Northern Hemisphere ice sheets and the climate system. *J. Geophys. Res.* **93**:11175–11191.

Oeschger, H., and C.C. Langway Jr., eds. 1989. The Environmental Record in Glaciers and Ice Sheets. Dahlem Konferenzen. Chichester: Wiley.

Overpeck, J.T., and J.E. Cole. 1990. The role of corals, varved sediments and models in understanding global environmental change. *EOS* **71(29)**:983–989.

Overpeck, J.T., L.C. Peterson, N. Kipp, J. Imbrie, and D. Rind. 1989. Climate change in the circum-North Atlantic region during the last deglaciation. *Nature* **338**:553–557.

Peterson, L.C., J.T. Overpeck, N. Kipp and J. Imbrie. 1991. A high resolution Late-Quaternary upwelling record from the anoxic Cariaco Basin, Venezuela. *Paleoceanog.* **6**:99–119.

Pollard, D. 1983. A coupled climate ice-sheet model applied to the Quaternary ice ages. *J. Geophys. Res.* **88**:7705–7718.

Prell, W.L., and J.E. Kutzbach. 1987. Monsoon variability over the past 15,000 years. *J. Geophys. Res.* **92**:8411–8425.

Raymo, M.E., W.F. Ruddiman, N.J. Shackleton, and D.W. Oppo. 1990. Evolution of Atlantic–Pacific $\delta^{13}C$ gradients over the last 2.5 m.y. *Earth Planet. Sci. Lett.* **97**:353–368.

Rind, D., D. Peteet, and G. Kukla. 1989. Can Milankovitch orbital variations initiate the growth of ice sheets in a general circulation model? *J. Geophys. Res.* **94**:12851–12871.

Santer, B.D., W. Brüggemann, U. Cubasch, K. Hasselmann, H. Höck, E. Maier-Reimer, and U. Mikolajewicz. 1993. Signal-to-noise analysis of time-dependent greenhouse warming experiments. Part 1: Pattern analysis. *Clim. Dynam.*, in press.

Schweingruber, F.H., K.R. Briffa, and P.D. Jones. 1991. Yearly maps of summer temperatures in Western Europe from A.D. 1750 to 1975 and Western North America from 1600 to 1982: Results of a radiodensitometrical study on tree rings. *Vegetatio* **92**:5–71.

Shackleton, N.J., A. Berger, and W.R. Peltier. 1990. An alternative astronomical calibration of the Lower Pleistocene timescale based on ODP Site 677. *Trans. R. Soc. Edin. Earth Sci.* **81**:251–261.

Shackleton, N.J., and N.G. Pisias. 1985. Atmospheric carbon dioxide, orbital forcing, and climate. In: The Carbon Cycle and Atmospheric CO_2: Natural Variations Archean to Present, ed. E. Sundquist and W.S. Broecker. *Geophys. Monog.* **32**:303–317. Washington, D.C.: American Geophysical Union.

Stuiver, M., T.F. Braziunas, B. Becker, and B. Kromer. 1991. Climatic, solar, oceanic and geomagnetic influences on Late-Glacial and Holocene atmospheric $^{14}C/^{12}C$ change. *Quat. Res.* **35**:1–24.

Thompson, L.G., et al. 1989. Holocene to late Pleistocene climatic ice-core records from Qinghai-Tibetan Plateau. *Science* **246**:474–477.

Author Index

Subject Index

Note: Figures and tables are indicated by *italic* page numbers

Index compiled by Paul Nash